Ladies in the Laboratory IV

Ladies in the Laboratory IV

Imperial Russia's Women in Science, 1800–1900

A Survey of Their Contributions to Research

Mary R. S. Creese
with contributions by Thomas M. Creese

ROWMAN & LITTLEFIELD
Lanham • Boulder • New York • London

Published by Rowman & Littlefield
A wholly owned subsidiary of The Rowman & Littlefield Publishing Group, Inc.
4501 Forbes Boulevard, Suite 200, Lanham, Maryland 20706
www.rowman.com

Unit A, Whitacre Mews, 26-34 Stannary Street, London SE11 4AB

British Library Cataloguing in Publication Information Available

Library of Congress Cataloging-in-Publication Data

Creese, Mary R. S., 1935–
Ladies in the laboratory IV : Imperial Russia's women in science, 1800–1900 : a survey of their contributions to research / Mary R. S. Creese ; with contributions by Thomas M. Creese.
pages cm
Includes bibliographical references and index.
ISBN 978-1-4422-4741-3 (hardcover : alk. paper) — ISBN 978-1-4422-4742-0 (ebook)
1. Women in science—Russia—History—19th century. 2. Women scientists—Russia—Biography. I. Creese, Thomas M. II. Title. III. Title: Imperial Russia's women in science, 1800–1900, a survey of their contributions to research.
Q141.C6934 2015
509.2'520947—dc23 2014040495

∞™ The paper used in this publication meets the minimum requirements of American National Standard for Information Sciences—Permanence of Paper for Printed Library Materials, ANSI/NISO Z39.48-1992.

Printed in the United States of America

Contents

Figures

Acknowledgments

From the start of this project the library staff of the University of Kansas provided important help without which this study would have been much the poorer. My thanks to them. I thank also Pam LeRow of the College of Liberal Arts and Sciences Digital Media Services who once again did an excellent job in manuscript reformatting. I wish to acknowledge the very considerable general assistance patiently given by Thomas Creese; he also prepared the maps and graphs and clarified certain calculations in medical statistics. I thank Alla Polekutina of the Moscow Academy of Science, who gave me much help and encouragement at the start of this long survey project, many years ago. It is a special pleasure to express my gratitude to Olga Valkova of the Institute of History of Natural Sciences and Technology, Moscow, for very valuable help with botanist Olga Fedchenko.

Introduction

This volume is the fourth and concluding part of a survey that has brought to light the contributions of about one thousand nineteenth-century women whose published work appeared in journals listed in the London Royal Society's *Catalogue of Scientific Papers, 1800–1900.*[1] The first three parts of the survey examined scientific work of American and British women, women from twelve western European countries, and women from four regions that were then British colonial territories.[2] Volume 4 concerns women authors from imperial Russia, who make up about 12 percent of all the women, including the British and North Americans, whose approximately 3,400 papers are listed in the *Catalogue*. Within the context of continental Europe, these women of imperial Russia dominate the picture (see fig. 1).

The great majority of the late nineteenth-century Russians had university training in the medical sciences, in contrast to their sisters in North America, Britain, and western Europe, among whom self-educated or independently educated women, active in the natural sciences, outnumbered medical women, according to work (in all fields) listed in the *Catalogue*.[3] This contrast reflected the marked differences between Western societies and tsarist Russia. By the middle of the nineteenth century, a revolutionary wave was moving many of Russia's intellectuals toward rejection of the old, established feudal order, under which most of the population was composed of landless serfs. During the same period Russian feminism was on the rise, fueled by the appearance of liberal women's magazines and literary works such as Nikolai Chernyshevskii's 1863 novel *Chto delat?* (*What Is to Be Done?*). Liberal-minded young women, inspired by the new idealism and eager to bring help to Russia's poverty-stricken masses, enthusiastically embraced the revolutionary doctrine, known as nihilism, within which they fashioned ways to find freedom from the traditional restraints and restrictions of paternal control. Education in the sciences was seen as the best preparation for meaningful participation in the work of social improvement; training in medicine in particular was one of their major goals. Russian university professors sympathized with the women and provided instruction whenever they could, but

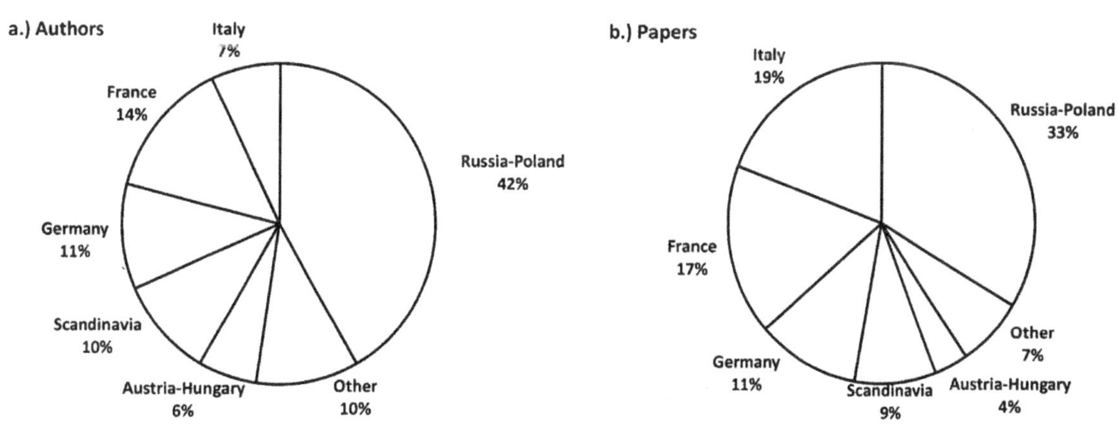

Figure 1. Distribution of continental European authors by country and by number of publications. Data from the Royal Society's *Catalogue of Scientific Papers, 1800–1990.*

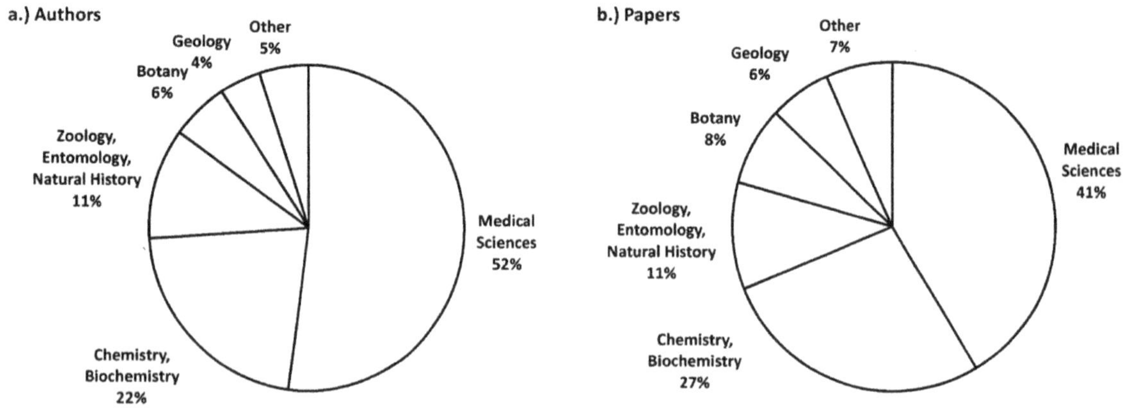

Figure 2. Distribution of Russian authors by field and number of publications. Data from the Royal Society's *Catalogue of Scientific Papers, 1800–1990.*

from time to time the tsarist government prohibited universities from accepting female students. As a result many sought higher education in the West, at first in Swiss universities.

Women in the medical sciences, chemistry, and zoology together make up about 85 percent of Russian women who published articles in scientific journals indexed in the Royal Society's *Catalogue*. Botanists and geologists, and particularly paleontologists, were also fairly well represented, producing respectively 8 and 6 percent of the papers. Mathematics and astronomy attracted fewer workers, and in archaeology the Royal Society listed one contributor who produced two papers. The medical women, many of whom carried out impressive doctoral research in specialized areas such as neurology, outnumbered those in the next most popular field, chemistry, by more than two to one (see fig. 2).

Sketches offered here of the careers of these daughters of imperial Russia vary considerably in length. Fairly full biographies of a few of them are now available,[4] but many more are without mention in western European and American literature on the history of science. The young Russians who formed the majority of that great wave of late nineteenth-century women who received doctoral degrees in medicine at Swiss universities, although capable and skillful research workers (as judged by the quality of their doctoral work), had no plans for careers in research. Nevertheless, when viewed as a whole, the large body of work they produced as students makes a notable contribution to the scientific research of the time. That so much of it came as a by-product of their eagerly sought-after physician's training makes it all the more remarkable. In their subsequent careers—the work of caring for their country's poor, particularly the women and children—they filled the lowest positions in Russia's medical profession, and their lives and experiences were rarely recorded. Perhaps their mention here will lead on to the rediscovery and recognition of a few. A small, select group, perhaps ten or twelve, did rise to full staff positions in well-regarded scientific institutions, where their research capabilities flourished and their productivity was notable.

The first five chapters of this volume discuss the contributions of women of imperial Russia, excluding those identified as Polish. Chapter 6 covers the sizable subgroup of women from the Russian region of what was then a partitioned Poland. An appendix discusses the careers of women from a number of other countries, including four Europeans, whose contributions are listed in the Royal Society's *Catalogue*.

Much of the historical importance of these nineteenth-century women lies in their demonstration of female competence in science-based work, a demonstration that the male establishment could no longer afford to ignore without arousing considerable social unrest as the century went on. Russians in impressive numbers led women's entry into modern science. Furthermore, the lives of some of them, where their stories are known in moderate detail, give fascinating glimpses into colorful personalities, at times working in exotic places; among them we find an intrepid botanist collecting in the foothills of the Pamirs in the late 1860s, a pioneering physician in military service in the Serbian army in the 1870s and later practicing in Muslim communities in Turkey and Egypt despite restrictive rules against female doctors, and a young zoologist leading a new marine-biology research station in the Crimea through its first years at the beginning of the twentieth century. These were courageous and enterprising adventurers as well as hardworking, dedicated scientists or medical women.

Transliterating Russian names into English presents some peculiar problems, and sometimes a choice had to be made. By and large the Library of Congress system was followed for both personal names and titles of publications.

NOTES

1. Royal Society, London, comp., *Catalogue of Scientific Papers, 1800–1900*, 19 vols. (Cambridge: Cambridge University Press, 1867–1925).
2. Mary R. S. Creese, *Ladies in the Laboratory? American and British Women in Science, 1800–1900* (Lanham, Md.: Scarecrow Press, 1998); Creese, *Ladies in the Laboratory II: West European Women in Science, 1800–1900* (Lanham, Md.: Scarecrow Press, 2004); Creese, *Ladies in the Laboratory III: South African, Australian, New Zealand, and Canadian Women in Science, Nineteenth and Early Twentieth Centuries* (Lanham, Md.: Scarecrow Press, 2010).
3. Creese, *Ladies in the Laboratory?* and *Ladies in the Laboratory II.*
4. See for instance Ann Hibner Koblitz, *A Convergence of Lives: Sofia Kovalevskaia: Scientist, Writer, Revolutionary* (Boston and Basel: Birkhaüser, 1983); Cordula Tollmien, *Fürstin der Wissenschaft: Die Lebensgeschichte der Sofja Kovalevskaja* (Weinheim and Basel: Beltz Verlag, 1995); Olga A. Valkova, *Olga Aleksandrovna Fedchenko* (Moscow: Nauka, 2006); Elena Roussanova, *Julia Lermontowa (1846–1919): Die erste promovierte Chemikerin* (Norderstedt: Books on Demand, 2003).

Chapter One

Russia's Early Medical Women

Nearly all of Russia's medical women identified in this survey received at least some part of their training at universities in western Europe, until 1873 largely at the University of Zurich. That year Tsar Aleksandr II promulgated a decree imposing severe penalties on women from his domains who continued studies at Zurich, a city that tsarist authorities, with good reason, considered a hotbed of revolution and conspiracy. Many moved to the University of Bern, which had opened to women in 1871, and Russian student numbers there increased steadily throughout the 1880s and 1890s. The universities of Geneva and Lausanne also trained and graduated a number of Russian medical women before 1901, and one Russian-Italian received a medical degree from the University of Naples in 1893.

In their own country the government's fear of change prevented the opening of medical education to women; since the tsar's councilors were reluctant to grant women the rights and privileges accorded to male physicians, their foreign medical degrees were not officially recognized, and consequently they were banned from government-controlled salaried positions. Training in Russian medical academies, particularly the Medical-Surgical Academy in St. Petersburg, run by the Russian War Department, was of a standard comparable to that in western Europe, but overall health care in the country remained very poor, simply because of a shortage of doctors; periodic epidemics could not be controlled, the syphilis rate was high, and infant mortality common.

As a result, despite its fears of allowing changes in the official status of medical women, the Russian government was glad enough to have them care for the sick; since salaries for physicians in rural areas were controlled by the boards of local district councils (*zemstva*), and were often very low, rural posts tended to be somewhat more readily available to women. Additional opportunities existed in territories to Russia's east and south, into which the empire was then expanding; there the inhabitants were Muslim peoples among whom women were permitted to receive medical care only from women.

The following sketches of late nineteenth-century Russian medical women are grouped according to the university at which they did most of their study or research. In the case of Bern University, because of the considerable numbers involved, they are further subdivided into groups according to the faculty member with whom they worked.

THE UNIVERSITY OF ZURICH

Following the established chronology in histories of women's entry into the medical profession in modern times, we begin with the story of NADEZHDA PROKOF'EVNA SUSLOVA[1] (1843–1918), later ERISMANN, and then GOLUBEV, now recognized as the first woman graduate in medicine in modern times. Thanks to her autobiographical notes published in 1900 and summarized in English in Tuve's 1984 work, fairly detailed biographical information is readily available.[2] A short account of her life is sketched below.

Born on 1 September 1843 in the village of Panino in the Nizhni Novgorod district along the Volga valley about two hundred miles east of Moscow, Nadezhda was the second daughter in the family of three children of Prokofii Suslov, a serf, and his wife, Anna. The family was rich by serf standards;

Prokofii Suslov could read and write, and he held the post of agent to the village's owner, Count Dmitrii Sheremetev, a member of the wealthy and powerful Sheremetev noble family with many generations of service to Russia's tsars. A good administrator, Suslov was moved from one to another of the count's hereditary estates, the first being Makarev on a tributary of the Volga, where the peasants were suffering severely from droughts and the epidemic diseases that periodically swept through the country. The hardships the people were enduring made a deep impression on the young Nadezhda, but they were hardly conducive to kindly feelings toward the count's agent. Consequently the Suslov family tended to keep to itself and became very close-knit; all the children learned to read and write, and the radical journal *Sovremennik* (Contemporary) was an important part of their reading.

In 1854, when Nadezhda was eleven, her father was transferred to Count Sheremetev's estate of Ostankino near Moscow, a property that had fallen into considerable disrepair; the magnificent late eighteenth-century Ostankino palace, which had been raided and occupied by Napoleon's generals in 1812, was then in sore need of restoration, a task assigned to Suslov. His family was given living quarters in the palace, and the Count got to know and like the three children. Using his influence, he was able to place both Nadezhda and her older sister, Apollinaria (Polina), in a boarding school for girls of the nobility in the ancient city of Tver. As children of a serf, the Suslov girls would not normally have been admitted to this school, known as "Madame Penigka's." Nadezhda was not very impressed by the institution. She later recalled that she felt out of place and that the education provided seemed to her dull and boring; little science was offered, although languages and history were well taught and she learned them thoroughly.[3]

Thanks to Prokofii Suslov's outstanding restoration of the Ostankino palace and estate, Count Sheremetev was able to invite recently crowned tsar Aleksandr II to stay at Ostankino during one of his visits to Moscow; organizing the necessary elaborate preparations once again brought out Suslov's talents. Impressed by his agent's managerial skills, Sheremetev transferred Suslov to St. Petersburg, where he had his permanent residence and many additional valuable properties. Suslov was appointed manager of all these holdings, and his family had its own living accommodations. Nadezhda and her sister were enrolled in one of the city's boarding schools for daughters of the nobility and also received private tuition. She completed studies to gymnasium level and received a governess diploma by the time she was sixteen. Her brother, Vasilii, also completed his high school education and went on to the university.

Nadezhda and her close school friend Mariia Obrucheva felt themselves very much part of the new generation of Russian women, committed to the revolutionary ideas of the nihilists and eager to take part in service to the people; the liberation of women and their acceptance into the universities, particularly to medical training, were priorities. Nadezhda's parents fully supported her educational ambitions, but Mariia Obrucheva's father, an army general, had a more conventional outlook; to break away from parental control Mariia had to resort to the tactic of fictitious marriage, arranged with a young physician, Petr Bokov, a friend of her brother.

In 1861, when several women were applying for admission to St. Petersburg Medical-Surgical Academy, Nadezhda and Mariia followed Petr Bokov's suggestion and asked the young physiology professor Ivan Sechenov for permission to attend his lectures as auditors. Sechenov, then at the start of his brilliant career, agreed; several other faculty members likewise accepted women auditors. Hardworking and enthusiastic, both Nadezhda and Mariia were given short research projects at the end of their first year; Nadezhda's investigation of the influence of light electrical stimulation of the skin was reported in two short papers in 1862 and 1863 (see bibliography).

Aleksandr II's emancipation proclamation of 1861 brought an end to serfdom in Russia, but many of the peasants were dissatisfied with their land allocations. Government suppression of their protests followed and student unrest intensified; secret societies sprang up and student demonstrations in which women took part were frequent. The universities were closed, and when they reopened in 1863 under a new government charter, women were denied admission; only those already in the St. Petersburg Medical-Surgical Academy were allowed to continue until the autumn of 1864.

By the early 1860s, however, Nadezhda Suslova was something of a marked woman, a major cause being the fictional pieces embodying a defense of nihilism and feminism that she published in *Sovremennik* magazine. Her brother and sister, and her closest friends, Petr and Mariia Bokov, were arrested for taking part in antigovernment protests; Nadezhda also came under police observation for a time. Following Sechenov's advice, she decided to continue her education in western Europe and went to Zurich in 1865. Her parents gave her as much financial help as they could, but family money was then short; after the required emancipation of his serfs, Count Sheremetev no longer employed Prokofii Suslov, who had retired to a small house in Nizhni Novgorod, where his son was a barrister.

Two years were to pass between Nadezhda Suslova's arrival in Zurich and the formal opening of the university to women, but the Zurich medical faculty accepted her as an auditor. She did well in her courses, which ranged across a broad list of subjects including pharmacology, obstetrics, and surgery; she was also happy to accept assignments in laboratories and clinics. Granted leave to carry out the research required for a medical dissertation, she initially planned to move to Paris and work in the city's obstetrical clinics. But before she left Zurich, Ivan Sechenov, her former St. Petersburg advisor, invited her to join him for doctoral research at the Institute for Physiology at Graz Medical University, where he was then working.

Sechenov was becoming well known in the new field of nervous reflexes; one of his particular interests was the relationship of the nervous system to muscular activity, live frogs being used as research animals.[4] Although neurological research was somewhat removed from the kind of practical, general-physician service she meant to take up, Suslova accepted Sechenov's invitation and carried out a program of careful, painstaking research correlating reflex muscle action with the functioning of the lymphatic hearts spaced along the nerve systems of frogs.[5]

In 1867, her dissertation research carried out and her studies almost completed, Suslova asked for permission to take the oral examination for a Zurich medical degree. This required an initial first step, namely a decision by the university senate and the cantonal government to accept women as full students. Thanks to the presence of several liberally minded men on the Zurich faculty and the movement toward liberal reform in the canton, the historic decision to do so was taken; Zurich became the first university in Europe to include women among its registered students. Suslova matriculated in February 1867 and undertook the defense of her dissertation the following December. Despite her shyness and much sharp questioning from surgeon Edmund Rose, physiologist Ludimar Hermann, and others of the medical faculty, she defended her views courageously in the presence of a large audience. Suslova's 1867 medical degree was the first conferred on a woman, following standard procedures, by a recognized Western university with high academic standards; the event, reported widely across Europe and America, ushered in a new era of medical education for women. Suslova's research findings were published in a German journal in 1868 (see bibliography).

While in Zurich she lived quietly and unobtrusively, avoided political activity, and worked hard. She had few friends and, being the only woman studying medicine in a class of generally unfriendly men, found life lonely and somewhat trying; indeed it seems that the experience nearly broke her.[6] "Hers was a quiet, serious nature, with deep feelings and a thoughtful, melancholy look from deep-set brown eyes."[7] However, after a few months she met Swiss medical student Friedrich Huldreich Erismann (1842–1915), one of the few Swiss students who at this early date favored the higher education of women. Through her Erismann became interested in the problem of the dismal living conditions of most of the Russian population, conditions that presented almost unlimited opportunities for public health workers. He became a convinced socialist, studied public health in Germany, and decided to join Suslova in work in that field; they were married in Vienna on 16 April 1868. Nadezhda then returned to St. Petersburg, and Erismann went back to his studies in Heidelberg.

After passing a special examination set by St. Petersburg Medical-Surgical Academy professors, Nadezhda Erismann was formally allowed to practice medicine in Russia; she was the first woman to be granted this right, although without the privilege in rank it brought to male physicians. In 1868 she opened a private practice in gynecology and pediatrics in St. Petersburg and soon had patients from a wide spectrum of society, from the indigent poor to middle- and upper-class women eager to consult a female doctor. Erismann, his enthusiasm for supporting the causes of workers' organizations further strengthened, arrived in St. Petersburg in 1869. After passing the necessary Russian qualification examinations, he joined his wife's practice and also began investigating health conditions in the city's slums, work in which Nadezhda also took part.[8] Most of the findings were published under Erismann's name only in the journal *Archiv sudebnoi meditsiny i obshchestvennoi gigieny* (Archive of forensic medicine and social hygiene). Suslova Erismann brought out a paper under her own name on the education of mothers in the rearing of children in their first years (see bibliography).

In 1872 Erismann returned to Zurich for further training in social health work and while there became more and more involved in Marxist circles. After Tsar Aleksandr II's 1873 proclamation prohibiting further study in Zurich by Russian women and more strengthening of measures against revolutionary activity, Erismann returned briefly to St. Petersburg. Within a short time, however, he moved to Munich, where he concentrated on study and research in public health. Suslova visited him there in the autumn of 1873, but they gradually drifted apart and their divorce was finalized on 18 August 1883 in Aarau district court.[9] Although both were dedicated to improving the living conditions of Russia's poor, they differed strongly in their ideas for bringing this about. Suslova believed in personal service, while Erismann saw the changing of structures and institutions as the fundamental requirement.[10]

For a period in 1873 Suslova was again under surveillance by the secret police, but her difficulties were soon resolved. Her brother having died of tuberculosis and her sister, Apollinaria, having lost touch with the rest of the family, she moved to Nizhni Novgorod to practice nearer her parents.[11] As time passed she became more and more conservative and traditional in her outlook and her way of life.

In 1885, when she was about forty-two, Nadezhda married Aleksandr Efimovich Golubev (1836–1926), a friend of Sechenov, whom she had known when she was a student in Graz. Golubev was at that time professor of histology at the Women's Medical Courses in St. Petersburg and also conducted a practice in Nizhni Novgorod. When he inherited a large fortune he gave up teaching and retired to Alushta, a beautiful region on the southeast coast of the Crimea. Nadezhda accompanied him to the villa he built on the lower slopes of Kastel Mountain overlooking the town of Alushta and the sea. She continued to practice, working among the local population, which included very poor Tartars; her services were given free of charge. In addition, she donated large sums of money to deserving groups and built an elementary school and a library for the poor on the Golubev estate. Golubev continued scientific work until he went blind from glaucoma, after which Nadezhda spent a great deal of time with him.

Her role in opening an important branch of higher education to women and her dedication to helping the poor is well recognized by historians of women in medicine, but she is said to have been far from satisfied with her life's work; she even told a young friend that the study of medicine led only to disappointment.[12] Her most important publication is considered to be her autobiographical account of her early years and her time as a student in Zurich and Graz;[13] here again she gave strong expression to her feminist views and her thoughts on the problem of male domination. Nadezhda Suslova died in her seventy-fifth year on 20 May 1918, in Alushta.

One Russian woman student who stayed in Zurich and finished her medical training there after the tsar's 1873 edict was JULIE SINCLAIR[14] (1849–1937), later SINCLAIR-EGLI, who received her medical degree in 1876. No publication by her is listed in the Royal Society's *Catalogue* since she did not report her doctoral dissertation research in any standard scientific journal. Nevertheless, her story is sketched here because she was one of the adventurous young Russian women who went to Zurich for medical training in the early 1870s, joining that small, multinational group, several of whose members were especially interesting characters about whom a modest amount of information is fairly readily available; Sinclair's life story has an international flavor, which perhaps adds to its appeal.

Born on 12 July 1849 in Yaroslav, southeast Poland, a few miles from the Russian border, she was the daughter of Georg Sinclair, an engineer of Scottish background. He, or one of his forefathers, in the long tradition of the restless Scot, had migrated east to the realm of the tsars and very successfully settled there.[15] By at least the 1870s Georg Sinclair had been knighted and held the position of privy councilor; his wife, Julie Sinclair's mother, Aleksandra Pokrowsky, came from one of Russia's noble families.

Julie received her early education in St. Petersburg and in 1871 traveled to Switzerland to enroll for the winter semester in the medical faculty of the University of Zurich; she was then twenty-two. Thanks to the memoirs of her close friend and only female classmate, the German student Franziska Tiburtius, we have a number of glimpses into her life during her student years. Despite the fact that several women had studied medicine since Nadezhda Suslova led their entry into Zurich's medical faculty in 1868 (see above), many Swiss male students were still strongly against having women in their midst. Sinclair and

Tiburtius were given a very noisy and intimidating reception on their first entry into the large lecture hall, and order was restored only by the arrival and intervention of the lecturer. But there were many pleasant times that left cherished memories; particularly notable were the social gatherings of a close group of about six women students and three men, one of whom later married Julie Sinclair.[16] By 1873 she had already decided that she would not return to Russia.

She completed eleven semesters of medical studies, concentrating on ophthalmology under the guidance of Friedrich Horner (1831–1886). Director of the university's eye clinic and an outstanding teacher, Horner was widely known for his successful work on eye surgery; he was the first to use antiseptic methods in the treatment of eye diseases, an advance that markedly reduced complications. For her doctoral research Julie Sinclair studied the formation of acquired cataracts covering the lens of the eye, one of several areas of great interest to Horner. Using calf, pig, ox, and sheep eyes extracted from dead animals, she carried out a long series of investigations to test the passage of fluids carrying various chemicals and cell nutrients into the occular tissues and the filtration capabilities of these tissues. Her first set of tests, in which she examined the composition and thicknesses of the layers of cataract material, indicated that cataract formation was related to inflammation-causing changes in the fluids and tissues surrounding the lens. A second set of tests used lenses removed immediately after death. Overall the research indicated that cataract formation resulted from a succession of inflammatory influences that disrupted the supply of essential nutrients to the lens and in advanced cases led to destruction of the lens. She received her degree in 1876, following presentation of her fifty-page dissertation to the Zurich University medical faculty.[17]

Shortly after graduating she married Theodor Egli (1844–1933), a Swiss who had received his medical degree two years before her and had already established a gynecological practice in Zurich. One of the close social group of Sinclair's student days and a particularly approachable and sympathetic person, Egli was someone to whom many of the young women turned for advice and help in times of stress. He had been a businessman before taking up medical studies, had spent five years in America, and had relatively wide experience of the world.

Following her marriage Julie Sinclair-Egli joined her husband's practice as assistant and junior partner. The first such husband-and-wife physician team in Zurich, they were notably successful, and their practice became extensive. They had two daughters, Alexandra Olga and Theodora, both of whom later married. Alexandra Facchetti-Egli moved to Milan, and Theodora von Scheven-Egli to Krefeld, a city on the west of the Rhine, south of Germany's Ruhr region. Tiburtius tells us that, over the years, Julie Egli's grandsons played during summer vacations on the grounds of the delightful Egli country house on the shore of Lake Zurich. In 1917, after more than forty years in practice, Julie and Theodor Egli retired to Milan. They lived there until 1932, returning to Zurich a few days before Theodor Egli's death, at the age of eighty-eight, in December of that year. During her widowhood Julie Egli moved around, from Zurich to Milan and then to the small town of Aricate, close to the Italian-Swiss border, before crossing the Atlantic to the United States. She died of pneumonia on 9 March 1937, in her eightieth year, at 1412 South Sixth Street, Louisville, Kentucky.[18] Why she went to the United States at an advanced age remains a question; possible reasons might include a visit to Egli relatives or old friends, or a trip to look into some continuing interest stemming from Theodor Egli's early business activities there. The remarkable qualities in Julie Egli about which there can be little doubt, however, are her youthful vitality and abundant energy, qualities that remained with her throughout her long life.

ADELÁIDA NIKOLAEVNA LUKANINA[19] (1843–1908), née RYKACHËVA, later PAEVSKAIA, did not receive her medical degree from the University of Zurich, but completed the bulk of her studies there. Perhaps most widely known as a writer of semibiographical fiction and memoirs, Lukanina was also one of Russia's earliest women physicians. She was born and spent her childhood years on her father's estate in Novgorod Province; growing up in that rural environment, which would be the setting for most of her literary works, she developed a deep appreciation for nature and concern for forests and wildlife.

She is thought to have spent several years at the old and respected St. Petersburg girls' school, the Smolny Institut. Planning to become a country schoolteacher, she studied pedagogy, but was also drawn toward science. About 1868, while attending science lectures at the St. Petersburg Higher Courses for Women, she was given much help by chemist and composer Aleksandr Borodin (1833–1887); under his guidance she began advanced studies in chemistry. She also sat in on lectures at the St. Petersburg Medical-Surgical Academy, where Borodin taught, although this institution was not yet open to women. By then she had freed herself from parental control by contracting a nihilistic-style marriage of convenience with Iulii Lukanin, a St. Petersburg bookstore owner; she became involved in the radical political activity sweeping the city's student community and in 1870 spent a few days in prison when the authorities tried to reassert their control. Anxious to escape further police notice, she moved to the semi-independent duchy of Finland and began studies at the University of Helsingfors, which had just opened its doors to women. While there she attended lectures at the medical school. Two years later, having decided that medicine was where her main interest lay, she enrolled at the University of Zurich.

From then on Lukanina focused on medical studies, turning aside from the possibility of further chemical research. Nevertheless, her early work, carried out under Borodin's guidance, constituted a creditable contribution to the field; her studies on the oxidation of albumins to produce urea, which Borodin discussed before the Russian Chemical Society in 1871,[20] and her investigation of the reaction of succinyl chloride with the aromatic ketone benzoin, reported in 1872 (see bibliography), were both noteworthy. Borodin is thought to have had a high opinion of her work.

Lukanina's involvement in student revolutionary politics continued after her arrival in Zurich, then one of the centers of antiestablishment activity and the headquarters of internationally known Russian radicals including Mikhail Bakunin, Mikhail Sazhin, and the philosopher and sociologist Pyotr Lavrov. Lukanina, one of the *Kosakenpferdchen* (little Cossack horses), as the Russian women students were called, enthusiastically attended the many political meetings and talks then taking place, and helped in some of the work being done. In 1873, when the tsar's government ordered Russian women students to leave Zurich or be banned from future professional employment in Russia, Lukanina chose to stay on and finish her studies. By 1875, with her courses completed but with a dissertation still needed to fulfill her Swiss degree requirements, she decided to move to America to try to get a medical degree by examination.[21]

She went first to Philadelphia, the leading medical center in the United States and the location of the degree-granting Women's Medical College of Pennsylvania. Along with recommendations from lecturers and certificates from classes she had attended at Zurich University, she also had a letter of introduction to Dr. Jacob Mendez Da Costa (1849–1900). America's leading specialist in clinical medicine, professor at Jefferson Medical College, and visiting physician at a number of other Philadelphia medical institutions, Da Costa had received much of his education in Europe. Lukanina hoped that, with his help and advice, she would be able to take the American degree examination without great delay and then find a position as a hospital house surgeon. But Da Costa was no supporter of medical education for women and somewhat abruptly referred her to the Women's Medical College. There Dean Rachel Bodley explained the necessary formalities and requirements, which would include attending clinical classes; furthermore, she warned Lukanina that assistant positions in hospitals were difficult to obtain.

Considerably disappointed by this setback, which somewhat shattered her idea about the position of women in the medical profession in America, Lukanina spent some time mainly occupied with studying American medical textbooks. But in June 1875 she received an offer of an assistant position at the New England Hospital for Women and Children in Boston, an institution founded and directed by Dr. Marie Zakrzewska. Zakrzewska had moved to America from Berlin in 1853 with qualifications in midwifery; three years later she was awarded a medical degree by Western Reserve Medical College in Cleveland, Ohio, which admitted a few women in the 1850s.

Lukanina accepted the opportunity Zakrzewska offered. She spent a busy summer at the hospital and attached dispensary, while also continuing her studies. By the autumn she was back in Philadelphia completing the required courses and hospital work; she received her degree in 1876 after passing the examinations without much difficulty. A short visit to Boston followed, so that she could renew her acquaintance with Marie Zakrzewska and deepen her experience of American life and customs, a subject in which she had an intense interest. Zakrzewska offered her a position in the New England Hospital, but she decided to return to Europe. Her American experiences were described in detail in a series of articles in the relatively liberal St. Petersburg journal *Vestnik Evropy* (European messenger), published after she was back in Europe.[22]

By 1877 she was in Paris but she still had eight years to wait until, in 1885, she was finally allowed to return home; by then Russian officialdom's suspicions of her radical political associations would appear to have abated. In the meantime she had to earn a living; money was in short supply. For help in finding a route to publishing a number of stories she had written, and in finding work translating and tutoring, she turned to Russian novelist Ivan Turgenev, one of the many Russian intellectuals who had moved to western Europe during the repressive rule of Tsar Nicholas I. Turgenev thought well of Lukanina's literary efforts and encouraged her writings; she in turn spent many hours at his side before his death at Bougival, on the outskirts of Paris, in 1883.

Lukanina's first story, "Liubushka," set in rural Novgorod province, appeared in *Vestnik Evropy* in 1878; her four-part cycle of stories "Starinnye dela" (Olden-day matters), published in the same journal between 1880 and 1887, combined fiction and memoir in its treatment of rural life. She had a strong concern for the environment, especially the northern forests whose gradual disappearance she deeply regretted. Other stories and articles she published include "Palata No. sto tretii" (Ward no. 103) based on her experiences as a medical assistant working among Zurich prostitutes (1879), and a fourteen-page paper describing lectures on various subjects she had attended at the Sorbonne.[23] Her major translations dealt with the lives and literary output of Victor Hugo and Sir Walter Scott; included in these were sections of interpretation and criticism. Both works were published after her return to Russia and under the name A. Paevskaia, adopted after her second marriage.[24] Nikolai Paevskii, a Russian psychiatrist living in Paris, had also turned to Turgenev for help in getting his writings published; he and Lukanina were most likely married in the late 1880s, although the date of the event is uncertain.

Lukanina returned to St. Petersburg in 1885, but her first opportunity to practice came only during the cholera epidemic of 1892. In this crisis women physicians previously barred from professional work were sent into the country's remote villages. Lukanina had a medical team of largely female staff working in villages along the Volga and in the Cherdaklinskii district of the Volga basin, east of Niznhi Novgorod; a region of wooded plains, deciduous forest, and lakes with a scattering of villages such as Cherdakly, Uren, and Turgenev, it was a landscape familiar to her since her youth. Her account of this trip, "Komandirovka na kholeru" (Journey to the cholera epidemic), published ten years later, dwelt heavily on her personal interactions with her peasant patients and with her staff rather than on medical or scientific matters, but it did perhaps help to once again bring to public attention the deplorable situation of the Russian peasantry in the 1890s.[25]

She continued to practice throughout the 1890s and also to write, publishing a number of stories for children as well as occasional articles on scientific and medical topics. Following the end of her second marriage, about 1894 or 1895, she opened a laundry, dry-cleaning, and dyeing business. This undertaking was successful enough to enable her to offer lodging, job training, and literacy instruction to her female employees, some of whom were also her patients; such work was very much in keeping with Lukanina's constant effort to improve the lives of Russia's working class women. She died on 25 December 1908, when she was about sixty-five, after a lengthy period of poor health.

Daria (Dasha) Schirman,[26] a daughter of M. A. Schirman, was born on 24 May 1874 in Rostov-on-Don, southern Russia. She attended a Rostov gymnasium, graduating in 1891, and a year later enrolled in the medical faculty of the University of Bern, where she studied during the winter semester of 1892–1893. She moved to the University of Zurich in the autumn of 1893 and completed her medical studies there over the course of the following five years.

Her doctoral dissertation research, carried out in the university's Anatomy Institute, was directed by Philipp Stöhr (1849–1911), professor of anatomy at Zurich from 1889 to 1897. Stöhr was well known for his work on embryological development, especially as it could be related to phylogenetic history, that is, the "ontogeny repeats phylogeny" concept. Daria Schirman's research was a microscopic examination of the fine structure of the mucous membrane layer of the lining of the large intestine of the guinea pig at the fetal stage of its growth. Her work extended earlier work by Stöhr on the fine structures in the mucous membranes of the human stomach.[27] Schirman observed that fine structures not present in the mucous membrane of an adult guinea pig were in fact present in the fetal form and later absorbed. Her results were published in the *Verhandlungen der Physikalische-Medicinischen Gesellschaft zu* Würzburg in 1899 (see bibliography); she received her degree from the University of Zurich in 1898.[28]

Ol'ga Vassilievna Leonova[29] (fl. 1890s–1910), later Leonova-von Lange, published at least ten studies on congenital abnormalities in humans; her particular interests were eye defects and limb deficiencies, conditions stemming from disruptions in central nervous system controls during early fetal development. No information about her family background or her later life has been uncovered.

Leonova's principal place of study and research during the twenty-year span of her work outlined here was most likely the Department of Neuropathology at the Imperial University of Moscow (now Moscow State University), founded in 1869 by Paris-trained Aleksei Yakovlevich Kozhevnikov (1836–1902). This department and those at several other Russian universities were places of great activity in early neuroscience research; international links were strong, and students were encouraged to spend time in western Europe.[30] Over the course of a number of years Leonova visited Zurich University for periods of work and study with Swiss neuropathologist Constantin von Monakow.

A pioneer in brain anatomy, von Monakow in 1885 set up a small laboratory in Zurich, which later developed into the Brain Research Institute of Zurich University. Despite its limited space and a bare minimum of equipment in its early years, this laboratory was a place of considerable research activity, visited by many outstanding medical men (including Americans) seeking further training. The general atmosphere was particularly pleasant; von Monakow described the sparsely equipped little laboratory as having an atmosphere of cheerful, comradely, vigorous work, a "kameradschaftlicher, arbeitsfrischer, frölicher Geist,"[31] where much mutual help was given. Leonova and Polish student Stanislawa Poplawska (see chapter 6) were most likely the first women to work there.

Leonova's research under von Monakow's guidance began in the early 1890s; much of the work was carried out in Moscow, but she is known to have spent the summer of 1895 in Zurich and the following winter in the United States. She may well have visited the University of Pennsylvania in Philadelphia, one of the earliest centers of American neurological research, since she would have met its faculty member Carl Seiler when he studied in von Monakow's laboratory.

Her research over several years was devoted to exploring nerve cell connections in the occipital lobe of the brain and the relationship of eye abnormalities to dysfunctions in the central nervous system. Although this area remained her primary interest, she later expanded her investigations to include congenital limb deformities, which she believed also resulted, at least in part, from disruptions in the central nervous system during fetal development. Her methods involved microscopic examinations of thin sections, sometimes preparations shared with colleagues and sometimes made by her from embryos or stillborn infants passed along to her from large obstetrical centers in Moscow. Central nervous system research was a fast-developing field at that time and Leonova an enterprising and enthusiastic contributor. By 1904 at the latest she had a doctoral degree and was fairly well known among workers in brain and spinal-cord research. At the 1906 Itinerant Meeting of the Society for Psychiatry and Neurology in Vienna she presented a lecture on her findings in an eye-abnormality investigation.[32] Her later publications, expanded discussions of monograph length, form part of a series of works by von Monakow's students brought out with his assistance.[33]

Only minimal information has been uncovered about early University of Zurich student ANNA KLEIMANN (KLEINMANN).[34] The daughter of wholesale merchant Leon Kleimann of Odessa, she enrolled in the medical faculty at Zurich in the winter semester of 1869, two years after Nadezhda Suslova (see above) received her degree there. Kleimann's first research, a joint study with Rosa Simonovich (see below) directed by physiologist Ludimar Hermann, concerned the effects of tartar emetic, using dogs as test animals. Results were published in Pflüger's *Archiv* in 1872 (see bibliography). Kleimann then went on to doctoral research with pediatrics specialist Oscar Wyss and investigated particular causes of death in young children in Zurich canton.[35] Her study was one of many statistical investigations in this area being carried out across Europe during the second half of the nineteenth century; the very high infant mortality rate and its causes had by then become a major concern for medical communities.[36] Kleimann saw respiratory diseases, and gastrointestinal problems associated with the substitution of cow's milk for mother's milk as being among the several possible causes being considered. Being very close to completing her work when the Russian government announced that Russian women students must leave Zurich by the end of 1873, she requested and received special permission to graduate early in 1874; her degree was conferred on 8 January of that year.

THE UNIVERSITY OF BERN

Between the mid-1870s and 1900 Bern University's faculty of medicine graduated about thirty-seven women students from imperial Russia whose research, much of it of notably high scientific quality, was

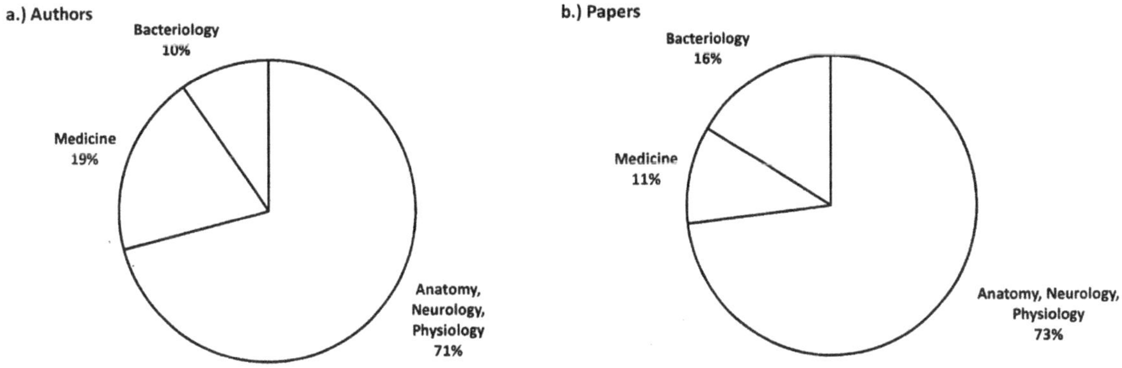

Figure 3. Distribution of Russian authors by medical subfield. Data from the Royal Society's *Catalogue of Scientific Papers, 1800–1990.*

listed in the Royal Society's *Catalogue* (see fig. 3). A great many of the following accounts concentrate on this research; biographical information about most of the women's subsequent careers was largely beyond the reach of this survey.

Twelve of these students specialized in morbid anatomy and neurology, carrying out their studies under the direction of the eminent German pathologist and anatomist Theodor Langhans (1839–1915). His fellow faculty member German physiologist Hugo Kronecker (1839–1914) guided seven Russian women through doctoral research; Kronecker, chairman of Bern's physiology department from 1885, built a new Institute of Physiology that rapidly acquired an international reputation for excellence and attracted many foreign students and research workers. Six of the women discussed here were students of anatomist Maximilian Flesch (1856–1943), leader of the Bern School of Veterinary Medicine from 1882 until 1885. Swiss surgeon Emil Theodor Kocher (1841–1917) mentored three more; a close colleague of Langhans, Kocher was known especially for his promotion of aseptic methods, and his success in greatly reducing mortality in thyroidectomies. Other women medical graduates discussed in this section were probably the sole female students with significant research publications who were mentored by various other Bern faculty members.

Students of Theodor Langhans

SERAFINA (SERAPHIMA) SCHACHOVA[37] was born in 1854 in Ekaterinoslav, southern Ukraine, a city whose surrounding region saw frequent peasant unrest during the years when she was growing up. Her family would not appear to have been among the impoverished since she was educated privately up to university entrance standard. In 1871, at the age of seventeen, she went to Switzerland and enrolled in the medical faculty of the University of Zurich for the winter semester of that year. Her studies there continued until August 1873, when the directive from the tsarist government required Russian women students to leave Zurich if they wished to return to future careers in their own country. While in Zurich Serafina Schachova lived in the "Obere Palmhof," the preferred boarding house for medical students since it was very close to the cantonal hospital and not far from the university.[38] Although the Obere Palmhof was the meeting place for women students' debating clubs, there is little readily available evidence to suggest that Schachova was particularly involved in the revolutionary activities that attracted a number of Zurich's Russian women students about that time. She concentrated strongly on her studies and carried out a creditable piece of work at the Zurich Pathological Institute. Using material from pigeons of age range a few days to adult, she examined a succession of microscopic sections, revealing the process of bone-tissue formation and the development of bone inner structure. The work was reported in a short paper in 1873 (see bibliography).

After she enrolled in the medical faculty of the University of Bern in the winter semester of 1873, she became one of the earliest students of Theodor Langhans, who had joined the faculty of the Bern medical school as professor of pathological anatomy the previous year. Her research constituted one stage in a long succession of fundamental investigations by Langhans and others into the structure of kidney tissue with its complex of convoluted urine drainage tubules and the changes these undergo in nephritic conditions. Working with kidney tissue derived from dogs that had suffered induced nephritis caused by poisoning with the toxic chemical cantharidin, she found extensive changes in the mucus membrane of the drainage tubules, a portion of which became known as "the spiral tube of Schachova."[39] She completed her research and graduated in July 1876.[40] Her findings on the kidney tubules were not published in a scientific journal at the time and as a result did not become widely known until a few years later when Langhans described her work in some detail, emphasizing the significance of her contribution to the understanding of kidney structure and function. Her work was cited in kidney research papers for many years.[41]

Within a year after receiving her degree she returned to Russia, where she passed the state licensing examinations for physicians in 1878. Of her subsequent career we know that from 1910 she practiced medicine in Kharkov, Ukraine. She married someone by the name of Slagin, about whom no information has come to light.

The work Schachova carried out during her three years in Bern was extended in scope and detail by another Russian woman student, IDA ELIASCHOVA.[42] Eliaschova was born in 1852 in Mogilev, Belarus, about one hundred miles south-southwest of Smolensk. During her childhood years the surrounding region was one where serfdom was extensive and serious peasant revolts were frequent until the 1861 emancipation act. Information about her early education has not been uncovered, but by her early twenties she had enrolled at the University of Bern, where she studied for twelve semesters during the years 1876–1882. She concentrated on pharmacology and medicinal chemistry, carrying out her required dissertation research under the direction of pathologist Theodor Langhans; she received her doctoral degree in 1883.[43]

Eliaschova's work, done in the university's Institute of Pathological Anatomy, concerned the effects of the toxic chemical cantharidin on the kidney's small, urine-conducting channels. Although Schachova reported that cantharidin induced changes only in these drainage channels in her 1876 dissertation, within a short time other workers found kidney damage caused by this chemical to be markedly more extensive.[44] Many points remained to be clarified, including such matters as the choice of animal subject, dose size, period of administration, and the effect of different methods of tissue preservation for the microscopic studies. Using only rabbits and single lethal doses, Eliaschova reaffirmed the earlier observations of damage in the walls of the small channels sufficient to slow or almost completely block urine flow; in addition she observed severe injury to the gomeruli, the ball-shaped structures in the kidney's interior cavity that play a major role in the filtration process. In short, cantharidin poisoning caused changes similar to those found in cases of acute nephritis.

Ida Eliaschova's career after her six years of study and creditable research in Bern has not been uncovered.

OLGA (VON) STOFF,[45] later SIEBOLD, the daughter of a court councillor, was born in St. Petersburg on 8 February 1848. She completed gymnasium studies in St. Petersburg, and in 1872, at the age of twenty-four, went to Zurich. While she was there one of her Stoff relatives (possibly her mother, Katharina Stoff) was resident in the nearby town of Hottingen, now part of the city of Zurich. Olga enrolled in the medical faculty of Zurich University, but the following year, to comply with Russian government directives, she moved to the University of Bern, where she completed her training. Her dissertation research concerned the physiological effect of aniline, then a topic of interest in pharmacology. Her degree was awarded in February 1875.[46] Along with her fellow Russian medical student and close friend Sophie Hasse (see below), she also undertook an investigation of the circulation rate in the spleen (see bibliography), a demanding piece of work done during the autumn of 1872 under the guidance of histologist and anatomist Heinrich Frey of Zurich University.

Olga Stoff was remembered as being a very able and very serious young woman who sat all evening in a corner of the sofa in her student living quarters smoking one cigarette after another,[47] a habit generally seen as typical of Russian students of nihilist outlook. Immediately after graduating she returned to Russia and in 1876 passed the Russian physicians' licensing examination. Free then to play her part in the work of improving the lives of women, she opened a practice in gynecology and obstetrics in St. Petersburg. She married Samuil Friedrichovich Zibol'd (Siebold), a prison doctor, whose outlook concerning service among the city's poor would appear to have matched hers.

MARIE ZIELINSKA,[48] from Warsaw, Poland, might well have been included in the chapter on Polish women, but for the purposes of this survey she fits equally well with the Russian students of Bern University pathologist Theodor Langhans.

Born on 26 November 1859, she studied in the medical faculty at Bern for twelve semesters, from 1887 until 1893.[49] Her doctoral research, carried out in the Bern Institute of Pathological Anatomy, was a contribution to the body of work of Langhans and several other eminent scientists on the structure and function of the thyroid gland, the same area in which her fellow Langhans student Nadine Ott was working at this time (see below). Zielinska examined the development of the gland's lymph nodes in dogs of ages ranging from newborn to adult. In addition she looked at examples of swollen thyroid glands in dogs; for these she reported only overall swelling and no discernible tumors. Her work also included an examination of congenital enlargement of the thyroid gland in humans who had died within their first three years; her hope was to indicate that the disease had its beginning in the period of fetal development.

This lengthy and painstaking piece of research was reported in Virchow's *Archiv* in 1894 (see bibliography). Bern University awarded Marie Zielinska her degree in 1893.[50]

Nadine Ott,[51] from St. Petersburg, was born on 18 August 1867. In 1888 she enrolled in the philosophical faculty of the University of Bern, but two years later, her interests having settled on biological topics, she switched to the medical faculty, where she studied for three years (1890–1893).

Her dissertation research, directed by Theodor Langhans, was one of a long series of investigations by Langhans and his coworkers on the morbid anatomy of goiter and cretinism. Ott examined changes in the peripheral nerve system and immediately surrounding tissue brought about in animals suffering thyroid disorders. The work involved microscopic study of nerve connections in representatives of several different animal species from which the thyroid gland had been removed before death and for which survival after the operation varied from two months (for a dog) to a few days (for a rabbit). Her data and her comparisons with known data for healthy animals, as well as with known nerve data from a deceased human who had suffered from cretinism, added to the gradually increasing stock of information on the numerous effects of thyroid deficiency. The work was published in Virchow's *Archiv* in 1894 (see bibliography); she received her degree from Bern University in 1893.[52]

Marie Vassilieva-Kleimann[53] (1861–1893), the daughter of wholesale merchant Leon Kleimann, was born in Odessa, Ukraine, on 1 January 1861. She studied medicine at the University of Bern for eleven semesters between 1879 and 1886, with an additional semester at the University of Zurich in the winter of 1883. Although she carried out doctoral-level research in the Bern Institute of Pathological Anatomy under the direction of pathologist Theodor Langhans, she did not proceed to a degree and dropped her student enrollment after completing a dissertation-style report by 1886. When the report was finally published in 1890 (see bibliography), Langhans added a brief footnote stating that events unrelated to the research had held up its appearance. These events may well have included Marie Kleimann's marriage about 1888 to St. Petersburg-born Nikolai Vassiliev, who had received his doctoral degree from the Bern University faculty of medicine in 1887; Vassiliev adopted Swiss citizenship in 1888.[54]

Kleimann's research was a contribution toward extending and clarifying knowledge of the function of a small group of follicles (glands) found in the small intestine. Known as Peyer's Patches, these small aggregations of lymphoid tissue act as part of a mammal's defense system by isolating and attacking toxic invaders. Working with material from rabbits that had died after being given subcutaneous injections of ox bile, a mixture whose components include readily visible colored materials, she noted, in particular, a buildup of yellow pigment in the Peyer's Patches. Also considered were the effects of other pigments such as carmine and india ink.

Marie Vassilieva-Kleimann died in 1893, when she was about thirty-two, probably at the Vassiliev home in Muri, a municipality near Bern.

Tatiana Vaschkevich[55] was born on 7 May 1873 in Voronezh, a city on the Don River about three hundred miles south of Moscow in what was then largely a Cossack region.

After initial studies at the University of Geneva, and a summer session in the medical faculty of the University of Zurich in 1895, she spent about two years, from the autumn of 1896 to the autumn of 1898, at the University of Bern; she also studied for an additional semester in Zurich in the summer of 1897. Concentrating on pathology, she carried out her doctoral research in the Bern Institute of Pathological Anatomy under the direction of Theodor Langhans.

Her work was an early investigation of the mammalian immune system response to infection—namely, the production of high numbers of large cells of unusual shapes (later named lymphocytes) by organs such as the spleen. The phenomenon had been observed in humans, particularly children, in cases of diphtheria, and was also known in a few other conditions. Pathology textbooks of the time had little to say on the subject, but it had been one of considerable interest to Langhans for several years. When Vaschkevich joined his group in 1896, he gave her the assignment of investigating the conditions under which these greatly increased numbers of irregularly shaped cells were produced.

Drawing on a large supply of suitable spleen material, she carried out an extensive study using cells obtained mostly from juveniles who had succumbed to diphtheria. Also included in her investigations was material from a number of adults and from victims of other infections, such as peritonitis following surgical operations and meningitis developed after head injuries; the greatly increased numbers of unusually shaped cells she observed in these cases came from the lymph nodes closest to the infection sites.

A comprehensive understanding of the working of the immune response came many decades after Vaschkevich's time, but her observations fit well with present-day ideas. Her report appeared in Virchow's *Archiv* in 1900 (see bibliography); she received her degree from Bern University in 1899.[56]

ANNA MOGILOVA,[57] later REINSTEIN-MOGILOVA (1866–1948), was born in Włocławek, Poland, about one hundred miles northwest of Warsaw, on 27 February 1866. When she was still a child her family moved to Skvyra, an ancient town in the Kiev Province, central Ukraine, where she grew up. In 1886, after completing her secondary education at a Kiev gymnasium, she enrolled in the medical faculty of Bern University; she studied there for seven semesters over the next four years (1886–1890) and spent an additional semester at the University of Zurich in the summer of 1888.

Her doctoral dissertation research, carried out in the laboratory of the Bern Institute of Pathological Anatomy and directed by Theodor Langhans, concerned the tissue layers surrounding developing mammalian ova at one to two months after implantation. Langhans was well known for important histological work on placenta and investigations of the cellular layers undergoing organization in the fast-growing ovum. Although the cells in the tissue layers surrounding the embryo were initially considered to be entirely of maternal origin, some doubt had arisen after Langhans's observations of some involvement of the cell layers originating in the ovum. Mogilova carried out microscopic studies on tissues from three aborted human embryos, promptly prepared and stained, with the aim of shedding further light on the question of involvement of cells of oval origin. Her results, published in Virchow's *Archiv* in 1891 (see bibliography), would seem to have reinforced Langhans's earlier suggestions. She received her degree in February 1891.[58]

The same year she married Boris Reinstein, a young Russian activist revolutionary then in Bern. They went to Paris but Reinstein was very soon arrested for his alleged involvement in a plot against Tsar Aleksandr III. Anna Reinstein reacted quickly; she moved to the United States and by late 1891 had settled in Buffalo, western New York State, where she opened a practice in gynecology and obstetrics. Her husband joined her about a year later, after his expulsion from France. Both became American citizens.

The family, which in time included two children, appears to have depended largely on Anna's income for financial support, Boris Reinstein's main concerns being his work on behalf of the labor movement and the recruitment of new members to the Socialist Party. He became very prominent in the party, traveled frequently to give lectures and talks, and sometimes spent short periods in jail.[59] He also studied pharmacy at Buffalo College, from which he graduated in 1898.[60] For several years he had a pharmacy in Buffalo, but with the outbreak of the 1917 revolution in Russia he returned to his home country, gave up his United States citizenship, and served in Lenin's government; he died in Moscow in 1947, when he was about eighty-one.[61]

Anna Reinstein continued her popular practice for more than five decades, almost until her death on 21 June 1948, at the age of eighty-two. She was one of the first female obstetrician-gynecologists in western New York State and was well liked for her sympathetic care of poor women, whom she treated regardless of whether they could pay. Her two children, Nadia and Victor, both became physicians, Victor joining her in her practice. Victor also acquired a large area of land in Cheektowaga County, including what is now the Reinstein Woods Nature Reserve.[62]

MARIIA SCHLEE,[63] from Simferopol, Crimea, was born on 14 November 1857. After receiving her earlier education in Simferopol, she enrolled in the medical faculty of the University of Bern, where she studied for eight semesters during the years 1881–1885. Concentrating on gynecology, she became a student of gynecologist Peter Müller, a man who had a particular interest in medical training for women, and physiologist Theodor Langhans, whose investigations included several on problems in gynecology.[64]

Her doctoral research, carried out under the guidance of Langhans and Müller, concerned the stretching and pulling apart during pregnancy and childbirth of the abdominal wall support muscles, the muscles that slope at a descending angle from each side of the abdomen, coming together at the *linea alba* (median line). She also investigated the structural changes in the skin at the sites of the wrinkles and scars that commonly appear during pregnancy due to the stretching of the skin fibers. These were matters that had previously received little serious attention from medical professionals.

Schlee studied fifty-five living patients at the cantonal lying-in institution and also used material from corpses held at the university's Institute of Pathological Anatomy. Her precise measurements included total expansion in both upper and lower regions of the abdomen and the width of the muscle-band split in the downward-sloping muscles; she included practical details on how these measurements could best be made. Her investigations of the skin wrinkles involved microscopic examination of the skin's supporting tissue fibers.

Mariia Schlee's findings were published in *Zeitschrift für Geburtshülfe und Gynäkologie* in 1886 (see bibliography); she received her degree from the University of Bern the same year.[65]

Barbara Kossovich,[66] the daughter of agriculturalist S. [Samson?] Kossovich, director of the Moscow Agricultural Academy, was born in October 1865. After early studies in Moscow, she enrolled in the medical faculty of the University of Bern in October 1887. As well as seven semesters of study there between 1887 and 1891, she spent an additional semester at Zurich University in the summer of 1890.

Concentrating on pathology, she carried out her doctoral research—a contribution to studies on the microcephalic condition—in the Bern Institute of Pathological Anatomy under the guidance of Theodor Langhans. She examined parts of the brain and spinal cord of an infant whose remains, after its two weeks of life, had been passed on to the institute, Langhans's particular interest in cretinism being widely recognized. The infant was the thirteenth child in a family with a history of neurological problems, although the parents and siblings were normal. Kossovich made a detailed examination of both facial features and skull dimensions, recording, for instance, such abnormalities as the size reduction of the frontal, or forehead, bones and the vertical slope of the occipital bone that forms the lower part of the back of the skull. In her work on the brain and spinal cord she noted, among several critically abnormal features, the poor and uneven development of the two hemispheres and the lack of normal nerve connection between them, as well as the stunted appearance of parts of the brain stem connecting with the spinal cord.

Commenting that these findings agreed well with those of other workers in the field, she went on to suggest that the cause of the skull deformities in the case she had examined was a mechanical impediment to growth, since it considerably increased pressure on the frontal part of the fetal head relative to that on the back. She concluded that this pressure difference had produced the deformities that both disrupted brain growth and caused the spinal cord abnormalities.

Barbara Kossovich reported her study in a substantial paper published in Virchow's *Archiv* in 1892 (see bibliography), and was duly awarded her University of Bern degree.[67]

Sophie (Sof'ia) Lubnitzky[68] was born on 18 September 1856. She grew up in Bialystok, northeast Poland, a region then part of imperial Russia. In the autumn of 1879, at the age of twenty-three, she enrolled in the medical faculty of the University of Bern, beginning five years of study there.

Concentrating on pathology, she carried out her doctoral research in the Bern Institute of Pathological Anatomy under the direction of Theodor Langhans. Her investigations concerned the process of clot formation of arterial blood immediately following injury of the artery. The topic was one occupying the attention of a number of medical scientists throughout the 1870s and 1880s, a time when microscopes giving improved clarity and brightness became available.

Research by several workers had indicated that clots are formed by the agglomeration of white blood cells bonded together to form a uniform mass over the wound site. But this view was not universally accepted; one modification proposed was that the clot was formed exclusively by a particular kind of bone marrow cell that developed fibrous characteristics.

Working with a fully grown rabbit, Lubnitzky made a small incision in the neck artery and followed the gradual growth of the resulting clot at times that ranged from twenty-one seconds after wounding to three days. In addition, using a state-of-the-art Zeiss microscope with an oil immersion lens, she studied the form and composition of clots that had developed five minutes after the artery opening and again after a lapse of three days. After five minutes she found red blood cells and a few white blood cells in both outer and inner parts of the clot; after three days the clot was rich in white cells bound together into a fibrous network to whose inner surfaces the red cells adhered—a general picture that comes close to that now held.

Sophie Lubnitzky received her degree from Bern University in December 1884;[69] her findings were published in *Archiv für Experimentelle Pathologie* the following year (see bibliography).

Sofia (Sophie) Hornstein,[70] born on 27 September 1869, was from the city of Kishinev (now Chişinău) in eastern Moldova, in her time the Russian province of Bessarabia. She enrolled in the medical faculty of the University of Bern in the winter semester of 1887 and continued her studies there for five years.

Concentrating on pathology, she carried out a very creditable study in that area for her doctoral research, done under the guidance of Theodor Langhans at the Institute of Pathological Anatomy. She investigated a case of congenital asymmetric growth in an infant, several parts of whose right side were noticeably larger than the corresponding parts on the left; the abnormal growth was clearly apparent in the ears, arms, legs, and feet, as well as in the fingers and toes. The child was the third born to a woman of thirty-seven who had a long history of periods of mental derangement for which she had been treated in the cantonal psychology clinic in St. Urban and in the Waldau psychiatric clinic in Bern.

The infant's deformity was considered to be basically due to genetic predisposition. Following its death at the age of two and a half months in the Bern Children's Hospital, its remains were deposited in the university's Institute of Pathological Anatomy. After noting various problems, such as lymph swellings at various locations, signs of diseased lung membranes, and severe discoloration of the skin, Hornstein went on to make precise measurements of the extent of asymmetry in ears, arms, hands, and fingers. Although the arm bones showed only slightly greater length on the right side, the right hand and digits, particularly the thumb, were bigger than the left. A similar size-variation pattern was found for the lower limbs, the right foot being bigger than the left in all dimensions, with the difference most pronounced in the first and second toes.

Hornstein's microscopic investigations included comparisons of thickness variation between right and left sides for ear cartilage and lobes, and variation in hair, skin, leg bones, and metatarsal bones of the feet. She also examined nerves of both central and peripheral systems, where she seems to have observed fairly small differences between the two sides; muscle thickness variation was greater, the increase being on the right side. For these studies she found it essential to use particular staining techniques in slide preparation and an up-to-date Zeiss microscope, as her compatriot Barbara Kossovich was doing (see above).

From all these observations she concluded that the asymmetry was due, for the most part, to differences in the soft tissues, particularly the connective tissues in the skin, the internal membranes, and the fatty tissue; bone-thickness variation played a lesser role. These findings are far from inconsistent with present-day ideas; later research has also shown that connective tissue disorders are the underlying cause of various conditions, such as the sarcoma type of cancerous tumor, that have a genetic component.[71]

In her general remarks she pointed out that, while there were many literature reports of excessive growth in fingers and toes on one side, and a lesser number recording asymmetric growth of an arm and a leg, there were to date fewer dealing with investigations of asymmetry of a whole side, from ear to toe.

Sophie Hornstein received her degree from Bern University in May 1893;[72] her report of her research was published in Virchow's *Archiv* the same year (see bibliography).

Students of Hugo Kronecker

Sof'ia Handler, Juliia Divine, and Pélagia Beschasnova were three of the students who took part in the investigations of heart function carried out by Hugo Kronecker in the Bern Institute of Physiology, the widely known and much-respected research facility that Kronecker built and led for about three decades. Although Kronecker's work covered many topics in different research areas, his main interest was cardiac physiology. Using isolated perfused turtle-heart preparations, he had introduced, by the early 1870s, a modification of earlier investigative techniques that allowed rapid exchange of the perfusion fluid; this modification may be said to have paved the way to establishing the effect of blood and its constituents on heart function, and thus to the understanding of cardiac metabolism. Kronecker and his coworkers (as well as other leading physiologists of the time) were active in this area for many years before the modern picture of cardiac metabolism was gradually made clear.[73] Handler, Divine, and Beschasnova all made modest contributions to the effort.

SOF'IA HANDLER (GANDLER)[74] from Kiev, Ukraine, was born on 28 September 1855. A student in the medical faculty of the University of Bern from 1881 to 1886, she concentrated on physiology and, following Kronecker's move to Bern in 1884, carried out her doctoral research under his guidance. She set out to determine the extent of the dependence of heart-muscle power on blood-oxygen supply, and in so doing faced the difficulty of finding a way to reliably measure both the amount of work performed by the heart muscle in the contracting-resting cycle and the quantity of oxygen consumed during that cycle. Several pages of her dissertation were devoted to describing an apparatus that permitted the oxygen content of the circulating blood to be measured spectroscopically. With help from Bern craftsmen, Kronecker had succeeded in improving an earlier version of this apparatus, previously described by his London physiologist friend Gerald Yeo;[75] it consisted of a fairly complex system of interconnecting tubes and reservoirs, one of the connections going to a mercury manometer.

Working with isolated frog and turtle hearts perfused with variously modified rabbit or calf blood, Sof'ia Handler concluded that the contracting power of the muscle remained strong for a time after the oxygen supply was reduced; that is, a continuous, steady supply of oxygen was not necessary for heart-muscle activity.

Handler received her degree in 1886;[76] her report appeared in *Zeitschrift für Biologie* in 1890 (see bibliography).

Of British family background, JULIIA DIVINE[77] was born in Moscow on 16 July 1868. She completed her course of study at a Moscow gymnasium for young women and enrolled in the medical faculty of the University of Bern in the winter semester of 1893–1894. Her studies continued until 1899, except for a break during the summer of 1897, when she spent a semester at the University of Zurich.

Divine's doctoral research on heart-muscle function under Kronecker's direction came a decade after the work of Sof'ia Handler, but followed the same general lines. Using isolated turtle and frog hearts, she found that a heart perfused with calf's blood free from or very poor in oxygen (saturated with hydrogen or carbon monoxide) performed just as well as it does with arterial blood, as measured by pulse volume. On the other hand, blood saturated with carbon dioxide quickly reduced the performance of the heart, although the heart recovered when treated with carbon dioxide–free blood.

These results, reported by Kronecker at the Fourth International Congress of Physiologists in Cambridge in 1898, were somewhat controversial and did not agree very well with the ideas of a number of other prominent physiologists. Several short reports of Divine's work, including Kronecker's presentation at the 1898 Cambridge Congress, appeared in print in 1898–1899 (see bibliography); her full dissertation was published in *Zeitschrift für Biologie* in 1906. She received her degree from Bern University in 1899.[78]

PĖLAGIA (DMITRIEVNA?) BESCHASNOVA (Mme),[79] from St. Petersburg, was born in 1860. In 1889, in her late twenties and by then the wife of St. Petersburg government official P. Beschasnov, she began medical studies during the summer semester at the University of Zurich. After this introduction to the subject she appears to have withdrawn from formal studies for five years, but by the summer of 1894 she had enrolled at the University of Bern; with further occasional periods of interruption, she attended lectures and took part in research there over the next eleven years.

Mme Beschasnova's studies of isolated frog-heart performance focused on the effects brought about by perfusion of the heart with calf's blood diluted with sodium chloride solution containing very small amounts of other salts. She found that the diluted blood gave a steady rise in heart pulse rate, in contrast to that observed when concentrated blood was used.

Her report, presented by Kronecker at the 1898 Cambridge Congress, also appeared in short communications in Swiss and German journals in 1898 (see bibliography). She remained at the University of Bern until 1905, when she left without graduating.

NADEZHDA (NADINE) LOMAKINA[80] also collaborated in Kronecker's heart-function investigations. Born in September 1873, she was from Moscow. At age twenty she enrolled in the medical faculty at Bern University and studied there for six years (1893–1899), except for the summer of 1897, which she spent in the University of Zurich's medical faculty.

Her research with Kronecker at the Bern Institute of Physiology concerned the very rich nerve networks of the mammalian heart; she focused on the nerves along the sides of the descending branch of the coronary artery and those on the outer edge of the left ventricle, all of them nerves that are capable of carrying powerful impulses. Her work involved the tying off of parts of the system and observing the resulting changes in pulse rate. She pointed out that work on the physiological importance of these nerves was still in its early stages and that more detailed research remained to be done.

Lomakina's study was one of the four reports of work by his students that Kronecker presented at the International Congress of Physiologists in Cambridge in 1898. Her full discussion appeared in a fifty-two-page paper in *Zeitschrift für Biologie* in 1900 (see bibliography); she was awarded her doctoral degree by the University of Bern the same year.[81]

LIUDMILLA SCHILINA[82] was born in 1872 in Krasnoyarsk, about four hundred miles from the Mongolian border. Once a place of exile for political dissidents, Krasnoyarsk is now one of Siberia's major cities. By the mid-nineteenth century it was the administrative center for the Yenesi district; gymnasia for both boys and girls were among the city's prominent institutions.

Schilina studied medicine at the University of Bern for eight semesters from 1893 to 1898, and for an additional semester at the University of Zurich during the summer of 1896. Concentrating on physiology under the guidance of Hugo Kronecker, she carried out her doctoral research in the famous Bern Institute of Physiology. Her work was a comparison of the performance of two physiological instruments that used graphical methods: Ludwig's kymograph, the first such instrument invented and in use since 1855, and Hürthle's more recently devised tonometer. Her goal was to find out how these compared in measuring and recording blood pressure over a given range. She concluded that each instrument had its own particular disadvantages and drawbacks, gave dependable results over limited ranges only, and, in the case of the tonograph, under carefully specified conditions.

Kronecker reported Schilina's results at the Fourth International Congress of Physiologists in Cambridge in 1898. The paper was timely, the use of graphical instruments in physiological research being one of increasingly wide interest among physiologists at the time. It fitted well with a presentation by Paris physiologist E. J. Marey, which focused on the urgent necessity, if progress was to be made in physiology, of establishing a standardized measurement scale and a standardized method of calibration to evaluate precision in instruments that used graphical recording instruments. Kronecker became a member of the international committee then set up to formalize the rules Marey recommended.[83]

Schilina published a detailed fifty-three-page report of her research findings in *Zeitschrift für Biologie* in 1899 (see bibliography), and was awarded her degree by the University of Bern the same year.[84]

SARAH AMITIN (AMITINE),[85] the daughter of Samuel Amitin, was born in Gomel, southeastern Belarus, in March 1869. She enrolled in the medical faculty of Bern University in the summer of 1890 and studied there for nine semesters, concentrating on physiology under the guidance of Hugo Kronecker; an additional semester was spent at Zurich University, where she was a student during the winter of 1892–1893. Her dissertation research, carried out in Kronecker's laboratory with assistance from then docent, later professor, Leon Asher, concerned the effect of temperature change on blood vessel tone, that is, the degree of tension in blood vessel walls and therefore the amount of constriction imposed on blood flow; results were published in *Zeitschrift für Biologie* in 1897 (see bibliography). Sarah Amitin received her doctoral degree in 1896.[86]

NADEZHDA (NADINE) POPOVA,[87] the daughter of R. Popov of Rostov-on-Don, was born on 1 October 1860 in Taganrog, a port city on the Sea of Azov. In 1881, following studies at a gymnasium in Rostov, she enrolled in the medical faculty at Bern University; she remained there until 1887, except for a summer semester (1885) when she attended lectures at Zurich University. Her Bern dissertation research on the formation of plasma albumin in the gut was published in a short preliminary note coauthored by

Kronecker in 1887, and as a full twenty-five-page paper under her name only in 1889 (see bibliography). The importance of the findings is emphasized by the inclusion of the paper in the serial *Reprints on Physiology*.[88] She received her degree from Bern University in December 1887.[89]

Students of Maximilian Flesch

Elena (Helene) Konev,[90] Anna Gitiss,[91] and Anna Kotlarevskaia,[92] three of the Russian women who received their medical degrees from the University of Bern in 1886 and 1887, form a distinct little subgrouping. For their doctoral research all three took part in an investigation of ganglia, the aggregations of neurons, or nerve cells, that lie along a vertebrate's nerve system and control nerve function. Their combined efforts constituted a significant contribution to the field and were noted in commentaries on its early development.[93]

Carried out in the Anatomy Institute at Bern's School of Veterinary Medicine, the research was directed by German anatomist Maximilian Flesch, leader of the institute from 1882 to 1887; it was one of the last investigations he directed before he returned to his native country and resumed work as a physician.

The basic problem Flesch set before these three students was that of finding the best available tools and methods with which to examine ganglia. The work concerned the peripheral nervous system exclusively—that is, the part that carries messages from the brain and spinal cord to other organs. Variation in ganglion form and structure, as well as dye-absorbing capacity, had already been observed, but whether this variation reflected differences between cells or resulted from their earlier treatment during the investigations was unclear.

Starting out from Helene Konev's initial work on sample preparation, including staining and hardening techniques, Anna Gitiss went on to investigate two particular ganglia, the spinal nerve ganglion and the trigeminal or Gasserian ganglion. She determined the most advantageous succession of preparation steps to allow maximum differentiation of ganglion structure for nine animal species, from rat and dog to pigeon, grass snake, and lamprey. Optimum procedures varied across these species. Also working with two particular ganglia, Anna Kotlarevskaia, the third member of the group, extended the search for the best methods of slide preparation to produce good pictures within a reasonable time after the test animal's death. She also gave special attention to the physiological reaction of the ganglion tissue to the stains used; this included trials with dye injection into living test animals. The studies of these three students led to the conclusion that there were two distinct cellular types in neural centers (chromophilic and chromophobic cells), which differed in chemical composition and which therefore had diverse functions.

Brief background notes follow, but no information about any later professional careers of Konev, Gitiss, or Kotlarevskaia has been uncovered.

HELENE KONEV, the daughter of St. Petersburg businessman O. Konev, was born in St. Petersburg in 1858 or 1859. When she was about twenty she enrolled in the medical faculty of Bern University and studied there for nine semesters, with an additional semester in the medical faculty of Zurich University in the summer of 1883. Her doctoral research on the peripheral nervous system was reported in the Bern *Mittheilungen der naturforschenden Gesellschaft* in 1886 (see bibliography); she received her degree the same year.[94]

ANNA GITISS was born on 29 October 1861 in Kiev, Ukraine, where her father owned a business. She enrolled in the medical faculty of Bern University in 1882 and studied there for eight semesters. Like Helene Konev, she also visited Zurich University, taking classes there during the 1885 summer semester. Her dissertation research on the comparative anatomy of the peripheral nerve system, part 1 of the "Communications on the Anatomy of the Nervous System" from the Bern Anatomy Institute, was reviewed promptly and in detail in *Zeitschrift für wissenschaftliche Mikrosopie*.[95] Her full report appeared in the Bern *Mittheilungen* in 1887 (see bibliography); she received her degree the same year.[96]

ANNA KOTLAREVSKAIA was born in Moscow on 21 August 1859. She studied in the medical faculty of Bern University for nine semesters and received her degree in 1886.[97] Her contribution to the ganglion investigations, her dissertation research, formed part 2 of the "Communications on the Anatomy of the Nervous System" from the Bern Anatomy Institute. Like Gitiss's dissertation, it was reviewed promptly and in detail in *Zeitschrift für wissenschaftliche Mikroskopie*.[98] Her full report appeared in the Bern *Mittheilungen* in 1886 (see bibliography).

REBEKKA (REBECCA) HALPERIN[99] was born on 9 July 1865 in Elizavetgrad (later Kirovograd), central Ukraine, then a fairly important trade center and a city of Ukrainian cultural activity. After gymnasium-level studies in Elizavetgrad, she enrolled in the medical faculty at Bern University and studied there for nine semesters from 1883 to 1887.

Her dissertation research, carried out in the Bern School of Veterinary Medicine and directed by Maximilian Flesch, concerned cases of spinal deformation observed at the embryonic stage or later in cattle and goats. She examined three preparations of tissues from embryos showing varying degrees of abnormal spinal curvature and was led to the conclusion that the deformations originated not in the development of the embryonic spine itself but from a local, external source—a congenital irregularity in part of the mother animal's abdominal wall. The work was reported in a German journal of veterinary medicine in 1889 (see bibliography), a year after Flesch had given up his professorship in Bern and moved to Heidelberg. Rebekka Halperin received her degree from Bern University the same year.[100]

VICTORIA FAMILIANT,[101] the daughter of Leon Familiant, a merchant in Simferopol, Crimea, was born in Simferopol on 24 July 1857. She had at least one sister, Dorothea Familiant (1854?–1931), who attended a gymnasium in Odessa and then studied in the medical faculties of both Zurich and Bern Universities. Victoria Familiant entered the University of Bern faculty of medicine in 1881, after completing her gymnasium education in Russia. Six semesters of study and a lengthy period of research brought her a doctoral degree in 1885;[102] she appears to have stayed on at the university until 1886.

Her research was carried out at the Anatomy Institute of the Bern School of Veterinary Medicine under the guidance of Maximilian Flesch. The institute having obtained from a "Bohemian Menagerie" the brain of a five-year-old lion that had died of bone disease, Victoria Familiant was assigned to carry out a comparison between the brain folds in this animal and those of other carnivores as well as those of primates. The brain folds of the more common carnivores, such as dogs, cats, and foxes, were already well documented in the literature, and maps of the human brain were also readily available; the anatomical details of a lion's brain were incomplete.

Familiant recorded a considerable amount of detail about the lion's brain and compared its convolutions with those observed in dog and cat brains; she was also able to compare the lion's brain features she had observed with the corresponding features in the brains of a fifty-year-old human suicide and a nine-year-old microcephalic boy. Her results were published in a lengthy paper in the Bern *Mittheilungen der naturforschenden Gesellschaft* in 1885 (see bibliography). Comparative anatomy, a field that was still in its initial stages in Familiant's time, has remained a subject of interest and activity as more adequate investigative tools and methods have been developed.[103]

EKATERINA IVANOVNA VON KOVALENSKAIA (VON KOVALENSKAIA-VON KOMPANEISKAIA)[104] was from Starobel'sk, Ukraine, and also had connections with Kerch in the Crimea. She was born on 7 December 1844. When she was about thirty-nine, having already completed preliminary studies at the University of St. Petersburg, she enrolled in the medical faculty of the University of Bern for the winter semester of 1883–1884. Specializing in anatomy, in particular the anatomy of the brain, she carried out a comparative study on the microanatomy of the cerebral cortex of man and a number of other mammals. Her studies form an early contribution to the science of brain mapping and the relating of particular physiological functions to specific regions of the brain.

Kovalenskaia worked under the guidance of anatomist Maximilian Flesch in the Bern School of Veterinary Medicine, continuing studies in the brain-anatomy field in which her fellow Russian student Victoria Familiant (see above) carried out dissertation research. Familiant received her degree a year before Kovalenskaia, but these two students of Flesch would most likely have worked in the Anatomy Institute about the same time and may well have shared ideas and observations.

Using microscope sections she had prepared from human, ape, cat, and dog brains, Kovalenskaia studied the folds of the occipital lobe, the part of the brain located at the posterior of the cortex that organizes an animal's perceived information about its surrounding environment. Her concluding remarks expressed the well-justified opinion that much work remained to be done in brain function research, still an area of great interest to twenty-first century neurologists.

Flesch presented her detailed report at a meeting of the Bern Naturforschenden Gesellschaft in 1886; her full paper appeared in the Bern *Mittheilungen* (see bibliography). Kovalenskaia received her degree in 1886.[105]

Students of Emil Theodor Kocher

SOF'IA HASSE (1847–1925),[106] later ERISMANN, was the daughter of Jacob Hasse, descendent of an old and distinguished German-Lithuanian family, then a resident of St. Petersburg. She was born in that city in 1847 and completed her gymnasium studies there. In 1872, in her midtwenties, she enrolled in the University of Zurich medical faculty, but by the winter semester of 1873, because of the tsarist government's edict concerning Russian women students, she had moved to Bern University faculty of medicine. She received her degree in July 1876, following three years of study and research.[107] Her investigations, studies of the causes of blood poisoning, were carried out under the guidance of the distinguished surgery professor Theodor Kocher, widely known for his promotion of aseptic methods well before the necessity of employing these precautions was generally accepted. She also coauthored a note on the rate of circulation in the spleen (see bibliography). This study, carried out with her fellow Russian student Olga Stoff (see above), was done in the laboratory of Zurich histology and anatomy professor Heinrich Frey during the autumn of 1872. As their contemporary Franziska Tiburtius later recalled, Hasse and Stoff were close friends and were habitually found together.[108] Frey noted that they worked with great perseverance on a difficult topic.

In 1884 Sof'ia Hasse became the second wife of Swiss public health doctor Friedrich Huldreich Erismann (1842–1915), the former husband of physician Nadezhda Suslova (see above). Sof'ia Hasse had first met Erismann in her student days during one of his visits to Zurich and Bern, and in the latter years of his largely incompatible and disintegrating marriage had corresponded with him frequently. Erismann, who had followed Suslova to Russia in 1870, was, by 1884, well established in his outstanding career as a pioneer of public health work in Russia. Throughout the 1870s he had held leadership positions on a number of government commissions, such as those investigating the deplorable sanitary conditions in the Russian army and the health conditions of factory workers in the Moscow region; the latter study resulted in a seven-volume report that came out under Erismann's name. In 1881 Moscow University awarded him an honorary doctorate, and three years later he was appointed professor of public health. In 1891 he received the honor of election to the presidency of the Pirogov Society of Russian Physicians.

During these years of remarkable success in Erismann's career, he and his family lived in Moscow. No indication has been found of Sof'ia Erismann having practiced medicine either before or after her 1884 marriage. But we do know that she had three children, and Tiburtius reports that she had a happy family life.[109] By 1896 the Russian authorities had come to feel that Erismann's activities were becoming too radical, and the resulting pressures made the family decide to leave the country. They settled in Zurich, where Erismann quickly rose to prominence in municipal government; he was particularly active in a wide range of public health matters and in efforts to improve the conditions of the city's working classes. Sof'ia Erismann became a naturalized Swiss citizen in 1903 and lived in Zurich until her death in 1925, ten years after that of her husband.

ANNA BÉGOUNE,[110] later ROUX, from Kherson in southern Ukraine, was born on 12 December 1859. She enrolled in the medical faculty of Bern University in 1878 and, concentrating on surgery, became a student of Theodor Kocher. Her dissertation research, a clinical investigation involving numerous patients, both infants and adults, was part of a long program carried out by Kocher and his coworkers to clarify the relationship of the thyroid gland to goiter, research for which Kocher received the 1909 Nobel Prize in Physiology or Medicine. Anna Bégoune was awarded her medical degree in November 1883;[111] her research findings were published in a lengthy paper in the *Deutsche Zeitschrift für Chirurgie* in 1884 (see bibliography). That year she married Cesar Roux, who had received his degree from Bern University's medical faculty in 1880 and was then working as a physician in Lausanne. Roux later became professor of surgery and gynecology at Lausanne University, a position he held until 1926. Whether or not Anna Bégoune Roux practiced in Lausanne has not been established.

Malka Sof'ia Gomberg,[112] later Tienes-Gomberg, from Elisavetgrad, now Kirovograd, in central Ukraine, was born on 1 October 1865. She studied in the medical faculty at Bern University from 1888 until 1894. Like her fellow Ukrainian Anna Bégoune (see above), she concentrated on surgery under the guidance of physiologist and surgeon Theodor Kocher, head of the university's surgical clinic. Her dissertation research, carried out at Kocher's suggestion with further direction from physiologist Hugo Kronecker, was an investigation of whether there was an advantage to using blood transfusions during surgical procedures carried out under chloroform anaesthesia. The subject was one of considerable interest to research workers at the time. Gomberg's studies, done in Kronecker's physiological laboratory, involved trials on sixteen test animals; results appear to have been varied. Her report was published in the *Deutsche Zeitschrift für Chirurgie* in 1894 (see bibliography). Gomberg was awarded her degree in July of that year.[113]

She was one of only twenty-six women physicians practicing in Germany who had qualified before 1900, most of them at the University of Zurich. Before 1910 she passed the German state licensing examinations and obtained full German physician certification (under her married name of Tienes-Gomberg). This freed her from restrictions to which physicians not licensed by the state were bound, although they were allowed to practice. By 1911 she had established a practice in Kassel, Hesse.

Other Medical Students at Bern

Rosa (Rosalie) Simonovich[114] occupies a notable position in that gradually increasing flow of Russian women who received medical degrees from Bern beginning in the mid-1870s. She was the first to be awarded the degree, conferred in 1874.

From the Ukrainian city of Odessa, Rosa was born on 20 June 1850, one of at least four children of Paul Simonovich of Odessa. She had a private early education, and, no doubt encouraged by the success of Nadezhda Suslova in obtaining medical training at Zurich University (see above), she enrolled in the medical faculty there for the winter semester of 1869. Accompanying her to Zurich was her brother, Mikhail Simonovich, then a student at the Novorossisk University in Odessa; in 1872 her sister Theolina followed. Both Mikhail Simonovich and Theolina also enrolled in the medical faculty, but Mikhail left Zurich in March 1870.[115]

It is perhaps of interest to note that Rosa Simonovich's first boarding house was the Fritschi Haus, so named after its Swiss landlady. This boarding house, where she lived from the autumn of 1869 through the summer of 1871, was the center of studies of socialist ideas, political economy, and the rise of workers' movements. The "Fritschi Circle" of ten to twelve students formed in 1871–1872 was the starting point for the All-Russian Social Revolutionary Organization, and a number of its student members, after they returned to Russia, were condemned to prison or exile in the government's roundup of young dissidents and revolutionary propagandists in 1877. Although there is no readily available record of Simonovich's taking part in Fritschi Circle discussions, she would at least have come into contact with those who did, and with their ideas. By the following winter semester she had moved to a different boarding house, and from the winter of 1872 until the autumn of 1873 she lived in yet a third, namely the "Bremerschlüssel."[116] The latter was the headquarters of supporters of the radical revolutionary Mikhail Bakunin. It also housed the Bakunin library of socialist books and other literature; Rosa Simonovich was one of the first members of this library. She and her sister also assisted revolutionary and writer Mikhail Sazhin (Armand Ross) in the completion of his 1873 work *The Historical Development of the International: Statehood and Anarchy*.[117]

Rosa Simonovich's studies in Zurich included joint research with fellow Odessa student Anna Kleimann (see above) carried out under the direction of physiologist Ludimar Hermann. Their investigation of the effects of tartar emetic on dogs was published in Pflüger's *Archiv* in 1872 (see bibliography). In 1873, when Russian women students were directed to leave Zurich, Simonovich and two other Russian women who were close to completing their studies petitioned St. Petersburg to grant them permission to remain in Zurich until 1874.[118] This was granted in order to let them take their doctoral examinations. Nevertheless, despite the dispensation, Rosa Simonovich moved to Bern University.

She enrolled in the Bern medical faculty in the autumn of 1873, two years after the institution opened its doors to women, and studied there for two semesters, carrying out research to fulfill doctoral degree requirements under the direction of Heinrich Dor, the first professor of ophthalmology at Bern University. Appointed in 1867, Dor filled the position for less than a decade, resigning in 1876 because of what he considered inadequate facilities at his clinic.

Simonovich's study concerned the importance of the alkaloid hyoscyamin, an isomer of atrophine and a possible alternative to atrophine for use in the temporary paralysis required in pupil dilation in ophthalmic work. Following Dor's suggestions that she look fairly closely into the possibilities for hyoscyamin, Simonovich, after careful review of the results of earlier workers in this area, carried out a long series of trials. Working with a number of human subjects but mainly with material obtained from various other sources, she measured the dilation caused with great accuracy but obtained inconclusive results. In a short footnote Dor stated his opinion that accurate chemical analysis of hyoscyamin samples available from different sources would be necessary before further ophthalmic trials could usefully be attempted. As matters stood, sample purity was questionable, and the influence of other active compounds present as contaminants could not be ruled out. Simonovich published her report in 1874 (see bibliography), the year she was awarded her degree.[119] Her graduation, a noteworthy event in the history of Bern University,[120] was noted in the liberal newspaper *Der Bund*, Switzerland's leading newspaper at the time, when the country's political and cultural environment was somewhat polarized. No information about her subsequent career has been uncovered.

Although she did not publish any article indexed in the Royal Society's *Catalogue*, MARIE SIEBOLD[121] (1847–1918?), who received a medical degree from Bern University in 1874, six months after Rosa Simonovich, nevertheless deserves inclusion in this survey. Siebold is notable not only for her early graduation but for her most eventful and remarkable later career; her story covers a span of more than four decades, until at least the first few years of World War I, when she carried out medical work in the Balkans, in Turkey, and in Eygpt, in whichever locations the authorities permitted her to practice.

A member of the many-branched German family of Siebold, she was one of eight children of Johann Friedrich Siebold from the city of Hitzerode, some twenty miles south of Gőttingen.[122] The name of her mother, wife of Johann Friedrich, is not known. In 1838, at the age of twenty-four, Johann Siebold immigrated to St. Petersburg, where he earned a living for twenty years by giving private music lessons. In 1858 he was appointed music instructor and organist for the St. Petersburg French Reformed Church; somewhat later he held the same offices in the German Reformed Church. Born on 18 January 1847, Marie Siebold attended the St. Petrischule, a German school attached to St. Peter's Lutheran Church. The Petrischule was the oldest school in St. Petersburg, founded about 1710, a few years after Peter the Great sanctioned the establishment of Protestant churches in that city. In 1858 the school became a gymnasium, which included a women's gymnasium section. Academic standards were high, the teaching of modern languages, sciences, and the arts being especially notable. The school was widely known in the West.[123]

After graduating from the Petrischule, Marie Siebold enrolled in the medical faculty of Zurich University, beginning in the autumn of 1870. Her studies there continued until 1873, when Tsar Aleksandr II's decree required Russian women students to leave Zurich or face serious future restrictions. The winter semester of 1873–1874 saw her enrolled in the medical faculty of Bern University, where one more year of study, and completion of dissertation research in gynecology and internal medicine, fulfilled her graduation requirements. Her work was carried out under the direction of the gynecologist Peter Müller (1836–1922), an outstanding teacher and the author of several gynecology textbooks; Müller also had a strong interest in the question of medical training for women.[124] Siebold's dissertation concerned the changes occurring during pregnancy, including the influence of the extent of initial uterine expansion on its rate of eventual shrinkage. No journal report of her findings was found, only a twenty-four-page manuscript pamphlet. She received her degree from Bern University on 2 December 1874.[125]

After returning to St. Petersburg in 1875, Marie Siebold was unable to establish a practice because her foreign degree was not recognized by Russian medical authorities; only in national emergencies were women allowed to take part in work that was normally the province of male physicians. Two years later, with the outbreak of the 1877–1878 war between Russia and Turkey, and the epidemic of cholera that accompanied it, such an emergency arose. Marie Siebold applied for and was granted a position as assistant physician in a military hospital. After satisfying surgery standards, she was designated chief physician with captain's rank in a Serbian military hospital—the Serbs, allies of Russia, being willing to accept the placement of male assistant physicians under a female chief although the Russians were not.

After the signing of the 1878 Treaty of San Stefano, which marked the defeat by Russia of the Turks in the Balkans, Marie stayed on in Belgrade, where she opened a private practice. This was possible only through the good offices of Queen Natalia, whose husband, King Milan I of Serbia, elevated Siebold to the ranks of the aristocracy by conferring on her the Order of Takovo, 5th Class, the class of this service medal that conferred a knighthood.[126] She was also awarded the medal of the Serbian Red Cross, an order recognizing exceptional merit in service to the Red Cross in both peace and war. After the introduction in 1880 of an ambulance service at the Belgrade public hospital, Serbia's first such city service, the numbers of sick arriving for treatment overwhelmed the available staff. Siebold was able to ease the difficulty and for three years gave up her private practice to join the three Serbian physicians at the hospital. While there she came to know Rasa Milosevic, husband of University of Zurich graduate and Serbia's first female physician Draga Ljocic.[127] Although she did not sympathize with Rasa Milosevic's radical political views, Siebold was firm in her support of fellow physicians. Her attempt to assist Milosevic when he was imprisoned resulted in her expulsion from Serbia, a punishment that was not reduced by her service in a military hospital during the disastrous Serbian campaign against Bulgaria in 1885–1886.

By 1888 Marie Siebold was in Constantinople. Since women could not legally engage in medical work, despite the recognized need for female practitioners for the Muslim population, her clientele was limited to Western women resident in the city, mainly people attached to embassies. But as her skills in surgery and childbirth assistance became known she was also called to help in the households of the Sultan's brother and that of the heir to the throne, service that led to her appointment as physician responsible for the health of the women of the Sultan's harem, a high honor. It is widely accepted that, after seventeen years in this position, Siebold became the victim of a harem intrigue, which, to the regret of many, led to her expulsion from Turkey about 1905. Various stories were circulated about her, including a claim that she performed an abortion to save a woman's life, which was illegal under Turkish law. None of the stories seem to have been proved. But, as was pointed out by the *Lancet* correspondent in Constantinople in 1908, her standing was not legal according to Turkish law, and only because of her outstanding work was her presence tolerated by the authorities for two decades.[128] Perhaps she was fortunate in being allowed to stay as long as she did.

Hoping to return to Belgrade and start a practice, she applied for Serbian citizenship. But the Serbian government denied the request, having heard through the German embassy in Constantinople that she had been accused of contravening medical ethics and other offenses. Finding herself without any cooperation in her efforts to continue her professional work in the Balkans, and running out of money to live on, she took a hospital position in Ramleh (now Ramle, Israel). Later she worked as a surgeon in Cairo, and in 1912 she became chief physician in the Children's Health Center in Tanta, about sixty miles north of Cairo, already then a city of one hundred thousand inhabitants. Conditions in the center were extremely primitive; much of the city's population was superstitious, mistrusted Western medicine, and held to their beliefs in wonder cures. The few Europeans stationed there went to Cairo to see a doctor. This left Siebold with few patients other than babies and small children, most of whom, after an initial successful visit, were never taken to see her again.

At the outbreak of war in 1914 she went back to Serbia and, although then about sixty-seven, volunteered for fieldwork. She was appointed head doctor in a military hospital in Albania, a region in a chaotic state, with Serbia controlling much of the north while the Greek military occupied the south. There was much to do, with sanitary conditions to bring under control and old wounds of patients to attend to. This work ended when, with the rest of the hospital, she was captured by Central Powers forces and imprisoned in Bulgaria.[129] Released before the end of the war, she was sent to Belgrade, where she is thought to have died in 1918, or, according to some statements, in 1938, when she would have been ninety-one.

Marie Siebold was well aware of the value and the pioneering quality of her work, work that was carried out in large part despite strong, powerful opposition. As she wrote from Egypt to her Zurich student friend Franziska Tiburtius, she regretted none of it, the bad experiences as well as the good; she had lived and experienced much, and she cherished all her memories. Tiburtius admired her not only for her skills as a surgeon, but for her deep and sympathetic understanding of others and her abilities in dealing with life's practical problems.

MARIE EKUNINA[130] was born in St. Petersburg on 14 June 1853. A fellow student of Nadine Ziber-Schumova (see below) in Marceli Nencki's Institute of Medical Chemistry at Bern University from 1874 to 1879, she carried out research on the acidic reaction that developed in animal tissue after death. After completion of this work and publication of her findings (see bibliography), she was awarded her medical degree in 1880.[131] No further scientific publications by her have been found, but she may have had a career in private practice.

STEFANIIA (FANNY) BERLINERBLAU,[132] later BERLIN (1850–1921), the daughter of Markus Berlinerblau, was born in Kherson, Ukraine, on 18 May 1850. Markus Berlinerblau later moved to Sevastopol. After completing her gymnasium education she went to Switzerland and by the winter semester of 1870 was enrolled in the medical faculty of the University of Zurich; her studies there continued for three years. The time was one of considerable revolutionary discussion and agitation in Zurich's sizable Russian student community; Fanny Berlinerblau is reported to have strongly sympathized with the radical revolutionary ideas of such men as writer, naturalist, and philosopher Prince Pyotr Kropotkin and writer Sergey Stepnyak-Kravchinskii.[133] She was one of the large group of Russian women students nicknamed *Kosakenpferdchen* (little Cossack horses), characterized by their spirited behavior, nihilist views, and bold, unconventional dress, but there is little evidence that her idealism led as far as action in the form of taking part in political agitation.[134] She was known to her friends as a quiet person, always ready to help others, and fully committed to pursuing the goal she had set herself with the persistence that was also a marked characteristic of these Russian women students.

In 1873, because of the tsarist government's decree that Russian women students leave Zurich or face future severe restrictions on their professional prospects, she moved to the University of Bern. There she quickly dropped all the outward appearance of the *Kosakenpferdchen* and turned herself into an elegant young woman. Her degree requirements still to be completed included dissertation research, work she carried out at the Bern Institute of Anatomy with guidance from anatomy professor Christoph Theodor Aeby (1835–1885). She investigated some of the details of blood circulation in mammals, including humans, by tracing movement of substances from arteries to veins and in this way demonstrating artery–vein connections. Using mostly rabbit-tissue preparations, she followed the pathways of dyes that she had injected into arteries and concluded that such substances did indeed move through the capillaries of the arterial system to tiny veins, and that a sharp boundary between arterial and venous systems should not be envisaged. Her eleven-page report was published in the Leipzig journal *Archiv für Anatomie und Physiologie* in 1875 (see bibliography); she received her degree in February of the same year.[135]

Russia at that point was going through a time of discontent in the universities and much student unrest;[136] Fanny Berlinerblau decided against going home and instead sought opportunities in the United States to further her medical experience. In 1877 she was appointed resident physician at the New England Hospital for Women and Children in Boston, then led by Berlin-born Marie Zakrzewska, who had been instrumental in establishing the institution fourteen years earlier. In 1879 she joined the hospital staff as one of four women surgeons.

Fanny Berlin, the name she adopted after moving to the United States, was one of ten women physicians who in 1878 founded the New England Women's Medical Society; another seven years were to elapse before women were accepted into the Massachusetts Medical Society. An example of her notably up-to-date surgical methods was described in her short paper in the *American Journal of Obstetrics* in 1881 (see bibliography). This reported her successful surgical correction in cases of severely prolapsed uterus using a procedure first described a few years earlier by the distinguished Paris surgeon Léon Le Fort, and applicable in the large number of cases in which pregnancy was no longer likely to take place.[137]

By then this pioneer of women surgeons in the United States was becoming well known for her skills. She soon rose to the position of chief surgeon at the New England Hospital for Women and Children, and she held the post until 1894. Still in her midforties, she was then appointed an honorary surgeon to the hospital, and for the next two decades had a large, private surgical practice in Boston. In the winter of 1911–1912 she made a long visit to Europe, staying for an extended period in Berlin with Franziska Tiburtius, her old friend from her student days in Zurich. Tiburtius reported that Fanny, despite her lengthy residence in the United States, had never come to feel that America was her home, and her long visit was probably a feeler toward a more permanent return to Europe; the city of Berlin attracted her strongly.[138] But the First World War broke out a few years after Fanny Berlin's return to the United States and American-European connections were severely disrupted. She retired completely in 1916, at the age of sixty-six, when her eyesight began to fail, and died five years later, on 4 September 1921, at her residence in the Roxbury district of Boston.

SOF'IA FRENKEL, née RABINOWITSCH,[139] was born on 25 November 1866 in Grodno, then a Russian city a few miles from the Polish border. She enrolled in the faculty of medicine at the University of Bern in 1884, when she was barely eighteen, and studied there until 1889, except for one semester at the University of Zurich in the summer of 1886. She concentrated on pharmacology and carried out her dissertation research under the guidance of Bern University physiologist and neurologist Hermann Sahli (1856–1933), one of whose special interests was blood flow rate, its measurement, and the factors that influence it.

Sof'ia Frenkel carried out a lengthy clinical investigation of the effects of a number of neurologically active compounds, such as caffeine, morphine, and digitalis, in the treatment of patients with a variety of serious health difficulties that in most cases included heart irregularities. One of the major effects she followed was blood pressure change, as measured with the Basch sphygmomanometer. She took some pains to make clear that this recently designed instrument, the first noninvasive and portable device for measuring blood pressure, had drawbacks, one being that it measured only one of the extremes rather than the average pressure. In addition it was difficult to use, and the numbers it produced varied in unexpected ways and could only be considered comparative. Nevertheless, despite its problems, the instrument was the best available for Sof'ia Frenkel's purposes.

Working with twenty or more patients of differing ages and basic fitness, she recorded a great deal of detailed data that included pulse rate, urine production and condition, and intensity level of the patient's major complaints, as well as blood pressure; the information was collected every few days over periods of up to two months during and after treatment with one or a succession of the above-mentioned chemicals. Her findings, published in the *Deutsches Archiv für Klinische Medicin* in 1890 (see bibliography), indicated that the effects of the drugs she used varied considerably and that the improvements to patients' conditions were temporary only.

Frau Frenkel received her degree in 1889.[140] No information about her subsequent life and activities has been uncovered; she may have returned to Grodno or to Warsaw, but she also had a family connection, Herr M. Frenkel, in Wiesbaden, Germany.[141]

DINA SANDBERG,[142] from Pinsk, Belarus, was born on 6 October 1864. She enrolled in the medical faculty of the University of Bern in the winter semester of 1887 and studied there for six semesters, concentrating on internal medicine. Her doctoral research, carried out in Bern University's Institute for Hygiene, later the Institute for Hygiene and Bacteriology, was a work in medical statistics. She examined death rates from tuberculosis for both males and females in regions of England that she classified as agricultural or industrial, the study being an attempt to shed light on possible causes of this common, devastating disease and point toward practical control measures.[143]

Selecting ten-year census data and official, attested death records for England and Wales (which had been collected regularly and systematically from the late 1830s),[144] she deduced death rates for both types of region. The census data provided population numbers and the fractions of the population employed in either agriculture or industry; annual death numbers, for both males and females, from all causes and from tuberculosis, were obtained from the official death records.

Her analysis (using the rates and proportions she had calculated) of the steady decline in deaths from tuberculosis across occupational groups for both sexes showed the following: (1) the decline in death rate had been marked and steady in all categories over the last three decades of the period studied, and the reduction of deaths from tuberculosis had been a leading cause of this decline; and (2) death rates were consistently higher in industrial districts than in agricultural districts, both overall and for tuberculosis. However, decreases were stronger where the death rates were higher; that is, stronger in the industrial districts and much stronger among women. By 1886 death rates among women in agricultural districts were comparable to those for men, and in industrial districts they were markedly lower. The effect was particularly prominent in deaths from tuberculosis.

These findings reflected changes in the work environment in the industrial trades over the time period covered; while the benefit to agricultural workers of employment in the fresh air remained constant, the industrial group, initially employed in closed, crowded buildings, was, from the early 1870s, benefiting from better ventilation in workplaces. The rapid decline in the death rate among females in agricultural districts in the five-year period 1873–1877 Sandberg attributed to the introduction in the early 1870s of state-mandated public health regulations, coupled with dampness control in dwellings by more adequate drainage provisions around building sites; females were affected more than males because the former spent more time inside their houses.

Dina Sandberg's calculations made good use of data available at the time, and her conclusions may well have helped emphasize and speed up the much-needed improvement in public health standards. Her report appeared in *Zeitschrift für Hygiene* in 1890 (see bibliography), and she was awarded her degree from Bern University the same year.[145]

OLGA KOVALEVSKAIA[146] was born in November 1869, in Odessa, Ukraine, where she received her gymnasium education. She enrolled at Lausanne University in 1886 as one of the institution's earliest women medical students, but transferred to Bern for the winter semester of 1889–1890. She became a student of surgeon and microscopist Ernst Tavel of the Bern Institute for the Study of Infectious Diseases, a colleague of surgeon Emil Theodor Kocher, who also carried out much research on infection control and prevention. Kovalevskaia's doctoral research was a basic study of blood infection; she examined the behavior of white blood cells and the process of chemotaxis, that is, the ability of a cell to "creep" toward the source of a diffusing chemical. She received her degree in 1896,[147] following publication of her lengthy paper in *Annales de Micrographie* (see bibliography).

THE UNIVERSITY OF GENEVA

Many Russian women studied midwifery at the University of Geneva from the early 1870s until the start of the First World War, but a research report by only one female medical graduate identified as Russian, KLAVDIA MARKOVA, was found listed in the Royal Society's *Catalogue*.[148]

A student of psychologist and neurologist Édouard Claparède,[149] then a *Privatdozent* at the University of Geneva, Markova carried out an investigation of the part played by tactile sensation in the essentially muscular process of small-object recognition. Working in the Laboratory of Experimental Psychology headed by Théodore Flournoy, she carried out a series of studies that attempted to distinguish between two types of sensation—muscular and tactile. Her findings suggested that touch plays a larger role than previously thought. The study contributed to Claparède's theories of childhood learning, especially the role of play and physical activity during early childhood, a subject of some controversy among specialists at the time.

Markova was awarded a doctoral degree by the University of Geneva in 1900,[150] following publication of her research in *Archives des Sciences Physiques et Naturelles* (see bibliography).

EKATERINA (CATHERINE) SCHIPILOVA[151] did not receive a degree from the University of Geneva, but she carried out several years of notable research there. One of the most productive Russian women neurologists-physiologists of the 1880s and early 1890s, she had wide interests that ranged over a number of the fields in which there was much activity at that time, from protein chemistry in muscle tissues to cranial nerve investigations and work on digestive enzymes.

Schipilova was born in St. Petersburg in 1859. By her early twenties she had moved to the University of Geneva to work and study under the direction of German physiologist Moritz Schiff. Schiff had carried out extensive research on the role of the central nervous system and had published an important textbook on the physiology of muscles and nerves, with which Schipilova was most likely familiar.[152]

Her first publication reported joint work on muscle biochemistry with Aleksandr Iakovlevich Danilevskii, a leading Russian contributor to protein chemistry and enzyme investigations, who spent some years in Switzerland and Germany in the late 1870s and early 1880s. This joint work served as Schipilova's introduction to studies on the nature and cause of rigor mortis, an area in which she made a lasting contribution (see bibliography). Her suggestion that muscle stiffening was brought about by the precipitation of proteins by lactic acid produced after death remained current, alongside one or two alternatives, for several decades. Only in 1943 was it shown that the loss of adenosine triphosphate from muscle after death was the major factor.[153]

During the later 1880s Schipilova carried out a number of extensive studies on the digestive enzymes; she examined the effects of bile and the secretions of Brunner's glands (duodenal glands) on the digestive action of pepsin, the major stomach enzyme. The results reported were somewhat at odds with the ideas of Geneva physiologist Aleksandr Aleksandrovich Herzen, and also with later findings.

Her major interest over a ten-year period was her neurological work on the central nervous system in frogs. She focused on cranial nerves, in particular those known as the VIIIth pair, the set that controls hearing, equilibrium, and sensation. Her 115-page monograph on the work was published in a Swiss journal in 1895 (see bibliography).

Ekaterina Schipilova also had interests in botany and spent some time in the laboratory of Geneva botanist Robert Chodat examining floral irregularities in leguminous plants (see bibliography). Her conclusion, a departure from earlier theories, was that the initial stages of development of the flower buds were quite regular, no one side of the bud developing faster than the other; the irregularities in the mature flower stemmed instead from periods of unequal growth in the various parts during the final stages of development.

After the death of Moritz Schiff in 1896, Schipilova left Geneva. No post-1900 publications by her were found in the usual indexes.

THE UNIVERSITY OF LAUSANNE

Lydiia Egon de Besser[154] was one of three Lausanne University Russian women medical students whose research reports were found listed in the Royal Society's *Catalogue* of pre-1901 scientific papers. The daughter of Aleksandr Egon de Besser, owner of lands in the Kovna (now Kaunas) district of Lithuania, she was born in 1868 in Penza, Russia, about four hundred miles southeast of Moscow; later her home was in St. Petersburg. She received her secondary education at a gymnasium in Riga, Latvia, and in the early 1890s went to Lausanne University, where she enrolled in the medical faculty. In the summer of 1892 she had a semester of study in the medical faculty of Zurich University.

Having a strong interest in anatomy, she carried out her doctoral research in the University of Lausanne anatomy laboratory under the direction of Édouard Bugnion (1845–1939), professor of physiology, anatomy, and embryology. Her work concerned joint and muscle function, one of many areas in which Bugnion made notable contributions.

De Besser investigated two distinct sets of muscles and tendons—those of the hand and wrist, and those of the lower limb, especially the Achilles tendon. For a number of bone positions she first determined the "natural" length of the muscle in question, that is, when neither extended nor retracted; her next step was to measure the degree of lengthening and shortening of the muscle fibers in extended and retracted states. In addition, she determined the extent of muscle retraction after the tendon was sectioned. Precise information about these factors was of more than theoretical interest; it was needed in surgical work involving the suturing of tendons in living animals, including man, after accidental cutting, and in the correction of abnormalities.

In the second part of her research she worked with dogs, rabbits, and frogs, investigating muscle retraction after cutting Achilles and triceps tendons, and tendons controlling flexor and extensor muscles in paws of dogs. In some cases observations were recorded on postmortem muscle movement following tendon sectioning. A four-page summary of de Besser's results from her lower-limb investigations appeared under Bugnion's name, with full acknowledgments, in 1895; she published details of both parts of her work in two papers in 1898 (see bibliography), and received her degree the same year.[155] She returned to Russia, probably to St. Petersburg, but no information about her subsequent life and work has been uncovered except that in 1906 an article that was probably by her appeared in the St. Petersburg Military-Medical Journal; it concerned bubonic plague in Mongolia and the Baikal region in the period 1889–1897,[156] which suggests that she may have had particular interests in public health statistics.

ANNA LUDKEVICH,[157] like her compatriot Lydiia Egon de Besser, was a student of Lausanne anatomy professor Édouard Bugnion. Her doctoral dissertation research, carried out in the late 1890s, was a study in comparative anatomy in which she examined the shoulder joint in man and several other species.

At that time cross-species comparisons of joints and their functioning under the requirements placed on them were still few in number, but Darwin's theories of evolution and species development were spurring interest in the area and providing new insights. Bugnion was already exploring the field fairly vigorously; Ludkevich's detailed investigations across eight species contributed to his studies.

Starting with the shoulder joint of man, and using both vertical and transverse cuts, she measured such features as size and curvature of socket and ball, and recorded cartilage coverage; amplitude of movements and the functioning of ligaments were also noted. Further work included the role of the surrounding bones, such as the shoulder blades, and the attached muscles. Finally she looked at joint movements, including rotation, for subjects of different ages and occupations.

Her horse study generally followed the same pattern, with notes emphasizing special features such as the marked difference in the shape and position of the horse shoulder blade compared to that in man. Similar measurements were made for ox, sheep, pig, dog, and tiger. The shoulder construction of the Barbary ape was found to be of intermediate type between that typical of quadrupeds and that of man—a most satisfactory finding from the evolutionist point of view. Various differences recorded included the strongly curved collarbone, the long, slender humerus, and dissimilarities in shoulder blade structure.

Ludkevich's sixty-two-page report was rounded off by an appendix added by Bugnion, which communicated the results of his examinations of shoulder joints in an orangutan from Borneo and a gorilla obtained from the university medical school collection. He also included observations on exceptional amplitudes of shoulder movement in humans engaged in particular occupations, such as that of boatman. The three sections (Ludkevich's dissertation, Bugnion's eleven-page introduction, and his appendix) together formed a substantial monograph published in the *Bulletin* of the Vaudois Natural Sciences Society in 1899 (see bibliography).

After completing her research, Anna Ludkevich received her degree from the University of Lausanne in 1899.[158]

VERA SOLOMON[159] (b.1872 or 1873) was from the western Ukraine city of Kamenetz-Podolskii, now a cultural center known especially for its annual Cossack Games. At least by the 1890s her mother, Caterina Solomon, had moved to Kichinev, Bessarabia (later Moldavia), then a heavily Jewish population center. Vera Solomon went to Switzerland in the early 1890s and appears to have studied in Geneva for a period before enrolling in the medical faculty of the University of Zurich for the winter semester of 1895. She left Zurich in June of the following year but returned for the 1896 winter semester, only to leave again in the summer of 1897.

She completed her medical studies at the University of Lausanne between 1897 and 1900, carrying out dissertation research in the department of hygiene and bacteriology then led by Italian bacteriologist Bruno Galli-Valerio (1867–1943); Galli-Valerio had only recently taken up the Lausanne professorship, moving from the University of Milan in 1897. Directed by him, Solomon undertook an investigation of sites into which the rabies virus might be introduced into laboratory animals, work whose ultimate aim was the provision of further information on reliable methods for detecting the organism's presence in suspected human cases. A number of workers in various laboratories, such as the Pasteur Institute in Paris, and Galli-Valerio himself had reported studies in this area in the early 1890s.

Transmission attempts had often failed or resulted in the death of test animals from other diseases that possibly originated in unhygienic procedures used, but some had been successful. These included the introduction of virus-containing solutions into a region of the brain, namely, the subdural space, a region that normally has connections with the cerebrospinal fluid. Although this required boring through the top of the skull and carried with it the possibility of infection, Solomon followed Galli-Valerio's suggestions and tried the method; she noted the additional difficulty of being sure she had actually succeeded in entering the subdural space with its connecting channels, rather than injecting into a muscle. Taking careful precautions against infection, she closed the wound with two sutures, washed the site with alcohol, and covered it with cotton soaked in strong acid solution.

The second method she tried involved the introduction of a virus-containing solution into the mucous membrane of the nose, entry being deep enough to bring the virus into contact with glands and nerves; this was ensured by a certain amount of abrasion.

Following up on the implications of the Chinese folk-medicine practice of prescribing the liver of a mad dog to humans at risk for rabies, Vera Solomon also studied the influence of the bile of infected rabbits on the rabies virus; she found that this bile did considerably reduce the effectiveness of the virus, but in her animal tests it only increased survival time by about seven days, an effect she labelled as antiseptic rather than antitoxic.

The work was carried out in the laboratory of the Lausanne Institute for Hygiene, which through Galli-Valerio then had close links with the Pathological Institute of the Veterinary College in Milan; the latter, a highly respected institution established in 1791, was one of Europe's oldest schools of veterinary medicine. Solomon's test animals were mostly rabbits; no mention of sedation was made, the animals simply being held in place by an assistant, an unpleasant reminder of how such research appears to have been conducted at that time.

By and large Solomon's results were good. In the method of introduction into the brain, she initially lost a few animals, but her subsequent reduction of the amount of fluid injected gave fairly reliable results; death from rabies occurred in six to seven days on average in twenty-one trials. When the virus was introduced into the nasal mucous membrane, seventeen trials produced a greater spread of survival time, ranging from four or five days to about two weeks.

Vera Solomon received her degree from the University of Lausanne in 1900;[160] her dissertation research was reported in the *Centralblatt für Bakteriolgie* the same year (see bibliography).

THE UNIVERSITY OF NAPLES

Sof'ia Mikhailovna Bakunina[161] (1870–1956) was the middle child of three of the Russian revolutionary and philosopher Mikhail Bakunin (1814–1876) and his wife, Antonia Kwiatkowska (c.1841–1887), the elder daughter of a Polish businessman, then a political deportee in Tomsk. Sof'ia was born in Locarno, Switzerland, not far from the Italian border, on 14 January 1870. The Bakunin household moved frequently in the course of Bakunin's revolutionary activities, but shortly after his death, in 1876, Antonia Bakunin and her three children settled in Naples, one of the places where they had previously lived. A short time later Antonia, still in her midthirties, married socialist lawyer Carlo Gambuzzi, one of Bakunin's Italian comrades who had followed him to Switzerland and with whom Antonia had had a close liason for some time. Indeed, it was widely believed that Gambuzzi was the father of Antonia's three children; even Bakunin himself is said to have thought so.[162] It should be remembered that the Bakunin-Kwiatkowska relationship may well have been a nihilist (platonic) marriage, common enough among ambitious young Russian women of the time who were eager to grasp this one chance open to them to escape paternal domination. Such relationships were not infrequent across all ranks of nineteenth-century society, and without DNA evidence, paternity questions stay unanswered. In any case, Bakunin's strong, fatherly affection for Antonia lasted throughout his life; he fully acknowledged the children and enjoyed their company during his last days.

Growing up in Naples from the age of six, Sof'ia most likely received her early schooling there. She then studied medicine at the University of Naples, graduating in 1893. Sometime after her graduation and before 1904 she married a well-known Naples surgeon, Giuseppe Caccioppoli (1852–1947), becoming his second wife.

Her research, reported in two papers and a talk at the 1894 International Medical Congress in Rome (see bibliography), was a lengthy, ambitious program of work that originated from a consideration of the then controversial but now almost forgotten "recapitulation theory" of German zoologist Ernst Haeckel (1834–1919). Haeckel's views were based on morphology, but Sof'ia Bakunina's interests were in function and activity, that is, how an organism operates. She set herself to establish an analogy between the life processes (functions) in the succession of embryonic stages in the development of a higher animal and the life functions of simpler organisms in their adult stage. Over the course of two years she examined the processes of secretion and respiration in developing chick embryos. For example, in her series of respiration studies that involved limiting oxygen supply, she followed the succession of recognizable stages of the embryo's growth and development and its accompanying gradual increase in oxygen demand; this allowed her to observe the changes in the mechanism of supply of oxygen throughout these stages. Initially the embryo received its oxygen directly through tissue, as happens in simple protozoa; then followed a stage involving a colorless, circulating liquid, and finally, by the time the chick was close to hatching, its oxygen needs had grown to the point of requiring blood vessels carrying hemoglobin, as is found in the adult form. The research was carried out in the Institute of Pathology at the Santa Maria del Popolo degli Incurabili, a Naples hospital that dated back to the sixteenth century and had always had a teaching vocation; for more than one hundred years it was the headquarters of the College of Surgeons and Physicians, and was the authority that conferred medical degrees.

Sof'ia Bakunina and her husband had two sons, Ugo and Renato. Renato Caccioppoli (1904–1959), an outstanding mathematician, became a faculty member at the University of Padova and later at the University of Naples.[163] Sof'ia Bakunina died in 1956, when she was about eighty-six.[164]

ST. PETERSBURG AND MOSCOW

Relatively few nineteenth-century Russian medical women whose contributions are listed in the Royal Society's *Catalogue* would appear to have carried out the major part of their research in St. Petersburg or Moscow. Two, Ekaterina Shumova and Maria Manasseina, are fairly well known; briefer accounts about the work of two others—Nadezhda Schultz and Praskoviia Tsiklinskaia—are offered, and a few notes on three others are added.

Ekaterina Olimpievna Shumova,[165] later Shumova-Simanovskaia (1852–1905), received her medical training and her degree at the University of Bern. On that basis she might have been grouped with the other Bern medical graduates in this survey, but her subsequent notable research career in physiology and neurology was in St. Petersburg institutions.

The elder of two remarkable sisters, the daughters of Olimpie Shumov, Ekaterina was born in 1852, probably in St. Petersburg.[166] The Shumovs were a middle-class family and comfortably situated. Ekaterina was taught at home during her early years and may have attended St. Petersburg Higher Courses for Women, as did her sister, Nadezhda (see chapter 2). In 1872, when Nadezhda accompanied her husband, economist Nikolai Ziber, to Zurich, Ekaterina went with them and enrolled in the medical faculty of Zurich University. She moved to Bern the following year, when Tsar Aleksandr II's edict warned Russian women students to leave Zurich or face severe future restrictions. After four years of study in Bern's medical faculty and completion of doctoral research on changes in urine, she was awarded her degree in 1877.[167] She returned to St. Petersburg the same year.

Ekaterina Shumova passed the required Russian physicians' licensing examination in 1879 and then joined the staff of the Medical-Surgical Academy; for a time she was also involved with courses for women doctors at the St. Petersburg Nikolaevski Military Hospital. Her early research was carried out in one of the smaller Medical-Surgical Academy physiology laboratories, probably one managed by Ivan Pavlov; she studied the physiological effects of various chemicals such as camphor and chloral hydrate, measuring the latter's influence on blood pressure. These investigations were recorded only as notes in St. Petersburg clinical publications and not in major journals.[168]

Sometime between 1877 and 1884 she married Nikolai Petrovich Simanovskii, then a *Docent*, later professor, at the Medical-Surgical Academy. Simanovskii's main area of interest was vocal function and the treatment of voice disorders, but he also joined Shumova in a lengthy study of the effects of alcohol and morphine on a physiological oxidation process. Their joint report, probably Shumova's first publication in an internationally important journal, appeared in 1884 (see bibliography).

She went back to Bern in 1882 for further research experience. During her two-year visit she studied fat synthesis in the laboratory headed by Polish physiologist and biochemist Marcellus Nencki. By then her sister, Nadezhda Ziber, had become one of Nencki's coworkers, having stayed on in Bern after receiving her degree there in 1880.

Shortly after returning to St. Petersburg, Shumova-Simanovskaia joined Ivan Pavlov in his research on the physiology of digestion. At that time the consensus among physiologists was that the process of mammalian digestion was initiated by the presence of food in the stomach; that is, secretion of gastric juices resulted from a mechanical effect in which the central nervous system played no role. The Pavlov-Shumova-Simanovskaia team took an approach to physiological research that was new and fundamentally different. They used live animals (dogs), rather than the poisoned animals employed in earlier work, and developed the new laboratory procedures and techniques that were essential for success. Shumova-Simanovskaia played a major role in working out the method known as esophagotomy—a procedure that prevented swallowed food from reaching the stomach.[169] Using a combination of esophagotomy and a previously devised technique for draining off gastric secretions, Pavlov and Shumova-Simanovskaia were able to demonstrate that appetite, a brain signal transmitted to the stomach via the vagus nerve, causes gastric secretion without any food being present in the stomach. These findings, brought out in two joint papers in 1889 and 1890 (see bibliography), aroused a controversy among physiologists that lasted for several years. But Pavlov's ideas were gradually accepted and in 1904 he won the Nobel Prize in Physiology or Medicine for the work.

Not only did the Pavlov-Shumova-Simanovskaia investigation radically change physiologists' ideas about mammalian digestive processes, it also provided a method for obtaining large quantities of pure gastric juice, from which the digestive enzyme pepsin could be extracted in clean form. Until then, crude preparations of pepsin, often largely ineffective, had been prescribed by physicians as a remedy for dyspepsia. By the late 1890s the new "dog-technology," developed further in Pavlov's laboratory after Shumova-Simanovskaia had moved on, was supplying clean gastric juice for medical use and for research in physiological chemistry.

Ekaterina Shumova-Simanovskaia was probably the first woman that Pavlov accepted as a research coworker. She was a skilled experimentalist whose talents and techniques he valued and depended on, but their relationship appears to have been somewhat strained at times. Shumova-Simanovskaia was extremely sensitive to anything she interpreted as a slight, and her strong reactions are said to have troubled Pavlov not a little.[170]

In 1891 she joined the chemistry department of the newly opened Institute of Experimental Medicine in St. Petersburg, which remained her place of work for the rest of her life. The department was headed by Marcellus Nencki, who had left Bern University to accept the position. With him came Nadezhda Ziber, by then a widow and one of Nencki's longest and closest coworkers.

Shumova-Simanovskaia's collaborative work with Nencki included a lengthy and detailed study on the halogens present in the animal body, published in 1895, and an investigation of the action of gastric juice on toxins; in the latter work, an example of the use of clean gastric juice in physiological research, Nadezhda Ziber also took part (see bibliography). The sisters continued joint work in this area for several years and also collaborated in studies on subjects such as the flora of the small intestine and variation of the extent of physiological oxidative processes.[171]

Like most of the Russian medical women of her generation, Ekaterina Shumova-Simanovskaia joined in the effort to bring medical care to the poor. With her own resources she built a free outpatient clinic in the Krestetsk district of Novgorod Province, to the south of St. Petersburg, and she spent most of the summer months there tending the sick and distributing medicine. She died suddenly, of a brain hemorrhage, on 3 July 1905, when she was about fifty-three. In her memory her husband and sister founded a school in the Krestetsk district for the education of the children of peasants and other people of the region.

Maria Mikhailovna Manasseina,[172] née Korkunova (1843–1903), has been described as "a founder of the new science of physiological chemistry (now biochemistry),"[173] but her very considerable international recognition came from her work in neurology and physiology. The daughter of a well-known St. Petersburg historian and archaeologist, and member of the Academy of Sciences, Professor Mikhail Andreevich Korkunov (1806–1858), she grew up in a strongly intellectual milieu. After an exceptionally good early education at home, she went on to become one of the first women to attempt studies in advanced medical education in Russia. Her initial award of the special degree of woman doctor (*zhenshchina vrach*) was later upgraded to doctor of medicine.

While still a student she married Poniatovskii, a fellow student, and together they actively participated in revolutionary circles. In 1861 Poniatovskii was arrested by tsarist police and sent into political exile, where he later died. Maria remarried in 1865, her second husband being Viatscheslav Avksentievich Manassein (1841–1901), who had taken part in student unrest in Dorpat and served a prison sentence before completing his education at the St. Petersburg Medical-Surgical Academy in 1866; he later became well known in the medical community, editor of the first Russian medical magazine (*Vrach*), and a professor at the Medical-Surgical Academy.

Early in 1870 the Manasseins went to western Europe for a period of further scientific training. During the six months Manasseina spent at the Polytechnic Institute in Vienna she studied the process of alcoholic fermentation in the laboratory of Julius Wiesner (1838–1916). She demonstrated that fermentation is due to chemical substances, at the time called "unorganized enzymes," produced by yeast cells and themselves capable of bringing about chemical reactions without the presence of the living cells that produced them. This discovery was one of great importance since it undermined Louis Pasteur's physiological theory of fermentation and confirmed the chemical hypothesis put forward by distinguished chemists Justus von Liebig, Marcelin Berthelot, and a number of others. Manasseina published her results in Russian, rather than in a Western journal, in 1871 in a short note and in a twenty-page detailed report (see bibliography); the work was presented in German the following year but only in Julius Wiesner's collection of reports of his own and his students' researches.[174] The eminent Justus von Liebig, who had himself recently published on the subject,[175] was prompted to invite her to join his research group in Giessen—a notable honor from that famous scientist; however, as she later stated, family matters obliged her to return to St. Petersburg.[176]

In 1877 Viatscheslav Manassein, who had held a teaching post at the Medical-Surgical Academy for some years, became head of the central clinic. Maria did not continue her biochemical research; although involved briefly with the recently established Women's Medical Courses, she turned her attention to problems in neurophysiology. Working with one of her husband's friends, Georgian prince Ivan Romanovich Tarkhanov, a professor at the Medical-Surgical Academy and an outstanding physiologist, she began investigations in sleep science, also one of Tarkhanov's special interests. This area of neuroscience became the field in which Manasseina made her most celebrated contributions; indeed, she was a pioneer of experimental sleep science.

Her seminal study of the effects of prolonged sleep deprivation on dogs aged two to four months showed that the observed effects (culminating in death within five days or less) originated in the brain. Dissections of the dead animals revealed small hemorrhages all over the cerebral cortex; loss in body weight was relatively slight. The results were very different from those obtained from comparable food-deprivation experiments on dogs, which resulted in death in twenty to twenty-five days. Manasseina's conclusion that sleep was more important for an organism than food was a hypothesis that was well ahead of her time, and one that had a considerable impact on sleep studies. Her report, presented at the International Congress of Medicine in Rome in 1894, appeared in an Italian journal the same year (see bibliography). These findings were a first step toward a scientific understanding of sleep.

Her lengthy book *Sleep* was published in 1889. Then revised and considerably expanded to include such studies as her work on dogs, it was tremendously successful in western Europe as well as in Russia, where it ran to several editions; a French version appeared in 1896, an English translation in 1897, and Swedish editions in 1895 and 1898.[177] American psychologist George T. W. Patrick gave the book a favorable review in an 1898 issue of the journal *Science*; he saw it as providing a comprehensive summary of "[t]he best that is now known about the physiology, pathology, hygiene and psychology of sleep," presented in a readily understandable style.[178] The book included Manasseina's research on dreams, an area in which she worked for about five years, collecting data on the dreams of thirty-seven people. She concluded that, on average, dream frequency depended on age and level of education, the young having more dreams than the elderly, and the well educated more than those with minimal education.

Although her international recognition was primarily for her work in sleep science, her interests over the course of her more than thirty-year career in neurology, physiology, and psychology also touched on general cultural ideas, pedagogical questions, and handwriting interpretation. Extremely hardworking nearly all her life, she was a prolific author, publishing some forty-eight monographs, reviews, and scientific articles, and translating about fourteen books from western European languages into Russian.[179] An honorary member of several Russian scientific societies, she made frequent visits to western Europe and attended a number of international scientific congresses. At the Rome congress in 1894 she presented not only her important findings in sleep science but also two fairly lengthy reports on brain research (see bibliography). In Paris in 1900 she reported results of experiments on the effects of different diets on the disposition of dogs; her conclusion, that dogs on a vegetarian diet were calmer and less quarrelsome than those fed meat, agreed with that of University of Paris physiologist Charles Richet.

In the late 1890s, more than a quarter of a century after her fermentation research was published, the German chemist Eduard Buchner (1860–1917), when examining yeast extracts from which live cells had been removed, found that these extracts were capable of causing fermentation of sugar. The papers he published in 1897 effectively destroyed the theory still held by most physiologists and a great many chemists that fermentation took place only in the presence of living organisms.[180] Buchner received the Nobel Prize for Physiology or Medicine in 1907 for his discovery of the chemical nature of fermentation, a major step in the history of modern biochemistry and a blow against the old doctrine of vitalism.

Although aware of Manasseina's earlier work, he made no reference to it in his initial reports and later remarked that her investigations were too inexact to be useful.[181] Nevertheless, they were sufficiently exact to allow her to correctly draw her all-important conclusions, the conclusions that Buchner definitively confirmed. He and Manasseina used different procedures for producing cell-free yeast extracts, and his were more in keeping with chemical rather than microbiological procedures of the time. However, Manasseina's work, carried out at the Polytechnic Institute in Vienna under the guidance of the well-known and respected chemist Julius Wiesner, was painstaking and careful, and her extracts were well checked for the presence of living cells. With surprising self-confidence for a woman of her time, she responded vigorously and attempted to establish some priority for her contributions. She promptly published two papers, one in 1897 in the widely known *Berichte* of the German Chemical Society (also Buchner's place of publication), and the other in 1898 in the first volume of the Moscow journal *Le Physiologiste Russe* (see bibliography). This effort does not appear to have elicited any notable reaction from the scientific community. It was essentially ignored, the well-established Buchner having already been accepted as the discoverer of cell-free fermentation. Manasseina's role in early biochemical studies was forgotten until 1950, when a paper in the Moscow journal *Biokhimiia* brought her contribution back to light.[182]

Maria Manasseina's outlook on social and political matters changed fundamentally as she grew older; an idealistic revolutionary in her youth, she became an ardent supporter of the tsar in her mature years. Her 1897 book *Sleep* and other works were dedicated to the tsar and his family, a public demonstration of loyalty and devotion that brought her considerable material reward. Aleksandr III granted her a lifelong pension in appreciation of her literary output, and his son Nicholas II, soon after his coronation, presented her with a gift of 10,000 rubles, a very great deal of money at that time.

About 1879 she left Viatscheslav Manassein and went to live with her colleague Ivan Tarkhanov. Her refusal to give Manassein a divorce prevented him from marrying his lover, a niece of the writer Fyodor Dostoyevsky, in the usual religious ceremony. However, it should be remembered that the position of a divorced woman, on her own without a supportive family, was difficult in nineteenth-century St. Petersburg society, especially if she had scientific ambitions. With the assistance provided by the imperial family, Manasseina appears to have fared fairly well. Through her public lectures that covered literary, psychological, and pedagogical topics, she was widely known and popular in St. Petersburg intellectual and aristocratic circles. Her death in St. Petersburg, after a long and painful illness, on 17 March 1903, when she was about sixty, was much regretted. Remembered as kindly, caring, and hardworking, she was one of the outstanding Russian women of her time.

NADEZHDA (NADIA) KARLOVA SCHULTZ[183] was from St. Petersburg, where her brother, Friedrich von Schultz, held a state councillor position. Born on 10 November 1839, she had a private education throughout her earlier years. She later decided to get a medical degree and enrolled at the University of Zurich for the winter semester of 1872–1873. A year later, to comply with the tsarist government's requirement that Russian women students leave Zurich, she transferred to the University of Bern. Her seven semesters of study in the medical faculty there included research carried out in the Bern Pathological Institute, guided mainly by professor of surgery Emil Kocher with further help and advice from pathologist Theodor Langhans, head of the institute.

The work concerned the process of artery healing following ligature application. By the 1870s there was general agreement among medical scientists that after an already established time lapse a certain amount of flexible, vascular connective tissue, containing normal cells, can be found at the site of the ligature, securely joined to the artery wall; opinions about the type and origin of the cells that formed the new tissue, and the stages in the healing process, were still a subject of much debate. Several workers considered that cells in the coagulated blood that resulted from the ligaturing reorganized. Which particular cells in the coagulant were involved remained to be decided; some investigators emphasized the importance of white blood cells, others saw a significant role for red blood cells, and yet others thought the process depended mainly on epithelial cells from the artery wall.

Following Kocher's suggestion, Nadezhda Schultz repeated some of the earlier investigations using a wide range of ligatured tissues with which she was supplied. In addition she studied tissues from new ligatured tissue over time intervals of a few hours after the operation to 155 days. The work was designed to reveal the details of tissue formation at the cellular level. Test animals were rabbits; carotid, brachial, and femoral arteries were among those used. After the animal's death she removed a strip of injured artery wall, along with surrounding fluids, at successive intervals, and subjected these to microscopic examination.

Her general conclusions were as follows. After the ligature application, the resulting damaged wall area is closed off by white blood cells, which come together to form a strong, homogeneous, fibrous network. At the same time the coagulant produced by the ligaturing begins to break down; cells are introduced by the gradually returning blood flow, and when the channels and crevices in the white-cell network begin slowly to widen, these cells flow into the artery's fibrous protective walls, which are gradually transformed into normal epithelial cells. These preliminary steps can take up to eighty days, after which healing can begin. One of the major points Schultz took care to emphasize was the importance of the controling role of the white blood cells. She prudently declined to speculate on the extent to which her results would apply to human subjects.

In addition to this creditable piece of research she followed microscopically the disintegration of catgut ligatures, the binding material still used in surgical procedures when stitches cannot be removed manually.

Nadezhda Schultz received her doctoral degree from the University of Bern in March 1877;[184] her research was reported in the *Deutsche Zeitschrift für Chirurgie* the following year (see bibliography).

By at least the later 1880s she had returned to St. Petersburg and was working at the new Institute of Experimental Medicine under the general direction of Mikhail Afanasiev. Here she carried out a number of bacteriological investigations of the diseases, epidemic and otherwise, that afflicted populations across extensive areas of the country. Among her reports about this time was a note on work with Afanasiev on actinomycosis, a bacterial infection of the eye, presented at the Third Congress of Russian Physicians, held in St. Petersburg in January 1889 (see bibliography).

A much longer paper described work she carried out with help from Marcelli Nencki, leader of the institute's chemical and biochemical section from 1891; this concerned the control of bubonic plague by disinfection procedures (see bibliography), a subject that was then occupying the attention of a number of investigators, including Nencki (see also the section on Nadezhda Ziber). Schultz hoped to more clearly establish the effectiveness of disinfection procedures and determine which chemicals offered the best control possibilities. Her interests centered principally on measures to be tried in dwellings housing infected persons and other closed-in areas. She collected and cultured bacterial samples, using both broth and other culture media including "Swedish Filter Paper" strips; the latter, which were employed as supports for an emulsion of the microbes, were considered especially reliable test material because of their neutrality to the disinfecting substances. Ten chemicals were tested for success in killing the bacterium, some of them fairly harsh and now considered dangerous to humans as well; they included mercuric chloride, formaldehyde, quicklime, sodium hydroxide, and chlorophenol, the latter also being tested about that time by Afanasiev's coworker Grigorii Karpov.[185] Schultz recorded minimal lethal doses at stated concentrations and the times required to wipe out the bacteria. Also investigated was a more readily obtained and cheaper agent, pine pitch extract, a material of special interest to Nencki; although hardly comparable to the more powerful mercuric chloride or phenol, it too was fairly effective. Results for disinfection of the contents of enclosed spaces were somewhat variable and depended to a considerable extent on the care taken in the work; Schultz recommended formaldehyde solutions for such items as furniture and clothes.

Whether or not Nadezhda Schultz's work in the biological sciences continued beyond 1900 has not been determined; by then she would have been about sixty years old.

Praskoviia Vasilievna Tsiklinskaia[186] (1859–1923), bacteriologist, educationalist, and probably Russia's first woman professor, was born on 16 June 1859. After completing gymnasium studies she taught for six years and then entered the science section of the Higher Courses for Women in St. Petersburg. In 1889, to further her scientific training, she went to Paris, where, with the help of Ilya Metchnikov, then at the Pasteur Institute, she succeeded in enrolling in a bacteriology course, and at the same time attended science lectures at the University of Paris. In her reminiscences about this period in her life she remarked on the low numbers of women in the sciences and the fact that at first they were often far from welcome; in her bacteriology class of twenty students she was the only woman.

Following these studies she began her first scientific research at the Pasteur Institute, where she had the general guidance of Metchnikov, a brilliant investigator and outstanding teacher; a fellow Russian student, Georgii Norbertovich Gabrichevskii, provided additional help. An invitation to join Gabrichevskii's newly established bacteriological laboratory in the therapeutics clinic at Moscow University came in the early 1890s. Starting as a laboratory assistant, she later held a lectureship and is reported to have been the first woman to hold a position of that rank at Moscow University.

Most of her research, both at the Pasteur Institute and in Moscow, concerned thermophilic bacteria, widespread in a great many habitats—soils, rivers, the sewers of Paris, and the bodies of man and other animals. Her five-page 1899 paper on thermophilic molds, which appeared in the *Annales* of the Pasteur Institute (see bibliography), is probably her most widely cited work. It reported her morphological studies on a number of thermophilic molds and fungi, and included a description of a new variety, *Thermoactinomyces vulgaris*, isolated from soil; *Thermoactinomyces vulgaris* Tsilinsky (1899) is the type species of the genus *Thermoactinomyces*, which in turn is the type genus of the family Thermoactinomycetaceae Matsuo et al., 2006.[187]

In 1902–1903 Tsiklinskaia spent a year in Robert Chodat's Botanical Institute at the University of Geneva, continuing her research on thermophilic microbes; she was awarded a doctoral degree in science in 1903.[188] That year also saw the publication in the Pasteur Institute's *Annales* of a twenty-three-page paper on the thermophilic microbial flora of the intestinal tract of man;[189] the work included several important studies on infants and children. Tsiklinskaia's findings confirmed the conclusions of Ilya Metchnikov, whose extensive research had revealed the beneficial effects of sour milk containing acid-producing lactobacilli, particularly *lactobacillus acidophilus*, and the possible severe negative effects of the common colon bacillus, *bacillus proteus*.

When French scientist, physician, and polar explorer Jean-Baptiste Charcot returned from his two-year expedition to the Antarctic in 1905 with a rich collection of scientific materials, Tsiklinskaia undertook the identification and analysis of the microflora; this included microbes isolated from polar soils and from the gastrointestinal tracts of Antarctic birds.[190]

In 1908 Praskovia Tsiklinskaia became head of bacteriology at the Moscow Higher Courses for Women and remained there, as the only woman professor, for the rest of her life. A very active member of various committees of the medical faculty of the Higher Courses for Women, and a member of the Society of Pediatricians, she was awarded an honorary doctor of medicine degree from Moscow University in 1917, a notable recognition of her contribution to studies of intestinal flora and human health.

Although by the outbreak of the First World War much of her time was going to teaching and committee work, she contributed a short monograph on skin-penetrating microorganisms to Semen Zlatogorov's extended work on microorganisms, published in 1916–1918.[191] For her teaching, at which she was considered outstanding, she was honored by having a scholarship established in her name—the P. V. Tsiklinskaia Scholarship. After the 1917 October Revolution she became a member of the Technical Committee and participated fully in its undertakings. Her scientific work continued throughout her life; she spoke at the Bacteriology Congress in Kharkov in 1921, two years before her death, on 25 December 1923, in her sixty-fifth year.

Mariia Raskin[192] (fl. 1880s–1911) carried out her research in the Laboratory of Clinical Bacteriology, then headed by Mikhail Ivanovich Afanasiev, in the Institute of Experimental Medicine, St. Petersburg; her materials came from the Oldenburg Children's Hospital, St. Petersburg.

Among the subjects she investigated were the sources of both scarlet fever infection and the secondary infections it commonly led to, such as diphtheria, pleurisy, and infections of the middle ear; included in her studies was the bacterial content of discharges produced.[193] Her brief note on results from a comparable bacteriological investigation of a fatal case of pyemia (septicaemia), following an unusually severe attack of scarlet fever, appeared in 1889; the work involved examination of the suppurative discharges from abscesses on the victim's kidneys (see bibliography). Aspects of Mariia Raskin's research, and her major conclusions concerning diseases caused by streptococcus and staphylococcus bacteria, were presented at the Third Congress of Russian Physicians in St. Petersburg in January 1889.

The development of improved laboratory procedures and techniques, including finding a better stain for diphtheria bacilli, was another interest; her papers on culture media for pathogenic microorganisms found in milk were published in 1887 and 1889 (see bibliography), and her reports on a new stain to facilitate the identifying of diphtheria bacilli in 1911 and 1912.[194] Raskin's work in the late 1880s concerning complications arising from scarlet fever infections constituted a noteworthy contribution to the understanding of this area of bacteriology.

MARIIA RECHTSAMER, née SCHMOULEVICH[195] (b. 1865), was from Moscow. She began her studies in the medical faculty of Bern University in the winter semester of 1884–1885, and continued at the University of Paris in the early 1890s. Her degree was awarded in 1895 following completion of her dissertation research on heart diseases that simulate heart enlargement during growth.[196] By then she had married, becoming Mme Rechtsamer. About this time she also carried out studies in Tiflis (Tbilisi) on the cholera bacillus (see bibliography), a subject that was being investigated then by Matvei Rechtsamer.[197] Mariia Rechtsamer's translation into German of the work of bacteriologist Nikolai Sacharov appeared in 1902.[198]

ROSALIE ZAÏDMANN[199] was born on 23 December 1872 in the then predominantly Jewish city of Berdichev, Ukraine, about one hundred miles southwest of Kiev. Little has been uncovered about her early education, but she appears to have studied in France and received the *licenciée ès sciences naturelles* qualification. She enrolled in the faculty of medicine at the University of Montpellier, most likely in the late 1890s, and was awarded a doctoral degree on 24 March 1900. Her dissertation research, carried out under the direction of microbiologist Alexandre Joseph Rodet, concerned the pathology of typhoid fever and coli bacteria, two areas of much investigation in the 1880s and 1890s, and of special interest to Rodet from the 1880s. Zaïdmann's work involved laboratory animal studies.[200]

Little information has been collected about MARIIA A. MIROPOLSKAIA (fl. 1880s) beyond the fact that she published two papers, one in 1884 and one in 1885, reporting research carried out in the medical faculty of St. Petersburg University. Miropolskaia studied at the St. Petersburg Higher Courses for Women and held an assistantship in the clinic headed by Kronid Slavianskii, professor of gynecology and obstetrics. She may have had a doctorate by 1885.[201]

Her work was carried out in the medical faculty's laboratories of experimental pathology and chemistry, and was directed by Slavianskii and Aleksandr Borodin, professor of chemistry. Using cats and rabbits as test animals, Miropolskaia investigated the possibility of the passage of inert solid particles from the bloodstream of the pregnant female into that of the fetus she carried. The part of the work requiring the use of chemical methods was done in the medical faculty's chemical laboratory under Borodin's guidance. Miropolskaia injected an aqueous suspension of finely ground pure cinnabar (red mercuric sulfide) into the jugular vein of the test animal and, after the animal's death, extracted the fetus and analyzed its tissue for the presence of mercury using standard chemical procedures.

Considerable effort was required to modify these standard procedures in order to avoid unforeseen sources of mercury contamination and ensure true and reliable results. Among the setbacks encountered was the discovery that commercial sulfuric acid, a necessary reagent, often contained traces of mercury. This led to Miropolskaia's second paper, which reported a method for the removal of mercury from commercial sulfuric acid; presented by Borodin at a meeting of the Russian Physico-Chemical Society in 1884, it was included in the "Correspondance russe" section contributed by Olga Davidova (see chapter 2) to the *Bulletin* of the Société Chimique de Paris in 1885 (see bibliography, chemistry section). The medical conclusion, that passage of a solid, inert powder from maternal to fetal blood does not take place, was reported in *Archives de Physiologie* and also noted in the *London*

Medical Record (see bibliography). No further contributions of a chemical nature by Miropolskaia were found in the period up through 1914.

IDA LEVIN (fl. 1880s) was a student of Filipp Vasil'evich Ovsiannikov (1827–1906), one of the founders of Russian histology and comparative physiology. From the mid-1860s until his death, Ovsiannikov was a central figure among forward-looking St. Petersburg biologists. Head of the subdepartment of anatomy and physiology at St. Petersburg University, he organized the first physiological laboratory at the St. Petersburg Academy of Sciences in 1864. Among his major interests was the comparative physiology of the nerve endings of the cornea, particularly their fine structure.

Ida Levin contributed to his cornea research, carrying out a detailed study of nerve endings and comparing these with the network of nerve endings in the tail of the tadpole. She found the tail network to be fully analogous to the pattern found in the cornea (allowing for the anatomical variation that determined the nerve pathways in the tail). Her study, presented at a meeting of the St. Petersburg Academy of Sciences in 1884, was published in the academy's *Memoirs* in 1885; it was one of the earliest in the biological sciences by a woman working at a major Russian institution that was accepted for publication in the institution's leading journal (see bibliography).

LIUDMILLA SOLLOGUB's report, presented at a meeting of the Physical-Mathematical Section of the Imperial Academy of Sciences in St. Petersburg in 1886, concerned her neurological studies of nerve fibers. She discussed in particular the Remak's fibers, or "gray fibers," that she had observed at the endings of nerves in frog muscle; these fibers were of special interest since they lacked the insulating myelin sheath more commonly observed along most of the length of nerve fibers (see bibliography).

NOTES

1. Dr. Hannay Rohner, *Die ersten 30 Jahre des medizinischen Frauenstudiums an der Universität Zürich, 1867–1893*, Zürcher Medizingeschichtliche Abhandlungen, Neue Reihe, no. 89 (Zurich: Juris Druck, 1972), 17–19, 78; matriculation register, Zurich University, http://www.matrikel.uzh.ch/active/static/27414.htm (July 2014); Janette E. Tuve, *The First Russian Women Physicians* (Newtonville, Mass.: Oriental Research Partners, 1984), 12–26; J. M. Meijer, *Knowledge and Revolution: The Russian Colony in Zurich, 1870–1873* (Assen, Netherlands: Van Gorcum, 1955); Olga Valkova, "The Conquest of Science: Women and Science in Russia, 1860–1940," *Osiris* 23 (2008): 136–65, on 146; Thomas Neville Bonner, "Rendezvous in Zurich: Seven Who Made a Revolution in Women's Medical Education, 1864–1874," *Journal of the History of Medicine* 44, no. 1 (Jan. 1989): 7–27; Joy Harvey and Marilyn Ogilvie, "Suslova, Nadezhda Prokof'evna (1843–1918)," in *The Biographical Dictionary of Women in Science: Pioneering Lives from Ancient Times to the Mid-20th Century*, ed. Marilyn Ogilvie and Joy Harvey (New York and London: Routledge, 2000), 2:1252–53; "Ostankino Estate," http://www.moscow-taxi.com/out-of-town/ostankino-estate.html (Oct. 2010).
2. Tuve, *First Russian Women Physicians*, 12–26; Nadezhda Suslova, "Iz nedavniago proshlogo" [From the recent past], *Vestnik Evropy* [European messenger] 35, no. 6 (1900): 625–73, on 626.
3. Tuve, *First Russian Women Physicians*, 13–14.
4. See, for instance, I[van] M[ikhailovich] Sechenov, *Ueber die elektrische und chemische Reizung der sensiblen Rückenmarksnerven des Frosches* (Graz: Leuschner & Lubensky, 1868).
5. The lymphatic hearts, or ganglia are aggregations of cells along an animal's nerve system that control responses to external stimuli.
6. Meijer, *Knowledge and Revolution*, 56.
7. Schweizerischen Verband der Akademikerinnen, *Das Frauenstudium an den Schweizer Hochschulen* (Zürich: Rascher, 1928), 59, quoted in Bonner, "Rendezvous in Zurich," 11 and note 10.
8. See for instance, Friedrich Erismann, "Ein Beitrag zur Entwickelungs-geschichte der Myople, gestützt auf die Untersuchung der Augen von 4358 Schülern und Schülerinnen," *Archiv für Opthalmologie* 17, Abth. 1 (1871): 1–79.
9. Suslova's entry in Zurich University matriculation register.
10. Bonner notes that Erismann was one of the pioneers in public hygiene in Russia, directing the first investigations into factory conditions and workers' lives. "At one time he held the highest position in public health in Russia." (Bonner, "Rendezvous in Zurich," 12.)
11. Apollinaria (Polina) Suslova (1839–1918) was for a time the mistress of writer Fyodor Dostoevsky and is considered to be the prototype for several of the female characters in his novels. She herself published short sto-

ries; her diary, letters, and an autobiographical story continued to attract considerable attention for decades after the first publication of the collection in 1928, undoubtedly because of the light it sheds on Dostoevsky's life and work. See A. P. Suslova, *Gody blizosti s Dostoevskim*, later title, *Chuzhaia i svoi: dnevnik, povest', pis'ma* (Minsk: Universitetskae, 1928, 1994), *Mes années d'intimité avec Dostoïevski*, trans. Luba Jurgenson, with *avant-propos* by Verena von der Heyden-Rynsch (in German), trans. Jean Launay (Paris: Gallimard, 1995). Some years after separating from Dostoevsky, Apollinaria married the controversial writer and philosopher Vasilii Rozanov; that stormy union was also of limited duration. She died in Sevastopol in 1918, at the age of seventy-eight.

12. Rohner, *Die ersten 30 Jahre*, 19.

13. Suslova, "Iz nedavniago proshlogo."

14. Matriculation register, Zurich University, http://www.matrikel.uzh.ch/active/static/27256.htm (July 2014), and for Theodor Egli http://www.matrikel.uzh.ch/active/static/5122.htm (July 2014); Rohner, *Die ersten 30 Jahre*, 23–24; Franziska Tiburtius, *Erinnerungen einer Achtzigjährigen* (Berlin: Schwetschke, 1923), especially 149, 150, 155.

15. See, for instance, Ian G. Anderson, *Scotsmen in the Service of the Czars* (Edinburgh: Pentland Press, 1990).

16. One of the other men in this small group of Zurich students and junior male faculty members was Friedrich Erismann, at that time married to Nadezhda Suslova.

17. Julie Sinclair, "Experimentelle Untersuchungen zur Genese der erworbenen Kapsel-Katarakt," dissertation, medical faculty, University of Zurich, March 1876.

18. Death announcement for Julia E. Sinclair, *Lousiville Courier-Journal*, 11 March 1937. The Zurich University matriculation record for Julie Sinclair reports her death in Louisville as being caused by a fire.

19. Mildred Davies, "Lukánina, Adeláida Nikoláevna," in *Dictionary of Russian Women Writers*, eds. Marina Ledkovskaia-Astman, Charlotte Rosenthal, and Mary Fleming Zirin, (Westport, Conn.: Greenwood Press, 1994); Meijer, *Knowledge and Revolution*, 48–49, 99, 176 note 6; Tuve, *First Russian Women Physicians*, 34–43, 59; Georgianna Kelly, "Adelaide Lukanina's 'A Year in America,'" MA thesis, University of Kansas, 1995; Anne C. Hughes, "Lukanina, Adelaida N. (1843–1908)," in *Biographical Dictionary of Women in Science*, ed. Ogilvie and Harvey, 2:810–11; Iusuf S. Musabekov, *Iulia Vsevolodovna Lermontova, 1846–1919* (Moscow: Nauka, 1967), 23–24; matriculation register, Zurich University, http://www.matrikel.uzh.ch/active/static/26344.htm (July 2014).

20. See the report by Borodin in *Zhurnal' Russkago Khimischeskago Obshchestva* 3 (1871): 127.

21. See also the career of Fanny Berlinerblau (below). Berlinerblau completed her degree at Bern University before going to America, but in many ways the early experiences of these two Russian women were similar.

22. A. L. [Adelaide Lukanina], "God v Amerike. Iz vospominanii zhenshchiny-medika," *Vestnik Evropy* 16, no. 8 (1881): 661–66; 16, no. 9 (1881): 31–78; 17, no. 4 (1882): 495–538; 17, no. 6 (1882): 503–45. See also Kelly, "Adelaide Lukanina's 'A Year in America.'" Interwoven into these vivid sketches of the life of a female medical student in Philadelphia and Boston in the 1870s are Lukanina's views on such matters as the miserable existence of Irish immigrants and American Negroes in the slums of Boston. She also had some critical comments about the quality of medical education in America compared to that in Europe. *God v Amerike* was republished in book form in 1892 (St. Petersburg: tip. I. N. Skorokhodova, 1892), 298 pp.

23. A. L. [Adelaide Lukanina], "Zakony bur—Iz lektsii v Sorbonne," extracts from lectures on a number of different subjects, *Vestnik Evropy* 15, no. 5 (1880): 390–404.

24. A. Paevskaia, *Viktor Giugo: ego zhizn' i literaturnaia deiatel'nost': biograficheskii ocherk A. Paevskoi* (St. Petersburg: tip. Obshchestvennia pol'za, 1890; 2nd ed., St. Petersburg, 1893). A. Paevskaia, *Val'ter Skott: ego zhizn' i literaturnaia deiatel'nost'* (St. Petersburg: tip. Obshchestvennia pol'za, 1891).

25. A. Paevskaia, "Komandirovka na kholeru," *Russkoe Bogatsvo* [Wealth of Russia] 7 (1903): 120–69; 8 (1903): 49–93.

26. Matriculation register, Zurich University, http://www.matrikel.uzh.ch/active/static/27095.htm (July 2014); archives, Bern University, http://apps.uniarchiv.unibe.ch/syscomm/images/mata/6878_6889.gif (July 2014); *Jahresverzeichnis der Schweizerschen Universitätsschriften, 1897–1898; Catalogue des Écrits académiques Suisse, 1897–1898* (Basel: Schweighauserische Buchdruckerei, 1898), p. 51, http://babel.hathitrust.org/cgi/pt?id=uc1.b4228802;view=1up;seq=61 (July 2014).

27. Philipp Stöhr, *Zur Kenntniss der feineren Baues der menschlich Magenschleimhaut* (privately printed, 1882).

28. Daria Schirman (aus Rostov am Don), "Ueber die Rückbildung der Dickdarmzotten des Meerschweinchens," dissertation, medical faculty, University of Zurich, 1897–1898.

29. Information in technical papers; List of prominent Vologodians, Constantin von Monakow, Vologda Oblast Government, http://site.vologda-oblast.ru/en/region/famous/list/index.php?id_15=953&type_id_15=14 (July 2014); Constantin von Monakow, *Vita Mea. Mein Leben*, ed. Alfred W. Gubser and Erwin H. Ackerkencht (Bern: Verlag Hans Huber, 1970), 205, 208, 225.

30. Alla A. Vein, "Jean-Martin Charcot at the Birth of Russian Neurology," *European Neurology* 65, no. 2 (2011): 75–81.

31. von Monakow, *Vita Mea*, 208.

32. See "Wanderversammlung des Vereins für Psychiatrie und Neurologie in Wien. 5–7 Oktober 1906. Sitzung, 6. Oktober 1906," *Monatsschrift für Psychiatrie und Neurologie* 21 (1907): 167–77, on 176. Leonova's presentation was entitled "Zur pathologischen Entwickelung des Centralnervensystem. Das Verhalten der Rinde der Sulci Calcarina in einem Falle von Microphthalmia und Amelia (Amputation spontanée)," later published in *Archiv für Psychiatrie und Nervenkrankheiten* 45 (1909): 79–91.

33. See also Ol'ga Leonova von Lange, "Zur pathologischen Entwickelung des Centralnervensystems (Neue Beiträge). Ein Fall von Cyclopie combinirt mit Mikro- und Arhinencephalie," *Archiv für Psychiatrie und Nervenkrankheiten* 38 (1904): 862–94; "Zur pathologischen Entwickelung des Centralnervensystems. Ein Fall von Amelia (Amputatio spontanea) (Neue Beiträge)," *Archiv für Psychiatrie und Nervenkrankheiten* 43 (1907): 1218–50.

34. Matriculation register, Zurich University, http://www.matrikel.uzh.ch/active/static/11433.htm (Sept. 2014). Anna Kleimann would seem to have been an older sister of Marie Kleimann (Vassilieva-Kleimann), who studied medicine at Bern University in the early 1880s and spent a winter semester at Zurich in 1883 (see note 53 on the next page). However, no confirmation of this conjecture has been found.

35. Anna Kleinmann, "Über die Ursachen der Kindersterblichkeit speziell im Kanton Zürich," dissertation, medical faculty, Zurich University, 1873 (Zurich: Druck von Orell, Füssli, 1873, 70 pp. manuscript).

36. Vincent J. Knapp, "Major Medical Explanations for High Infant Mortality in Nineteenth-Century Europe," *Canadian Bulletin of Medical History/Bulletin canadien d'histoire de la médecine* 15 (1998): 317–36.

37. Barbara Bachmann and Elke Bradenahl, "Medizinstudium von Frauen in Bern, 1871–1914," doctoral dissertation, University of Bern, 1990 (Bern: Medizinhistorisches Institut der Universität Bern, 1990); archives, Bern University, http://apps.uniarchiv.unibe.ch/syscomm/images/mata/2866_2876.gif (July 2014); matriculation register, Zurich University, http://www.matrikel.uzh.ch/active/static/27025.htm (July 2014).

38. Verein Feministische Wissenschaft Schweiz, *"Ebenso neu als kühn": 120 Jahre Frauenstudium an der Universität Zürich*, eds. Katherina Belser, et al. (Zurich: eFeF-Verlag, 1988), 140–41. While in Zurich Schachova had the company of her relative, probably her mother, Frau Katerina Schachova (see matriculation register, Zurich University).

39. See for instance [John Ernest Ady], "The Kidney," in *Studies in Microscopical Science*, ed. Arthur Charles Cole, vol. 1 (London: Ballière, Tindall & Cox, 1883), 109–40, on 124, and *Anatomy: Descriptive and Applied*, ed. Henry Gray, 18th ed., revised by Edward Anthony Spitzka (Philadelphia and New York: Lea & Febiger, 1910), 1353.

40. Seraphina Schachowa, "Untersuchen über die Nieren," dissertation, medical faculty, University of Bern, 1876. It is perhaps worth noting that the first Russian women to graduate from the Bern medical school received their degrees only two years earlier (Marie Siebold from St. Petersburg and Rosa Simonovich from Odessa). Schachova was one of eight who graduated in 1876 (see Bachmann and Bradenahl, "Medizinstudium von Frauen in Bern").

41. Theodor Langhans, "Ueber die Veränderungen der Glomeruli bei der Nephritis nebst einigen Bemerkungen über die Entstehung der Fibrincylinder," *Archiv für pathologische Anatomie und Physiologie und für klinische Medicin* 76 (1879): 85–118, especially 108–9. For a later citing of Schachova's work, see, for instance, Dennis Brown and Sylvie Breton, "Mitrochondria-rich proton-secreting epithelial cells," *Journal of Experimental Biology* 199 (1996): 2345–58.

42. Bachmann and Bradenahl, "Medizinstudium von Frauen in Bern"; archives, Bern University, http://apps.uniarchiv.unibe.ch/syscomm/images/mata/3295_3307.gif (July 2014).

43. Ida Eliascheff, "Ueber die Wirkung des Cantharidins auf die Nieren," dissertation, medical faculty, University of Bern, 1883 (see bibliography).

44. See for instance, Victor Cornil, "Sur les lésions du rein et de la vessie dans l'empoisonnement rapide par la canthridine," *Comptes Rendus Hebdomadaires des Séances de l'Académie des Sciences* 90 (1880): 188–91, 536–39.

45. Bachmann and Bradenahl, "Medizinstudium von Frauen in Bern"; archives, Bern University, http://apps.uniarchiv.unibe.ch/syscomm/images/mata/2894_2906.gif (July 2014); matriculation register, Zurich University, http://www.matrikel.uzh.ch/active/static/27375.htm (July 2014).

46. Olga Stoff, "Ueber die Wirkung des Anilins auf den thierischen Organismus," dissertation, University of Bern, 1875.

47. Tiburtius, *Erinnerungen einer Achtzigjährigen*, 149.

48. Bachmann and Bradenahl, "Medizinstudium von Frauen in Bern"; archives, Bern University, http://apps.uniarchiv.unibe.ch/syscomm/images/mata/5627_5638.gif (July 2014).

49. Marie Zielinska should not be confused with Antonina Zielinska, who studied medicine at the University of Bern from 1892 to 1898 (with a summer session at the University of Zurich in 1896). Antonina Zielinska was also from Warsaw and had completed her secondary education at a Warsaw girls' gymnasium. Her doctoral dissertation, a bacteriological study on the diphtheria bacillus and the difficulties of distinguishing it from other pathogens that produce similar clinical symptoms, does not appear to have been published in a scientific journal indexed by the Royal Society. She received her Bern medical degree in 1898. Attempts to establish a family connection between her and Marie Zielinska did not succeed. See Bachmann and Bradenahl, "Medizinstudium von Frauen in Bern"; archives, Bern University, http://apps.uniarchiv.unibe.ch/syscomm/images/mata/8104_8115.gif (July 2014); matriculation register, Zurich University, http://www.matrikel.uzh.ch/active/static/27831.htm (June 2014); and Antonina Zielinska, "Über die Differentialdiagnose zwischen den Diphtherie- und den Pseudodiphtheriebacillus," dissertation, medical faculty, Bern University, 1898.

50. Marie Zielinska, "Beiträge zur Kenntniss der normalen und strumösen Schilddrüse des Menschen und des Hundes," dissertation, medical faculty, Bern University, 1893.

51. Bachmann and Bradenahl, "Medizinstudium von Frauen in Bern;" archives Bern University, http://apps.uniarchiv.unibe.ch/syscomm/images/mata/5755_5766.gif (July 2014) and http://apps.uniarchiv.unibe.ch/syscomm/images/mata/6436_6447.gif (July 2014; these archival entries differ on Ott's year of birth, the former giving 1867, the latter 1865).

52. Nadine Ott, "Ueber peri- und endoneurale Wucherungen in den Nervenstämmen einiger Thierspecies," dissertation, medical faculty, Bern University, 1893.

53. Bachmann and Bradenahl, "Medizinstudium von Frauen in Bern"; archives, Bern University, http://apps.uniarchiv.unibe.ch/syscomm/images/mata/3808_3818.gif (July 2014), see also entry column for 4102_4113; matriculation register, Zurich University, http://www.matrikel.uzh.ch/active/static/26040.htm (Sept. 2014); Bernard Degan, "Nikolai Wassiljewitsch Wassiljew: Arbeitersekretär," in *Käser, Künstler, Kommunisten: Vierzig russisch-schweizerische Lebensgeschichten aus vier Jahrhunderten*, eds. Eva Maeder and Peter Niederhäuser (Zürich: Chronos, 2009), 175–80. Bern archives and the Zurich matriculation register differ on the year of Marie Kleimann's birth; dates vary from 1860 to 1862.

54. Nikolai Vassilievich Vassiliev (1857–1920?) came from a distinguished St. Petersburg family; his maternal grandfather, Ivan Simonov, was rector of the University of Kazan and his father, Vassili Pavlovich Vasiliev, a much-respected scholar of Chinese culture, taught first at the University of Kazan and then in St. Petersburg. Vassili Pavlovich's democratic outlook, which put him much at odds with the constraints of the tsarist government, was fully absorbed by his son. Nikolai Vassilievich began his medical studies in Bern in 1879, interrupted these for a time to write for the German socialist publication *Sozialdemokraten*, but graduated from Bern University in 1887. After several years of medical work he became very active in the Bern Workers Union, taking a prominent part in a number of disputes. From the position of secretary of this union he went on to serve as a member of the Bern cantonal parliament for three years, ending in 1905. In the spring of 1906 he returned to St. Petersburg along with his second wife, Nadine Vassiliev-Kononovicha, whom he had married in 1897, and his five-year-old daughter, Natalia. Vassiliev died in St. Petersburg about 1920, after more than a decade of activity in Russian socialist politics.

Nadine Vassilieva-Kononovicha (1874–1956), from Odessa, also had an eventful life and would seem to have had a fairly successful career as a physician, although she had several interruptions. She studied in the medical faculty of Bern University from 1892 until at least 1896, and after her 1897 marriage to Vassiliev completed her courses at the University of Basle, where she passed the Swiss licensing examinations for physicians. She then started a private practice while also working toward a doctoral degree, which was conferred following presentation of her research on the problem of blood clotting after childbirth (Nadine Wassilieff-Kononowitsch, "Uber Thrombophlebitis in Wochenbett insbesondere über das sogenannte Mahler'sche Zeichen," dissertation, medical faculty, University of Basle, 1904–1905). Whether or not she practiced during the years 1906 to 1922, when she and her family were in St. Petersburg, has not been determined, but in 1922 she returned to Switzerland along with her daughter. After a short stay at her former home in Muri, near Bern, she moved to Basle and established a successful ear, throat, and childhood diseases practice, which lasted for close to three decades. She retired in 1952 and went to live with her daughter in Geneva, where she died on 25 January 1956. (See Degan, "Nikolai Wassiljewitsch Wassiljew," and, for Nadine Wassilieff-Kononowitsch, archives, Bern University, http://apps.uniarchiv.unibe.ch/syscomm/images/mata/6914_6925.gif (July 2014) and http://apps.uniarchiv.unibe.ch/syscomm/images/mata/7670_7681.gif (July 2014).

55. Bachmann and Bradenahl, "Medizinstudium von Frauen in Bern"; archives, Bern University, http://apps.uniarchiv.unibe.ch/syscomm/images/mata/8116_8126.gif (July 2014); matriculation register, Zurich University, http://www.matrikel.uzh.ch/active/static/27608.htm (Sept. 2014).

56. Tatiana Waschkewitsch, "Ueber grosszellige Heerde in den Milzfollikeln bei Diphtheritis und anderen Affectionen," dissertation, medical faculty, University of Bern, 1899.

57. Bachmann and Bradenahl, "Medizinstudium von Frauen in Bern"; archives, Bern University, http://apps/.uniarchiv.unibe.ch/syscomm/images/mata/5885_5896.gif (July 2014); matriculation register, Zurich University, http://www.matrikel.uzh.ch/active/static/26494.htm (July 2014).

58. Anna Reinstein-Mogilova, "Ueber die Betheiligung der Zellschicht des Chorion an der Bildung der Serotina und Reflexa," dissertation, medical faculty, University of Bern, 1891.

59. Stephen Powell, quoting from *Buffalo Courier*, 30 June 1906, "B. Reinstein scores one on arch-enemies," *The Buffalonian*, http://www.buffalonian.com/hnews/1906borisreinstein.html (July 2014)

60. "University of Buffalo, 26 April 1898: Graduates in Pharmacy,"*American Druggist and Pharmaceutical Record* 32, no. 9 (May 1898): 267.

61. See Michelle Kearns, "Reception to Mark Reopening of Reinstein Family Collection," *Buffalo News*, 16 October 2010; Charity Vogel, "Women to Remember," *Buffalo News*, 13 November 2005.

62. Kearns, "Reception to Mark Reopening of Reinstein Family Collection."

63. Bachmann and Bradenahl, "Medizinstudium von Frauen in Bern"; archives, Bern University, http://apps.uniarchiv.unibe.ch/syscomm/images/mata/3985_3996.gif (July 2014).

64. For Müller's interest in medical training for women, see Peter Müller, *Ueber die Zulassung der Frauen zum Studium der Medizin: Aus einem Cyklus von Populär-wissenschaftlichen, vom Lehrkörper der Berner Hochschule gehaltenen Vorträgen* (Hamburg: Verlagsanstalt und Druckerei Actien-Gesellschaft, 1894), 43 pp.

65. Marie Schlee, "Ueber die Dehnung der Bauchwand während der Schwangerschaft," dissertation, medical faculty, University of Bern, 1886.

66. Bachmann and Bradenahl, "Medizinstudium von Frauen in Bern"; archives, Bern University, http://apps.uniarchiv.unibe.ch/syscomm/images/mata/5569_5578.gif (July 2014); matriculation register, Zurich University, http://www.matrikel.uzh.ch/active/static/26123.htm (July 2014).

67. Barbara Kossowitsch, "Untersuchungen über den Bau des Rückenmarks und der Medulla oblongata eines Microcephalen," dissertation, medical faculty, University of Bern, 1892.

68. Bachmann and Bradenahl, "Medizinstudium von Frauen in Bern;" archives, Bern University, http://apps.uniarchiv.unibe.ch/syscomm/images/mata/3764_3774.gif (July 2014).

69. Sophie Lubnitzky, "Die Zusammensetzung des Thrombus in Arterienwunden in den ersten fünf Tagen," dissertation, medical faculty, University of Bern, 1884.

70. Bachmann and Bradenahl, "Medizinstudium von Frauen in Bern"; archives, Bern University, http://apps.uniarchiv.unibe.ch/syscomm/images/mata/5604_5614.gif (July 2014).

71. For a general discussion of connective tissue, see, for instance, Janet L. Hopson and Norman K. Wessells, *Essentials of Biology* (New York: McGraw-Hill, 1990), 302, 515–16.

72. Sophie Hornstein, "Ein Fall von halbseitigem Riesenwuchs," dissertation, medical faculty, University of Bern, 1893.

73. See Heinz-Gerd Zimmer, "Hugo Kronecker and the Dependence of Heart Function on Blood Supply," *Clinical Cardiology* 28 (2003): 489–90, "Profiles in Cardiology" section.

74. Bachmann and Bradenahl, "Medizinstudium von Frauen in Bern"; archives, Bern University, http://apps.uniarchiv.unibe.ch/syscomm/images/mata/4090_4101.gif (July 2014). Sof'ia Handler's name is rendered "Sara Gandler" in these archives.

75. Gerald F. Yeo, "An Attempt to Estimate the Gaseous Interchange of the Frog's Heart by Means of the Spectroscope," *Journal of Physiology* 6, no. 3 (1885): 93–121.

76. Sophie Handler, "Ueber die Reduction des Hämoglobins im Herzen," dissertation, medical faculty, Bern University, 1886.

77. Bachmann and Badenahl, "Medizinstudium von Frauen in Bern"; archives, Bern University, http://apps.uniarchiv.unibe.ch/syscomm/images/mata/7091_7102.gif (July 2014); matriculation record, Zurich University, http://www.matrikel.uzh.ch/active/static/25134.htm (July 2014). The name Divine is rendered "Divain" in both Bern and Zurich archives. A short note on Divine is included in the discussion of British women physiologists in Mary R. S. Creese, *Ladies in the Laboratory? American and British Women in Science, 1800–1900* (Lanham, Md.: Scarecrow Press, 1998), 151, 152.

78. Julia Divine, "Über die Atmung der Herzen von Kröten und Fröschen," dissertation, medical faculty, Bern University, 7 June 1899; *Zeitschrift für Biologie* 47 (1906): 335–78.

79. Bachmann and Bradenahl, "Medizinstudium von Frauen in Bern"; archives, Bern University, http://apps.uniarchiv.unibe.ch/syscomm/images/mata/7304_7315.gif (July 2014); matriculation register, Zurich University, http://www.matrikel.uzh.ch/active/static/24828.htm (July 2014). Beschasnova is listed by Bachmann and Bradenahl under two different family names, Beschasnoff and Bescharnon. The latter may be an error.

80. Bachmann and Bradenahl, "Medizinstudium von Frauen in Bern"; archives, Bern University, http://apps.uniarchiv.unibe.ch/syscomm/images/mata/8392_8402.gif (July 2014); matriculation register, Zurich University, http://www.matrikel.uzh.ch/active/static/26326.htm (July 2014).

81. Nadine Lomakina, "Über Verlauf und Bedeutung der Herznerven," dissertation, medical faculty, Bern University, 1900.

82. Bachmann and Bradenahl, "Medizinstudium von Frauen in Bern"; archives, Bern University, http://apps.uniarchiv.unibe.ch/syscomm/images/mata/7150_7161.gif (July 2014); matriculation register, Zurich University, http://www.matrikel.uzh.ch/active/static/27075.htm (July 2014).

83. See "The Triennial International Congress of Physiologists," *Nature* 58, no. 1507 (1898): 481–86, on 481; E. J. Marey, "Nécessité de créer une commission international pour l'unification et le contrôle des instruments inscripteurs physiologiques," *Journal of Physiology* 23 (1898–1899): [6]–[7].

84. Ludmilla Schilina, "Vergleich von Ludwigs Kymographen mit Hürthle's Tonographen," dissertation, medical faculty, Bern University, 1899.

85. Bachmann and Bradenahl, "Medizinstudium von Frauen in Bern"; archives, Bern University, http://apps.uniarchiv.unibe.ch/syscomm/images/mata/6331_6339.gif (July 2014); matriculation register, Zurich University, http://www.matrikel.uzh.ch/active/static/24658.htm (July 2014).

86. Sarah Amitin, "Ueber den Tonus der Blutgefässe bei Einwirkung der Wärme und der Kälte," dissertation, medical faculty, Bern University, 1896.

87. Bachmann and Bradenahl, "Medizinstudium von Frauen in Bern"; archives, Bern University, http://apps.uniarchiv.unibe.ch/syscomm/images/mata/4268_4279.gif (July 2014); matriculation register, Zurich University, http://www.matrikel.uzh.ch/active/static/26780.htm (July 2014).

88. E. F. W. Pflüger, H. J. Hamburger, et al., [*Reprints on Physiology IX*], originally collected by the Bibliothek des Instituts für Pharmakologie und physiologische Chemie zu Rostok, 1888–1890.

89. Nadine Popoff, "Ueber die Bildung von Serumalbumin im Darmkanale," dissertation, medical faculty, Bern University, 1887.

90. Bachmann and Bradenahl, "Medizinstudium von Frauen in Bern"; archives, Bern University, http://apps.uniarchiv.unibe.ch/syscomm/images/mata/3863_3874.gif (July 2014); matriculation register, Zurich University, http://www.matrikel.uzh.ch/active/static/26095.htm (July 2014). Bern records are unclear on the date of Konev's birth.

91. Bachmann and Bradenahl, "Medizinstudium von Frauen in Bern"; archives, Bern University, http://apps.uniarchiv.unibe.ch/syscomm/images/mata/5162_5173.gif (July 2014); matriculation register, Zurich University, http://www.matrikel.uzh.ch/active/static/25486.htm (July 2014). Bern and Zurich records differ (by about a year) on date of birth of Gitiss.

92. Bachmann and Bradenahl, "Medizinstudium von Frauen in Bern"; archives, Bern University, http://apps.uniarchiv.unibe.ch/syscomm/images/mata/4385_4396.gif (July 2014).

93. See Santiago Ramón y Cajal, *Texture of the Nervous System of Man and the Vertebrates,* trans. Pedro Pasik and Tauba Pasik, 3 vols. (New York: Springer, 1999–2002), 1:185. This is an annotated and edited translation of the original Spanish *Textura del sistema nervioso del hombre y de los vertebrados*, with the additions added by Cajal to the French version, 1909–1911.

94. Helene Koneff, "Beiträge zur Kenntniss der Nervenzellen in den peripheren Ganglien," dissertation, medical faculty, Bern University, 1886.

95. C. Nörner, "Gitiss, Anna, 'Beiträge zur vergleichenden Histologie der peripheren Ganglien,'" *Zeitschrift für wissenschaftliche Mikroskopie und für mikoskopische Technik* 4 (1887): 385–86.

96. Anna Gitiss, "Beiträge zur vergleichenden Histologie der peripheren Ganglien," dissertation, medical faculty, Bern University, 1887 (Bern: Haller, 1897). This work was also included in the mainly reprints collection *Articles on Neuritis and Neuroses*, put together by Moses Allen Starr and collaborators for the years 1885–1891.

97. Anna Kotlarewsky, "Physiologische und mikrochemische Beiträge zur Kenntniss der Nervenzellen in den peripheren Ganglien," dissertation, medical faculty, Bern University, 1886 (Bern: Haller, 1887). This work was also included in Moses Allen Starr's reprint collection (see note 96 above).

98. C. Nörner, "Kotlarewsky, Anna, 'Physiologische und mikrochemische Beiträge zur Kenntniss der Nervenzellen in den peripheren Ganglien,'" *Zeitschrift für wissenschaftliche Mikroskopie und für mikroskopische Technik* 4 (1887): 386–88.

99. Bachmann and Bradenahl, "Medizinstudium von Frauen in Bern"; archives, Bern University, http://apps.uniarchiv.unibe.ch/syscomm/images/mata/4670_4681.gif (July 2014).

100. Rebecca Halperin, "Uber die abnorme Krümmung der Wirbelsäule bei congenitaler Spaltbildung der Leibeswand," dissertation, medical faculty, University of Bern, 1889.

101. Bachmann and Bradenahl, "Medizinstudium von Frauen in Bern"; archives, Bern University, http://apps.uniarchiv.unibe.ch/syscomm/images/mata/4114_4125.gif (July 2014); see also archives, Bern University, entry for Dorothea Familiant, http://apps.uniarchiv.unibe.ch/syscomm/images/mata/2857_2862.gif (July 2014) and matriculation register, Zurich University, entry for Dorothea Familiant, http://www.matrikel.uzh.ch/active/static/25272.htm (July 2014).

102. Victoria Familiant, "Beiträge zur Vergleichung der Hirnfurchen bei den Carnivoren und den Primaten im Anschlusse an die Untersuchung eines Löwen-Gehirnes," dissertation, medical faculty, Bern University, 1885.

103. For an example of twenty-first century work in this area, see Eliot C. Bush and John M. Allman, "The Scaling of Frontal Cortex in Primates and Carnivores," *Proceedings of the National Academy of Science* 101, no. 11 (2004): 3962–66. Also available online at http://www.pnas.org/content/101/11/3962.full (July 2014).

104. Bachmann and Bradenahl, "Medizinstudium von Frauen in Bern"; archives, Bern University, http://apps.uniarchiv.unibe.ch/syscomm/images/mata/4562_4573.gif (July 2014).

105. Catharina Iwanowa von Kompaneiskaja von Kowalenskaja, "Beiträge zur vergleichenden mikroskopischen Anatomie der Hirnrinde des Menschen und einiger Säugethiere," dissertation, medical faculty, Bern University, 1886.

106. Bachmann and Bradenahl, "Medizinstudium von Frauen in Bern"; archives, Bern University, http://apps.uniarchiv.unibe.ch/syscomm/images/mata/2866_2876.gif (July 2014); matriculation register, Zurich University, http://www.matrikel.uzh.ch/active/static/25670.htm (July 2014); Rohner, *Die ersten 30 Jahre*, 19; Karin Huser, "Friedrich Erismann: Sozialmediziner im Zarenreich—Politiker in Zürich," in *Käser, Künstler, Kommunisten*, eds. Maeder and Niederhäuser, 115–18.

107. Sophie Hasse, "Septicaemi und ihre Ursache," dissertation, medical faculty, Bern University, 1876.

108. Tiburtius, *Erinnerungen einer Achtzigjährigen*, 149.

109. Tiburtius, *Erinnerungen einer Achtzigjährigen*, 150.

110. Bachmann and Bradenahl, "Medizinstudium von Frauen in Bern"; archives, Bern University, http://apps.uniarchiv.unibe.ch/syscomm/images/mata/3598_3609.gif (July 2014); Mariama Kaba, "Roux, César," *Dictionnaire historique de la Suisse*, http://www.hls-dhs-dss.ch/textes/f/F14606.php (July 2014).

111. Anna Bégoune, "Ueber Gefässversorgung der Kröpfe mit besonderer Berücksichtigung der Struma cystica," dissertation, medical faculty, Bern University, 1883.

112. Bachmann and Bradenahl, "Medizinstudium von Frauen in Bern"; archives, Bern University, http://apps.uniarchiv.unibe.ch/syscomm/images/mata/5687_5698.gif (July 2014), Gomberg's entry here is given under the name Malka Omberg; Eva Brinkschulte, *Women Doctors: The Fight for Professionalization, 1875–1954* (Berlin: University Press, 1997); "Dokumentation: Ärztinnen im Kaiserreich: Sammlung von etwas 1000 Biografien" (Berlin: Institut für Geschichte der Medizin, Freie Universität Berlin, 2009), http://web.fu-berlin.de/aeik/HTML/rec01038c1.html (July 2014); Beate Vogt, "Erste Ergebnisse der Berliner Dokumentation: Deutsche Ärztinnen im Kaiserreich," in *Weibliche Ärzte: Die Durchsetzung des Berufsbildes in Deutschland*, ed. Eva Brinkschulte (Berlin: Edition Hentrich, 1993), 159–68, on 160.

113. Sophie Gomberg, "Ueber die Belebungsversuche beim Chloroformtode," dissertation, medical faculty, Bern University, 1894. Since in 1947 the *Deutsche Zeitschrift für Chirurgie* merged with *Archiv für klinische Chirurgie* to form *Langenbecks Archiv für klinische Chirurgie*, Gomberg's 1894 paper may be considered the first by a woman from Bern to appear in that journal, an internationally recognized Berlin publication that published extended reports on new surgical procedures.

114. Bachmann and Bradenahl, "Medizinstudium von Frauen in Bern"; archives, Bern University, http://apps.uniarchiv.unibe.ch/syscomm/images/mata/2933_2944.gif (July 2014); matriculation register, Zurich University, http://www.matrikel.uzh.ch/active/static/27253.htm (July 2014); Pietro Scandola and Franziska Rogger, "Early Women at the University of Bern," in "The History of the University," Archives, Bern University, http://www.unibe.ch/eng/pdf/history.pdf (July 2014). See also Semen Maksimilian Dionesov, "Russkie tsiurikhskie studentki" [Russian students in Zurich], from [History of the medical education of Russian women] in *Sovetskoe Zdravookhranie* 30 (1971): 68–72.

115. Theolina Simonovich was born in Odessa in 1854. After completing her high school studies at a gymnasium in that city, she enrolled in the medical faculty of Zurich University for the summer semester of 1872, but left Zurich in August 1873, in compliance with the tsarist government's requirements. No indication of further study by her was found. See matriculation register, Zurich University, http://www.matrikel.uzh.ch/active/static/27252.htm (July 2014).

116. For Rosa Simonovich's residences in Zurich, see Verein Feministische Wissenschaft Schweiz, *"Ebenso neu als kühn,"* 140–44. For a comprehensive discussion of the Fritschi see Amy Knight, "The Fritschi: A Study of Female Radicals in the Russian Populist Movement," *Canadian-American Slavic Studies* 9, no. 1 (1975): 1–17.

117. Mikhail Petrovich Sazhin (1845–1934) used the name Armand Ross during the year he spent in the United States after escaping from his banishment from St. Petersburg for revolutionary activities. By 1870 he was back in Europe and once again becoming involved in revolutionary work in Zurich. For references to Rosa Simonovich's activities in Zurich's revolutionary circles, see Meijer, *Knowledge and Revolution*, especially 180 note 92, 199 note 92, and 204 note 52.

118. Tiburtius, *Erinnerungen einer Achtzigjährigen*, 154.

119. Rosa Simonowitsch, "Ueber Hyoscyamin und dessen Bedeutung für die Augenheilkunde," dissertation, medical faculty, University of Bern, 1874.

120. See Scandola and Rogger, "History of the University."

121. Bachmann and Bradenahl, "Medizinstudium von Frauen in Bern"; archives, Bern University, http://apps.uniarchiv.unibe.ch/syscomm/images/mata/2877_2883.gif (July 2014); matriculation register, Zurich University, http://www.matrikel.uzh.ch/active/static/27239.htm (July 2014); Werner Dettelbacher, "Dr. med. Marie Siebold—eine der ersten Ärztinnen in Serbien, in der Türkei und in Ägypten," *Münchener Medizinische Wochenschrift* 45, no. 115 (1973): 2051–53 (port.); Paulette Meyer, "They Met in Zürich: Nineteenth-Century German and Russian Women Physicians," in *Women Healers and Physicians: Climbing a Long Hill*, ed. Lilian R. Furst (Lexington: University Press of Kentucky, 1997), 151–77, on 161, 166. Marie Siebold's date of death is uncertain; sources give both 1918 and 1938 (see especially the Dettelbacher reference).

122. Marie Siebold was the third woman of this Siebold family to qualifiy as an obstetrician, two of her forebears, Josepha von Siebold (1771–1849), and Josepha's daughter, Marian Theodore Charlotte Heiland (Charlotte Heiland-Heidenreich von Siebold) (1788–1859), having been awarded degrees in obstetrics more than half a century before Marie. Josepha received her honorary doctorate of midwifery from the University of Giessen in 1815, and was the first woman to obtain this qualification from a German university. She then worked as a midwife and assistant to her second husband, Damian von Siebold, a man from a well-known family of physicians; he himself was a municipal practitioner in the city of Darmstadt, Hesse, and also served the surrounding country areas.

Charlotte von Siebold, Josepha's eldest child by her first husband, government advisor Georg Heiland, was formally adopted by Damian von Siebold. Interested in medicine from an early age, she benefited considerably from instruction from both her parents, and in 1811 went to Göttingen, where for two years she had an opportunity to listen to private lectures given by obstetricians Johann Friedrich Osiander and Konrad Johann Langenbeck of the Göttingen University medical department. In 1814, after satisfying the requirements of the Grand-ducal Medical College in Darmstadt, she was granted permission to practice obstetrics. Her ambitions did not stop there, and three years later, after further studies, she was awarded a doctorate in obstetrics by Giessen University after she presented and publicly defended a comprehensive, twenty-three-page work on extra-uterine pregnancy ("Ueber die Schwangerschaft ausserhalb der Gebärmutter und über Bauchhöhlenschwangerschaft insbesondere"). A notable feature of von Siebold's presentation was her documentation and critical examination of the case of a farm woman of twenty-three who had died following an abdominal operation and whose remains had undergone a postmortem examination; von Siebold's conclusions occasioned considerable discussion.

She then joined her parents' practice in Darmstadt and the surrounding area, where she worked tirelessly to help the poor, carried out a considerable amount of charitable work, and tried to improve the city's provision of obstetrical assistance; after Josepha von Siebold's death, Charlotte took on her mother's patients in addition to her own. Over the years her skills as a midwife became widely known far outside the state of Hesse; she was often called to assist at births in princely families, including that of the future Queen Victoria (child of Victoria, Duchess of Kent), and also that of Queen Victoria's future husband, Prince Albert, son of Duchess Louise-Saxe-Coburg-Gotha. Both these births were in 1819, two years after her graduation.

After Charlotte von Siebold's death in 1859, a number of landmarks in Darmstadt were named after her, one of them the Sieboldstrasse. A considerable amount of material has now been published on these early Siebold medical women; for Josepha, see, for instance, Jost Benedum, "Zur Geschichte der Medizinischen Fakultät," first published in 1982 in *Festschrift der Universität anlässlich ihres 375-jährigen Bestenens*, now available online at http://fss.plone.uni-giessen.de/fss/fbz/fb11/dekanat/geschichte/Facultas Medica/geschichte_fakultaet.doc/file/geschichte_fakultaet.doc (July 2014), and for Charlotte, see Ulrike Enke, "Geburtshelferin der englischen Königin," *Hessisches Ärzeblatt* 8 (2009): 525–26, online at http://www.laekh.de/upload/Hess._Aerzteblatt/2009/vollst._Ausgaben/HAEBL_08_2009.pdf (July 2014). In 2006 the University of Göttingen started a program promoting women in science, one section of which told the story of Charlotte Heidenreich von Siebold; see Universitätsmedizin Göttingen, http://www.med.uni-goettingen.de/de/content/forschung/182.html (July 2014). Both Josepha and Charlotte von Siebold are mentioned in *The Northeastern Dictionary of Women's Biography*, comp. and ed. Jennifer S. Uglo, 3rd ed., revised by Maggy Hendry (Boston: Northeastern University Press, 1998), 496–97.

123. *St. Petersburg Encyclopaedia*, s.v. "St. Petrischule," http://www.encspb.ru/en/article.php?kod=2804012124 (July 2014).

124. See note 64 above.

125. Marie Siebold, "Ueber die Bestimmung der Zeitdauer der Schwangerschaft nach der äusseren Untersuchung, sowie über den Einfluss der Ausdehnung des Uterus auf den Gang der Involution," dissertation, medical faculty, Bern University, 1874 (Bern: Rieder & Simmen, 1874).

126. The Order of the Takovo Cross for exceptional service and merit in war or peace was instituted in 1865 by the then new Serbian principality as its first decoration. It became obsolete in 1903, when a new dynasty took power.

127. See Mary R. S. Creese, *Ladies in the Laboratory II: West European Women in Science, 1800–1900* (Lanham, Md.: Scarecrow Press, 2004), 177–78.

128. Correspondent, Constantinople, *Lancet* 2 (1908): 1338.

129. Bulgaria also was involved in the occupation of Albanian territory after the scattering of Serb forces by the Central Powers during World War I.

130. Bachmann and Bradenahl, "Medizinstudium von Frauen in Bern"; archives, Bern University, http://apps.uniarchiv.unibe.ch/syscomm/images/mata/3182_3194.gif (July 2014).

131. Marie Ekunina, "Ueber die Ursache der sauren Reaction der thierischen Gewebe nach dem Tode," dissertation, medical faculty, Bern University, 1880.

132. Bachmann and Bradenahl, "Medizinstudium von Frauen in Bern"; archives, Bern University, http://apps.uniarchiv.unibe.ch/syscomm/images/mata/2894_2906.gif (July 2014); matriculation register, Zurich University, http://www.matrikel.uzh.ch/active/static/24810.htm (July 2014); Carl Alpert, "Berlin, Fanny," in *The Universal Jewish Encyclopedia*, ed. Isaac Landman (New York: Universal Jewish Encyclopedia, 1940), 2:216; Judith Chasin, "Fanny Berlin," in *Jewish Women: A Comprehensive Historical Encyclopedia*, 1 March 2009, Jewish Women's Archive, http://jwa.org/encyclopedia/article/berlin-fanny (July 2014); Rohner, *Die ersten 30 Jahre*, 23; Tiburtius, *Erinnerungen einer Achtzigjährigen*, 115, 155–56; death notice in "Obituary" section of *British Medical Journal* (19 November 1921): 2:870, online at http://www.bmj.com/content/2/3177/870.3 (July 2014).

133. Alpert, "Berlin, Fanny." Alpert's 1940 article contains a number of identifiable errors and should be treated with caution.

134. The name *Kosakenpferdchen* for that early generation of Russian women students is by no means forgotten and can be found in stories and in historical accounts of late nineteenth-century Zurich. See, for instance, "Kosakenpferdchen an der Platte," in *Quartierverein Fluntern* (2011), http://www.zuerich-fluntern.ch/index.php?nav=21 (July 2014).

135. Fanny Berlinerblau, "Über den directen Übergang von Arterien in Venen," dissertation, medical faculty, Bern University, 1875.

136. See Alexander Vucinich, *Science in Russian Culture, 1861–1917* (Stanford, Calif.: Stanford University Press, 1970), ch. 2.

137. Léon Le Fort, "Nouveau procédé pour la guérison du prolapsus utérin," *Bulletin Général de Thérapeutique Médicale et Chirurgicale* 92 (1877): 337–44 (noted briefly in the United States weekly periodical *Medical Record* [New York], 31 January 1880). Léon Clément Le Fort (1829–1893) was remembered especially for this procedure, known as "Le Fort's Operation," for correcting uterine prolapse. Professor of surgery at the University of Paris medical faculty from 1873 and surgeon at the Hôtel-Dieu, he was also known for advances in ankle and foot surgery.

138. Tiburtius, *Erinnerungen einer Achtzigjährigen*, 155–56.

139. Bachmann and Bradenahl, "Medizinstudium von Frauen in Bern"; archives, Bern University, http://apps.uniarchiv.unibe.ch/syscomm/images/mata/5162_5173.gif (July 2014), see also http://apps.uniarchiv.unibe.ch/syscomm/images/mata/5687_5698.gif (July 2014); matriculation register, Zurich University, http://www.matrikel.uzh.ch/active/static/25361.htm (July 2014). Bern and Zurich records differ in some of the details concerning Sof'ia Frenkel.

140. Sophie Frenkel, "Klinische Untersuchungen über die Wirkung von Coffeïn, Morphium, Atropin, Secale cornutum und Digitalis auf den arteriellen Blutdruck, angestellt mittelst des v. Basch'schen Sphygmomanometer," dissertation, medical faculty, Bern University, 1889.

141. See matriculation register, Zurich University (in note 139 above).

142. Bachmann and Bradenahl, "Medizinstudium von Frauen in Bern"; archives, Bern University, http://apps.uniarchiv.unibe.ch/syscomm/images/mata/5639_5650.gif (July 2014).

143. See also the earlier statistical work of Zurich student Anna Kleimann on infant mortality rates (above).

144. The London General Register Office began collecting data in 1837. In 1840 it was also given responsibility for the census, beginning with the census of 1841. See Edward Higgs, "The Early Development of the General Register Office," Online Historical Population Reports, UK Data Archive, University of Essex, http://

www.histpop.org/ohpr/servlet/View?path=Browse/Essays (by kind)&active=yes&mno=2002 (July 2014). For the first forty-two years its chief statistician was William Farr (1807–1883), a medical practioner, epidemiologist, and ardent reformer, considered one of the founders of medical statistics. See John M. Eyler, "Farr, William (1807–1883), Statistician and Epidemiologist," *Oxford Dictionary of National Biography* (Oxford: Oxford University Press, 2004), 19:99–101. Sandberg attributed all the data to him although he was never Registrar General; he retired in 1879 and died before the end of her study period. The death data she used appeared in *Annual Report of the Registrar-General of Births, Deaths, and Marriages in England* (London: Her Majesty's Stationary Office, 1859–1887).

Note also the statistical work of Florence Nightingale, who collected and analyzed data on deaths from diseases in the British Army. With help from William Farr and other medical colleagues she produced graphical presentations illustrating the impact of disease on the army and the effects of improved sanitary conditions. Her work led to improvements in sanitation that dramatically reduced army mortality rates from the 1860s on. See Creese, *Ladies in the Laboratory?*, 174–77.

145. Dina Sandberg, "Über die Abnahme der Lungenschwindsucht während der drei letzten Decennien nach Beruf und Geschlecht," dissertation, medical faculty, Bern University, 1890. Sandberg's report was also published in 1891 in St. Petersburg, with the title "Die Abnahme der Lungenschwindsucht in England während der drei letzten Decennien nach Beruf und Geschlecht" (Buchdruckerei der Kaiserlich Akademie der Wissenschaft).

146. Bachmann and Bradenahl, "Medizinstudium von Frauen in Bern"; archives, Bern University, http://apps.uniarchiv.unibe.ch/syscomm/images/mata/6180_6191.gif (July 2014); Mélanie Lipinska, *Histoire des Femmes Médecines depuis l'Antiquité jusqu'à nos Jours* (Paris: G. Jacques, 1900), 408.

147. Olga Kovalevskij, "Beiträge zur Kenntniss der Leukocytose und Chemotaxis," dissertation, medical faculty, Bern University, 1896.

148. It is perhaps worth restating here that the Royal Society did not index reports published in journals devoted primarily to clinical medicine.

149. Swiss neurologist and child psychologist Édouard Claparède (1873–1940), professor of psychology at the University of Geneva from 1915, became widely known for his extensive studies in memory and in infant psychology. In 1901, along with Théodore Flournoy, he founded and edited the much-respected *Archives de Psychologie*. With his international standing in the teaching of modern childhood education, his presence at the University of Geneva did much to establish Geneva as a center for studies in that field.

150. Klavdia Markova, "Contribution à l'étude de la perception stéréognostique," inaugural thesis, medical faculty, Geneva University, 1900 (Geneva: Eggimann, 1900). *Jahresverzeichnis der Schweizerischen Universitätsschriften, 1899–1900* (Basel: Schweighauserische Buchdruckerei, 1900), 37.

151. Information in technical papers; Alphonse Rebière, *Les Femmes dans la Science: Notes recueilles*, 2nd ed. (Paris: Nony, 1897), 251; "Schepiloff (Catherine)," in John Briquet, *Biographies des botanists à Genève de 1500 à 1931*, ed. François Cavilliet (Geneva: Albert Kundig, 1940), 428–29, in the series *Berichte der Schweizerischen Botanischen Gesellschaft*; "Schépiloff, Mlle Catherine," in *Section Genevoise, Publications des Members de la Société de Physique et d'Histoire Naturelle de Genève*, 1er suppl. (1896), 140.

152. Moritz Schiff, *Lehrbuch der Physiologie des Menschen: 1, Lehrbuch der Muskel- und Nervenphysiologie* (Lahr: M. Schauenburg, 1858–1859).

153. E. C. Bate-Smith and J. R. Bendall, "Changes in Muscle after Death," *British Medical Bulletin* 12, no. 3 (1956): 230–35. Bate-Smith and Bendall refer to Schipilova as the originator of the lactic acid theory (p. 230).

154. Matriculation register, Zurich University, http://www.matrikel.uzh.ch/active/static/25197.htm (July 2014); *Jahresverzeichniss der Schweizerischen Universitätsschriften, 1898–1899* (Basel: Schweighauserische Buchdruckeri, 1899), 39, online at http://babel.hathitrust.org/cgi/pt?id=uc1.b4228802;view=1up;seq=115 (July 2014).

155. Lydie Egon de Besser (de St. Pétersbourg), "De l'action mécanique des muscles des doigts et du poignet. II. De la rétraction des muscles après la section de leur tendon," dissertation, medical faculty, Lausanne University, 1898.

156. Besser, (L. V.), [Brief note on bubonic plague in man, observed in Mongolia and beyond Baikal in 1888–97], *Voyenno-meditsinskii zhurnal* [Military-medical journal], St. Petersburg, 215 (1906): 279–93.

157. *Jahresverzeichnis der Schweizerischen Universitätsschriften, 1899–1900* (Basel: Verlag der Universitätsbibliothek, 1900), 42, online at http://babel.hathitrust.org/cgi/pt?id=uc1.b4228802;view=1up;seq=178 (July 2014); *Catalogue des Étudiants de l'Université de Lausanne, 1895–1896* (Lausanne: Couchoud, 1895), 16.

158. Anna Ludkewitch, "L'articulation de l'épaule: Étude d'arthrologie comparée," dissertation, medical faculty, Lausanne University, 1899/1900.

159. Matriculation register, University of Zurich, http://www.matrikel.uzh.ch/active/static/26990.htm (July 2014). Solomon's year of birth is given as 1872 in her 1895 registration record and 1873 in that of 1896. These Zurich records give Salomon as an alternative spelling for her name.

160. Vera Solomon, "Experimentelle Untersuchungen über Rabies," dissertation, medical faculty, University of Lausanne, 1900. Listed in *Jahresverzeichnis der Schweizerischen Universitätsschriften, 1900–1901* (Basle: Schweighauserisch Buchdruckei, 1901), p. 52, entry 33, online at http://babel.hathitrust.org/cgi/pt?id=uc1.b4228802;view=1up;seq=258 (July 2014).

161. Faber Renato Fabbris, "Biography of Renato Caccioppoli," *MacTutor History of Mathematics Archive*, School of Mathematics and Statistics, University of St. Andrews, http://www-history.mcs.st-andrews.ac.uk/Biographies/Caccioppoli.html (Sept. 2014); Mark Leier, *Bakunin: The Creative Passion* (New York: St. Martin's Press, 2006), especially 239–41, 251–52; E. H. Carr, *Michael Bakunin* (1935; New York: Octagon Books, 1975); Francesca Patuelli, "Bakunin, Maria," *Scienza a due voci*, http://scienzaa2voci.unibo.it/biografie/5-bakunin-maria (July 2014); Vladimir Keidan, "Marija Michajlovna Bakunina," *Russi in Italia: Dizinario*, http://www.russinitalia.it/dettaglio.php?id=476 (July 2014).

162. Leier, *Bakunin*, 251–52.

163. Fabbris, "Biography of Renato Caccioppoli."

164. For a sketch of Sof'ia Bakunina's younger sister, chemist Maria Bakunin, see chapter 2.

165. Matriculation register, Zurich University, http://www.matrikel.uzh.ch/active/static/27186.htm (July 2014); Bachmann and Bradenahl, "Medizinstudium von Frauen in Bern"; archives, Bern University, http://apps.uniarchiv.unibe.ch/syscomm/images/mata/3019_3031.gif (July 2014); D. G. Kvasov and A. K. Fedorova-Grot, *Fiziologicheskaia Shkola I. P. Pavlova: Portrety i kharakteristiki sotrudnikov i uchenikov* [The physiological school of I. P. Pavlov: Portraits and characteristics of his coworkers and students] (Leningrad: Isdatel'stvo Nauka, 1967), 280–81; George Windholz, "Three Researchers in Pavlov's Laboratories," *NWSA Journal* 1, no. 3 (1989): 491–96; Daniel P. Todes, *Pavlov's Physiology Factory: Experiment, Interpretation, Laboratory Enterprise* (Baltimore: Johns Hopkins University Press, 2002).

166. Student records for Zurich and Bern universities differ by one year for Shumova's date of birth, and the Zurich record lists an alternative place of birth, Samara, the city at the confluence of the Volga and Samara rivers.

167. Catharina Schumoff, "Untersuchung über die Abhängigkeit des Übergangs in den Harn von der Verschiedenheit der Einverleibungsarten," dissertation, medical faculty, Bern University, 1877 (Bern: Buchdruckerei G. Michel, 1877), 36 pp.

168. See, for instance, E. O. Shumova, [On the question of the physiological effect of camphor], *Ezhenedel'naia klinicheskaia gazeta* [Weekly clinical gazette] 3 (1883): 177, 201. This periodical was edited and published by Sergei Botkin, professor at the Medical-Surgical Academy and head of its research laboratories.

169. For a full description of the procedure, see Todes, *Pavlov's Physiology Factory*, 95–96.

170. See Windholz, "Three Researchers," 492.

171. See, for instance, N. O. Sieber and E.-O. Schumoff-Simanowskaja, "De l'action de l'érepsine et du suc intestinal sur les toxines et sur l'abrine," *Arkhiv Biologicheskikh Nauk* (St. Petersburg) 10 (1903): 1–11; E. O. Shumova-Simanovskaia and N. O. Ziber, [Influence of bacterial toxins and the virus of rabies on the oxidizing function of the animal organism], *Arkhiv Biologicheskikh Nauk* (St. Petersburg), 11 syph. (1904): 68–78.

172. Obituary, *Niva* 13 (1903): 257–58; S. I. Vinokurov and R. V. Chagovets, "Mar'ia Manasseina i ee rol' b otkrytie beskletochnogo brosheniia" [Maria Manasseina and her role in discovering cell-free fermentation], *Biokhimiia* 15, no. 6 (1950): 558–62; John Lagnado, "Was the First Biochemist a Woman?," *Biochemist* 14, no. 5 (1992): 21–22; Athel Cornish-Bowden, "The Origins of Enzymology," *Biochemist* 19, no. 2 (1999): 36–38; Vladimir M. Kovalzon, "Some Notes on the Biography of Maria Manasseina," *Journal of the History of the Neurosciences* 18 (2009): 312–19; Ingrid Kästner, "'Leider haben mich Familienverhältnisse abgehalten . . .': Leben und Werk der Ärztin Maria Michailowna Manasseina," in *Geschlecterverhältnisse in Medizin, Naturwissenschaft und Technik*, eds. Christoph Meinel and Monika Renneberg (Bassum: Verlag für Geschichte der Naturwissenschaft und der Teknik, 1996), 62–68; S. Finger, "Women and the History of the Neurosciences," *Journal of the History of the Neurosciences* 11, no. 1 (2002): 80–86.

173. Kovalzon, "Some Notes on the Biography of Maria Manasseina," 313.

174. Maria Manassein, "Beiträge zur Kenntniss der Hefe und zur Lehre von der alkoholischen Gährung," in Julius Wiesner, *Mikroskopische Untersuchen, ausgeführt im Laboratorium für Mikroskopie und technische Waarenkunde am k. k. Polytechnischen Institute in Wien* (Stuttgart: Maier, 1872), 116–28.

175. Justus Liebig, *Ueber die Gährung, und die Quelle der Muskelkraft und über Ernährung* (Munich: Straub, 1870), in *Sitzungsberichte der Königl. Bayerischen Akademie der Wissenschaften zu München* 2, no. 4 (1869): 393–482; Louis Pasteur and Justus Liebig, *Note sur un mémoire de M. Liebig, relative aux fermentations* (Paris: Gauthiers-Villars, 1871).

176. Marie von Manasseïn, "Zur Frage von alkoholischen Gährung . . . ," *Berichte der Deutschen Chemischen Gesellschaft* 30 (1897): 3061–62, on 3061 (see bibliography). It is perhaps also worthy of note that by 1870 Manasseina already had a number of other strong research interests, particularly in education and child psychol-

ogy (see later discussion); there is little hard evidence to assume that her brief period of biochemical research, outstanding though her findings were, would have led her to rearrange her research priorities in favor of further work in that area.

177. Marie de Manacéine, *Le sommeil, tiers de notre vie: Pathologie, physiologie, hygiène, psychologie*, trans. Ernest Jaubert (Paris: Masson, 1896), 358 pp.; *Sleep: Its Physiology, Pathology, Hygiene, and Psychology* (London: W. Scott, 1897; New York: Scribner's, 1899; Washington, DC: American Psychological Association, 2009), 341 pp.; *Sömnen och dess betydelse för lif och hälsa: dess hygien och psykologi: Öfvers. från den genomsedda och tillökade engelska uppl. af O.H.D.* (Stockholm: Geber, 1895, 1898), 300 pp.

178. G. T. W. Patrick, review of *Sleep: Its Physiology, Pathology, Hygiene, and Psychology*, by Marie de Manacéine, *Science*, n.s., 7, no. 162 (4 February 1898): 175–76. George Patrick and his colleague, J. Allen Gilbert, both from the University of Iowa, had recently published their results from an experiment on sleep deprivation in humans. See G. T. W. Patrick and J. A. Gilbert, "Studies from the Psychological Laboratory of the University of Iowa: On the Effects of Loss of Sleep," *Psychological Review* 3 (1896): 469–83. The investigation, clearly suggested by Manasseina's work on dogs, did not inspire further research on human subjects until many years later. See Thomas Fuchs and Jeffrey Burgdorf, "Replication and Pedagogy in the History of Psychology IV: Patrick and Gilbert (1896) on Sleep Deprivation," *Science and Education* 17 (2008): 511–24.

179. These publications included the following: Maria Mikhailovna Manassina, *O vospitanii detei v pervye gody zhizni* [Child rearing during the earliest years] (St. Petersburg: tip. IAkova Treia, 1870); Marie de Manacéine, *Le surmenage mental dans la civilization modern: Effets, causes, remèdes*, trans. E. Jaubert, with a preface by Charles Richet (Paris: Masson, 1890), Russian original published 1886, and German translation by Ludwig Wagner, with revision, published as *Die geistige Überbürdung in der modernen Kultur* (Leipzig: Barth, 1905); Marie de Manasséine, *L'Anarchie passive et le Comte Léon Tolstoï: Le salut est en vous* (Paris: F. Alcan, 1895), a work in which she criticized the political views of Leo Tolstoy, who in his later years devoted much thought to political questions, such as agrarian reform, and Russian social conditions. The following additional titles, from Konovalov, "Notes on the Biography of Maria Manassina," 314, have not been verified: Maria Manasseina, [Principles of education from the first years of life to the end of university studies], five issues (St. Petersburg, 1894–1902); *Ob ustalost'* [On fatigue] (St. Petersburg, 1892); [Something about arts], 1895; [On consciousness], 1896.

180. Eduard Buchner, "Alkoholische Gährung ohne Hefezellen: Vorläufige Mittheilung," *Berichte der Deutschen Chemischen Gesellschaft* 30 (1897): 110–13, 117–24, 335.

181. Eduard Buchner and Rudolf Rapp, "Alkoholische Gährung ohne Hefezellen," *Berichte der Deutschen Chemischen Gesellschaft* 31 (1898): 209–17. See also Eduard Buchner, Hans Buchner, Martin Hahn, Ludwig Geret, and Rudolf Rapp, *Die Zymasegärung. Untersuchungen über den Inhalt der Hefezellen und die biologische Seite des Gärungsproblems: Aus dem Hygienischen Institut der Kgl. Universität München und dem Chemischen Laboratorium der Kgl. Landwirtschaftliche Hochschule zu Berlin* (Munich and Berlin: R. Oldenbourg, 1903), especially 17.

182. Vinokurov and Chagovets, [Maria Manasseina and her role in discovering cell-free fermentation].

183. Bachmann and Bradenahl, "Medizinstudium von Frauen in Bern"; archives, Bern University, http://apps.uniarchiv.unibe.ch/syscomm/images/mata/2933_2944.gif (July 2014); matriculation register, Zurich University, http://www.matrikel.uzh.ch/active/static/27177.htm (July 2014).

184. Nadezhda Schultz, "Ueber die Vernarbung von Arterien nach Unterbindungen und Verwundungen," dissertation, medical faculty, Bern University, 1877.

185. Grigorii Karpov, [L'action désinfectante des monochlorophénols et de leurs éthers salicyliques, et leurs métamorphoses dans l'organisme], *Archives des Sciences Biologiques* 2 (1893): 304–25.

186. N. V. Vorotyntseva and I. S. Vintovkina, "Tsiklinskaia, Praskovia Vasilievna (1859–1923): [On the 100th anniversary of her birth]," *Pediatriia* 24, no. 6 (1959): 94; Rebière, *Femmes dans la science*, 278; information in technical papers.

187. The spelling Tsilinsky, rather than Tsiklinsky, for the type species appears to arise from a misspelling of Tsiklinskaia's name in her original 1899 publication. The thermoactinomyces (genus and species) have remained subjects of interest from Tsiklinskaia's 1899 report until the present.

188. Prascovia V. Tsiklinskaia, "Recherches sur les microbes thermophiles; thèse presentée à la faculté des sciences de l'Université de Genève pour obtenir le grade de docteur ès sciences, par Prascovie Tsiklinsky . . ." (Geneva: H. Kündig, 1903), 91 pp.; also published in *Bulletin de la Société Impériale des Naturalistes de Moscou* (Moscow: typ. de l'Université Impériale, 1903).

189. P. V. Tsiklinski, "Sur la flore microbienne thermophile du canal intestinal de l'homme," *Annales de l'Institute Pasteur* (1903): 217–40.

190. P. V. Tsiklinski, "Flore microbienne," in *Expédition antarctique française 1903–1905, commandée par le Dr Jean Charcot: Sciences naturelles: Documents scientifiques,* eds. Jean-Baptiste Charcot and Louis Joubin (Paris: Masson, 1906–1908), 34 pp.

191. P. V. Tsiklinskaia, [A contribution on penetration of the skin by microorganisms], in Semen I. Zlatogorov, *Uchenie o mikroorganizmakh* [Studies on microorganisms], parts 1–3 (Petrograd: Ivd-vo Prakticheskaia meditsina, 1916–1918), 202–54.

192. Information in technical papers.

193. For a contemporary reference to this work, see Frank R. Blaxall, "A Bacteriological Investigation of the Suppurative Ear Discharge Occurring as a Complication in Scarlet Fever," *British Medical Journal* (21 July 1894): 116–18, on 117.

194. Marie Raskin, "Eine neue einzeitige Doppelfärbungsmethode für die Polkörperchen der Diphteriebazillen," *Deutsche Medizinische Wochenschrift* 37 (1911): 2384; Marie Raskin, "Ergänzung der Mitteilung über die einzeitige Doppelfärbungsmethode der Diphteriestäbchen," *Deutsche Medizinische Wochenschrift* 38 (1912): 609; see also the report in *American Journal of Medical Sciences* 143, no. 5 (1912): 906–7.

195. Bachmann and Bradendahl, "Medizinstudium von Frauen in Bern"; archives, Bern University, http://apps.uniarchiv.unibe.ch/syscomm/images/mata/4790_4801.gif (July 2014); information in publications.

196. Mme Marie Rechtsamer, née Schmoulevich, "Des troubles cardiaque simulant l'hypertrophie cardiaque de croissance," thesis, faculty of medicine, University of Paris, 1895, from *Catalogue des thèses et écrits académique, Part 3*, Ministère de l'Instruction Publique, Années scolaires, 1894–1899.

197. Matvei Rechtsamer, *Materialy k voprosu ob etiologii kholery* (St. Petersburg, 1895), reprinted in Dutch as *Bouwstoffen voor de etiologie der cholera* (St. Petersburg, 1895).

198. Nikolai Alekseevich Sacharov, *Das Eisen als das thätige Prinzip der Enzyme und der lebendigen Substanz*, trans. Marie Rechtsamer (Jena: Fischer, 1902).

199. *Nouveau Montpellier Médical, 1900*, eds. Delord-Boehm and Martial (Montpellier, 1900).

200. Rosalie Zaidmann, "Contribution à l'étude expérimentale du pouvoir pathogène de Bacilles d'Eberth et coli. Injections intra-spléniques," thesis, faculty of medicine, University of Montpellier, 1900. The work was reported in a short, joint note with Rodet (see bibliography). Eberth's bacillus, which causes typhoid fever, is now known as *Salmonella typhi*.

201. Although Mariia Miropolskaia's name is sometimes written with the title Mme, the *London Medical Record* for 1885 refers to her as Dr. M. A. Miropolskaia, as does the *Archive de Physiologie* paper of 1885.

Chapter Two

Opening Laboratory Doors

Chemists and Biochemists

Among the Russian women whose contributions to chemical research are listed in the Royal Society's *Catalogue*, two stand out as especially notable, Nadezhda Ziber-Shumova and the Russian-Italian Maria Bakunin. Five more were very able research workers, notable in various ways and productive to varying degrees—Anna Volkova, Juliia Lermontova, Evdokia Fomina-Zukovskaia, Vera Bogdanovskaia-Popova, and Zenaida Kikina. Most of the remaining ten typically authored or coauthored two or three papers reporting research carried out during student years at the Higher Courses for Women at St. Petersburg or Moscow Universities; a few also worked at laboratories in the medical and agricultural academies in St. Petersburg and Moscow.

The sum of the contributions to scientific knowledge of these ambitious young women was modest, unavoidably so since they were pioneers of nineteenth-century women's push for entrance to equal scientific training and career opportunities. They were notably fortunate in some ways, since many of them studied under the guidance of the outstanding Russian chemists of the time; their teachers and mentors included Mikhail Butlerov, Dmitrii Mendeleev, Vladimir Markovnikov, and Aleksandr Borodin. These men were active supporters of the cause of higher education for women, well ahead of their colleagues in nearly all the scientifically advanced Western countries at the time (Switzerland excluded). Most of the leading Russian academic chemists willingly made places for women in their laboratories.

The research efforts of the women chemists have one or two rather striking features. First, there is the relatively large proportion whose activities were centered around Mikhail Butlerov, the most influential Russian organic chemist of his generation, famous not only for his groundbreaking theoretical contributions to structural theory in organic chemistry but also for his mentoring of a sizable school of outstanding students. Much of the absorbing and rewarding work of verifying Butlerov's theories and the predictions these led to was carried out by his students, male and female; most of the women, with the exception of Lermontova, were minor figures in the present story, but nevertheless their modest contributions, made during their student years, are worth noting.

Also eye-catching is the participation of one or two women in the petroleum research being carried out by Vladimir Markovnikov at Moscow University. Both Juliia Lermontova and Evdokia Fomina-Zukovskaia collaborated with Markovnikov in his program of systematic studies of Caucasus petroleum deposits, studies of international interest in this early period of exploration and development of oil resources worldwide. Although neither Lermontova nor Fomina-Zukovskaia worked in petroleum research for very long, their contributions constitute an interesting note in the story of women's first steps into industrial research.

Nadezhda Olimpievna Shumova, later Ziber-Shumova[1] (1856–1914), daughter of Olimpie Shumov, was born on 18 May 1856, most likely in St. Petersburg. The family was middle class and fairly comfortably situated. Along with her older sister, Ekaterina (see chapter 1), Nadezhda was taught at home in her early years and then attended lectures at the St. Petersburg Higher Courses for Women. She concentrated on chemistry, which at that time was being taught by some of the most distinguished scientists in the field, including Mendeleev, Butlerov, and Wagner. While still in her teens she married Nikolai

Ivanovich Ziber (1844–1888), an economist from Kiev, with whom she went to Switzerland in the summer of 1872, when he audited classes at the University of Zurich.[2] The Zibers were accompanied by Nadezhda's sister, who registered in the medical faculty. The 1873 order from the tsarist government that Russian women students must leave Zurich or suffer severe restrictions on any future professional work in Russia disrupted the sisters' plans, but by 1875 Nadezhda was enrolled in the University of Bern medical faculty. Her major subjects were microbiology and physiological chemistry. For her dissertation research she carried out a study of mold-causing fungi, directed by Polish physiological chemist Marcellus (Marcelli) Nencki. She received her degree in 1880.[3]

Her long association with Nencki, a pioneer in the chemical approach to the study of microorganisms, began in 1877, her second year at the University of Bern. Nencki (1847–1901) had that year been appointed full professor and director of Bern's Institute of Medical Chemistry, where he remained until 1891. One of Nencki's coworkers for almost a quarter of a century, Nadezhda Ziber-Shumova authored or coauthored more than thirty papers in mainstream journals during this period, a record that most likely makes her the most productive woman research chemist of the late nineteenth century. Her work covered a variety of biochemical, chemical, and bacteriological topics including *in vivo* oxidation, bacterial fermentation and decomposition processes, enzyme studies, and sugar hydrolysis, as well as basic chemical studies such as the zinc chloride–catalyzed acylation of phenols known as the "Nencki reaction," published in 1881.[4] Particularly notable was her work on the chemical constitution of the blood pigment hemin (see bibliography), part of a long series of systematic investigations of this large molecule, mainly by degradation followed by identification of the resulting fragments; Nencki's group worked on the problem over many years and was followed by a number of other distinguished chemists.[5] Among the findings that emerged from these studies was the close chemical relationship between hemin and the plant pigment chlorophyll.

Nencki left Bern in 1891, having been invited, along with physiologist Ivan Pavlov, to organize the newly founded Institute of Experimental Medicine in St. Petersburg; Nencki became head of the Department of Chemistry and Biochemistry. Nadezhda Ziber-Shumova, a widow from 1888, followed Nencki to St. Petersburg and joined the institute's chemical department staff; the Polish chemist and health worker Szymon Dzierzgowski, Martin Hahn, and Ziber-Shumova were Nencki's three senior coworkers.

During the periodic epidemics and famines that ravaged vast areas of the country throughout the 1890s, the authorities often proved unable to handle the problems. Russian professionals in public health and sanitation worked to introduce improvements and provide immediate assistance; *zemstvo* doctors, a number of whom were women, were especially active in the work. When the cholera epidemic of 1892 began, Ziber-Shumova joined Nencki and Dzierzgowski in the effort to organize clinics and field observational units in affected regions. Their search for cheap, readily accessible antiseptics led to a detailed investigation of the chemical composition of pine pitch and a study of its disinfecting properties. Other bacteriological work carried out about this time included studies of pathogens causing epidemics in fish populations and efforts to combat protozoan infections of the skin. Especially notable was the work on *Rinderpest* (cattle plague). Commissioned by the founder of the Institute of Experimental Medicine, Prince Aleksandr Petrovich Ol'denburgskii, to carry out field trials involving serum immunization, Nencki and his team made several trips to affected regions in the late 1890s. Ziber-Shumova took part in expeditions to the Caucasus in 1895 and 1898 (see bibliography).[6] The work resulted in two stations being established, one in the Tiflis (now Tbilisi) region in the Caucasus, and the other in Siberia, near Chita, about five hundred miles east of Lake Baikal on the then partially completed Trans-Siberian Railway. The results were so favorable that herdsmen in both regions had no hesitations about bringing their animals for immunization; thousands were successfully treated.

Although German bacteriologist Robert Koch is the most widely recognized biologist for the successful investigation of livestock immunization against cattle plague by serum injection, it is worth noting that the work of Nencki and colleagues in the Caucasus was done before Koch's studies in East Africa. Indeed, Koch used the Nencki team's findings without ever making any reference to the latter group's contributions to the field. Nencki is said to have remarked philosophically that "C'est toujours comme ça!"[7] The incident is reminiscent of another Western scientist's ignoring the fact that his work was in effect a repetition of much earlier research by a Russian, Maria Manasseina, whose career is outlined in chapter 1.

Ziber-Shumova's joint studies with Nencki included examinations of gastric juice and its enzyme content, an extension of Nencki's chemical investigations with Ekaterina Shumova-Simanovska on the nature of the digestive enzyme pepsin.[8] Other papers, both joint and single-author, reported research on putrefaction, fermentation processes, the role of intestinal bacteria, and enzyme studies, some of the latter with her sister, Ekaterina.[9]

After Nencki's death in 1901, Nadezhda Ziber-Shumova became head of the institute's chemical department. She was the first female department head and with that rank received the right to full membership of the institute. In 1902 and 1904 she brought out two biographical articles on Nencki's scientific work.[10]

Over the following decade her research included a joint study with S. Metalnikov on the food and digestive processes of the greater wax moth, *Galleria mellonella*, and several other areas of interest including collaborative investigations with her colleague W. Dzierzgowski on lung tissue.[11]

In 1906, along with her brother-in-law, N. P. Simanovskii, she founded "Zorka" (Dawn), a school for girls in Novgorod Province. The school, established in memory of her sister, Ekaterina Shumova-Simanovskaia, Simanovskii's wife, was dedicated to the education of the children of peasants and country people in the region. Nadezhda Ziber-Shumova, one of the most productive women in the chemical and biochemical sciences of the late nineteenth century, died when she was about sixty, probably in 1914.[12]

MARIA (MARUSSIA) MIKHAILOVNA BAKUNIN[13] (1873–1960), the third child of Russian philosopher and revolutionary Mikhail Bakunin and his wife, Antonia Kwiatkowska, was educated in Italy and worked there throughout her whole career. Nevertheless, because of her parentage, she has a place among Russian women chemists.

Maria Bakunin was a noteworthy figure in the history of early twentieth-century chemistry in Italy; her life's work is now well recorded, most recently in the writing of her biographers Marco Ciardi and Miriam Focaccia, but already widely available through detailed commentaries and articles by her student and then coworker, the distinguished chemist Rodolfo Alessandro Nicolaus.

She was born in Krasnoiarsk, Siberia, on 2 February 1873. Her mother, Antonia, was the elder daughter of a Polish businessman exiled to Siberia for his political activities, as was Bakunin at that time. The family, following Bakunin's escape, had moved to Italy well before Maria's birth, but on the death of Antonia's last surviving brother she journeyed to Krasnoiarsk to visit her relatives in June 1872. Maria was brought to Switzerland a month after her birth.[14] The family moved frequently in the course of Bakunin's revolutionary activities, especially during his last years, but after his death in Bern in 1876, Antonia and her three children settled in Naples, one of the places where they had previously lived; here their lodging was in the Villa Capodimonte, near the great National Museum of Capodimonte. About a year later Antonia married socialist lawyer Carlo Gambuzzi, one of Bakunin's Italian comrades who had followed him to Switzerland and with whom Antonia had already had a close relationship for some years. Indeed it was widely believed among their friends that the Bakunin-Kwiatkowska marriage was from the beginning a nihilist (platonic) marriage and that, although the three children were fully acknowledged by Bakunin, Gambuzzi was their father.[15] From her second marriage Antonia had a fourth child, Tatiana Gambuzzi.

Growing up in Naples at a time when ideals of liberty were very much in the air and the general atmosphere was one that provided great intellectual stimulation, Maria developed into a strong-minded young woman. Along with her brother, Carlo, she attended the classical lycée Umberto I, a high school with a fine reputation, which had been established in 1862. The death of her mother when she was fourteen reinforced Maria's determination to do well in life, and after completing her lycée course she enrolled at the University of Naples in the chemistry degree course. In 1890, while still an undergraduate of seventeen, she was appointed to a position as *preparatore* (laboratory assistant) in the Institute of Chemistry. By then she had already developed a deep (and ultimately lifelong) interest in the details of molecular structure, the special relationships of the atoms in a molecule; in particular she focused on the fairly new area of stereoisomerism, or geometric isomerism, predicted in the classic 1875 paper by the Dutch physician-chemist J. H. van't Hoff.[16] The phenomenon was demonstrated and conclusively verified by the 1880s, the experimental work of several chemists having produced examples. Maria Bakunin became one of the contributors to this area.

Her dissertation research, guided by Agostino Oglialoro-Todaro, director of the Institute of Chemistry, concerned investigations of geometric isomers in cinnamic acid derivatives, the subject of her first paper published jointly in 1890 with Oglialoro-Todaro in the Naples journal *Rendiconto dell' Accademia delle Scienze Fisiche e Matematiche* (see bibliography). She received her degree from Naples University in 1895, the youngest and most outstanding student of the year.

Maria Bakunin's subsequent research covered a wide range of topics, but two subjects in particular held her attention throughout her whole career, one being stereoisomerism in the cinnamic acid series, and the other the mechanism of the Perkin condensation. Her entry in a competition set by the Accademia delle Scienze, Fisiche e Matematiche in 1900 was a collection of six papers on the stereochemistry of phenylcinnamic acid, phenylnitrocinnamic acid, and phenyloxycinnamic acid, plus a number of their derivatives (see bibliography). Requiring difficult and exacting experimental work that was carried out over the course of several years in the late 1890s, her entry was judged more worthy than that of the already well-established chemist Emil Erlenmeyer (Jr.?);[17] Maria Bakunin received the 1,000 lira prize. The work has the added interest of having engendered an international exchange, via journal articles, between the Bakunin-Oglialoro group and the young Belgian chemist Albert Joseph Jacques Vandevelde, then starting his long career at the University of Gand, who was also investigating condensation methods for preparing pure phenoxycinnamic acid and its derivatives.[18] In all these stereochemical studies Oglialoro was also much involved, and the close working relationship between him and Maria developed into a lasting marriage.

The subjects Maria went on to investigate included extensive research on substituted indones, begun in her student years, when she was focusing on geometric isomerism, and continued throughout much of her career; among these were studies of photochemically induced rearrangements. Her interest in reaction mechanisms resulted in several notable papers on condensation reactions, including her report on the development of a technique to allow temperature lowering and control in esterification and cyclization reactions, a method involving the use of catalytic amounts of phosphoric anhydride in an inert solvent; the procedure was employed industrially in Britain for the preparation of aspirin. Studies on the catalyzing effects of colloidal solutions in organic syntheses were additional contributions. Many of these investigations were carried out with students and collaborators.[19]

One of her studies, carried out with G. Dragotti, is of particular note; this was her investigation of the dark skin pigments found in most living animals, including man, a major interest being the relationship of these large polymeric structures to biological function, namely their photoprotectant properties.[20] Several decades later this subject was developed into a major research program at the Naples University Institute of Organic Chemistry, largely under the leadership of Maria Bakunin's student Rodolfo Nicolaus. Also of interest from the biological standpoint is Bakunin's early work on the toxic effects of morphine.[21]

She was one of a number of scientists who observed the 1906 eruption of Vesuvius from the town of Torre Annunziata, on the mountain's southern slopes. Perhaps the sight of this dramatic event stimulated her interest in rocks. Over the years she carried out analytical work on the bituminous schists of Sicily and the Giffoni Valley in the southern Apennines.[22] These investigations had considerable commercial significance; dry distillation of this sulfur-rich oil shale led to the isolation of a product of natural origin containing a high percentage of organically bound sulfur, namely, ammonium bituminosulfonate or ichthyol, which is used in medicine for treating some skin diseases and has bactericidal and fungicidal properties. For almost twenty years, beginning about 1911, she consulted for companies working the shale deposits of the Giffoni Valley. Among other studies were two analytical investigations of the waters of the mineral springs of the Pompei Valley, one a joint article that emphasized the rich variety of minerals in these springs and the relationships between the minerals and the bacteria present.[23]

In 1909 she began teaching applied chemistry at the Scuola Superiore Politecnica in Naples, and three years later became chair of applied technological chemistry at that institution. Throughout the years of World War II she headed the Department of Organic Chemistry at Naples University, where she remained for the rest of her career.

Maria Bakunin made a large and important contribution to chemical education in Naples institutions. She was very strict with her students, and her high standards resulted in a generation of fine chemists, a number of whom went on to outstanding careers. Among them were Francesco Giordani and Rodolfo Nicolaus. Giordani, professor of electrochemistry at the Naples School of Engineering, directed the new National Committee for Nuclear Research in the 1950s and 1960s, and was one of the "Three Wise Men" called on in 1956 to draft the plan for the Euratom Project to assess Europe's future needs for nuclear energy. Nicolaus, professor of organic chemistry at Naples University, became well known for his wide-ranging studies, including structure determination and synthesis of compounds of biological interest, particularly pigments.

Thanks to her energy and initiative, Maria Bakunin did much to ensure the survival of the university's chemical institute throughout the German reprisals and the destruction caused by Allied occupation in World War II. When her own home was burned down in one of these reprisals she moved into an empty room in the university and there continued to provide a meeting place and the hospitality she had offered in her home to those who came to her. After the war she helped to rebuild the university library; all her books and papers were placed there.

From the 1920s on she served on a number of Italian chemistry associations and commissions. A corresponding member of Lincei Accademia, ordinary member of the Society of Arts and Letters of Accademia Pontaniana, president of the latter from 1944 to 1949, and of the Physical Science and Mathematics Section of the Society for Arts and Letters from 1932 to 1952, she was well acquainted with some of Italy's finest scholars. Among them were chemist Stanislao Cannizzaro, with whom she often corresponded, and philosopher and politician Benedetto Croce, the person who nominated her for the distinction of president of the venerable Accademia Pontaniana.[24] In 1935, under the auspices of the Ministry of Agriculture, Industry, and Commerce, she visited Switzerland and Belgium to study university teaching in these countries; in 1934 she was an official delegate to the centenary celebration in Leningrad of Mendeleev's birth.

Her family connections were strong and she had a special affection and regard for her nephew Renato Caccioppoli (1904–1959), the son of her older sister, Sof'ia Bakunina (see chapter 1). An outstanding mathematician, Caccioppoli became a faculty member at the University of Padua in 1931 and from 1934 taught at Naples University. Outspoken in his opposition to the founder and leader of the Fascist Party, Benito Mussolini, he spent some time in jail for his political activities, a predicament from which his aunt Maria went to great effort to free him.[25]

A person of courage and charm, wide academic and cultural interests, and notable talent and ability in many branches of her field, it was Maria Bakunin's good fortune to begin her career at a time when the Italian chemical sciences were going through a period of vigorous development. Her contributions were remarkable, particularly for a woman of her time. She died in 1960, at the age of eighty-seven, in her home on Mezzocannone Street, not far from the university's chemical institute.[26]

The five women in the chemical sciences who are ranked here just below the most outstanding two in terms of contributions to scientific knowledge all had short careers, either because of early death or because of unsatisfactory work opportunities and competing claims on their time. Of the five, Juliia Lermontova is the most widely known; initially this was in large part through her long and close relationship with the celebrated Sofia Kovalevskaia, but more recently because of growing recognition of her own achievements. However, her predecessor Anna Volkova published seven very creditable papers in Russian and German journals before Lermontova received her University of Göttingen doctorate. Compared to Lermontova, Volkova had a quiet career, contained entirely within the St. Petersburg chemical community and without the added color and wide interest of the international element in Lermontova's training and work; but she too has left what might well be considered a remarkably distinctive record.

ANNA FEDOROVNA VOLKOVA[27] (d. 1876), most likely Russia's first woman chemist, had no formal university degree, but she succeeded in making a noteworthy contribution to the chemical community in the few years of her short life; her entire chemical career, from student to productive research worker, lasted no more than a decade.

Her family background and date of birth remain obscure, but she probably started her studies in the heady days of the 1860s, when agitation for higher education for women was at a high point in Moscow and St. Petersburg. The St. Petersburg Higher Courses for Women began only in 1878, but a program of "Systematic Public Lectures," open to both men and women, was instituted in 1869. Teaching staff for these lectures included several of St. Petersburg University's leading chemists, men of international standing, and at least from that year, Volkova was able to attend lectures by Aleksandr Butlerov and Dmitrii Mendeleev.

In 1869 she also started original investigations in the laboratory of the St. Petersburg Institute of Agriculture and Forestry, then led by chemist and agronomist Aleksandr Engel'hardt; the following year she moved to the St. Petersburg Institute of Farming (renamed Institute of Forestry in 1877). Throughout all her work she had steady encouragement and help from several senior chemists, including Engel'hardt, Pavel Lachinov (at the Institute of Farming), Butlerov, and especially Mendeleev.

Almost all her published papers were reports of her structure and properties studies of arylsulfonic acids and their derivatives, particularly the acid amides, an area then being extensively investigated by Engel'hardt and his coworkers. With guidance from Mendeleev, who was then teaching at the Institute of Farming, Volkova extended studies in this area. Her two papers on toluenesulfonic acids and acid amides in *Zeitschrift für Chemie* in 1870 were probably the first reporting chemical research by a woman from a modern chemical laboratory. After their publication she was invited to join the Russian Chemical Society, founded two years previously, as its first woman member.[28] About this time Nikolai Menshutkin enlisted her help in editing and abstracting for the society's journal.

Despite her very short career, ended by poverty-related illness and premature death in 1876, Volkova's research contribution is striking, especially for a woman of her time. During the period 1870–1873 she prepared and studied several new acid amides (see bibliography), but the work for which she is best remembered is the preparation of ortho-toluene sulfonic acid amide by an improved, efficient method (1870); this compound became a key intermediate in the manufacture of saccharine. Well regarded by her colleagues, and well integrated into the chemical community, she was the first and only woman to attend the Third Congress of Russian Naturalists and Physicians in Kiev in 1871; the first two congresses (1867–1868 and 1869) had been exclusively male gatherings.[29] Not only did Volkova present two papers at the Kiev congress but, most remarkably, she was elected chair of one of its chemistry sessions. Compounds she synthesized were among the new materials prepared by Russian chemists that were exhibited at the World Industrial Exhibition in London in 1876—the year of her death.

The story of Juliia Vsevolodovna Lermontova[30] (1846–1919), the first woman to be awarded a doctorate in chemistry by a German university and one who went on to several years of noteworthy research, was brought back to light for Russian readers by Soviet historian Iusuf S. Musabekov in 1967; her introduction into the Western literature on women in science came only sixteen years later, when Ann Hibner Koblitz published her biography of Sofia Kovalevskaia, with its many references to Lermontova. More recently two German-language works have provided valuable contributions toward a fuller picture of Lermontova's life and work; these are Cordula Tollmien's essay on the intertwined careers of Kovalevskaia and Lermontova, which presents a full explanation of the complicated circumstances surrounding Lermontova's doctoral examination, and chemist Elena Roussanova's 2003 biography, *Julia Lermontowa*. Roussanova's work, based on a master's thesis (*Diplomarbeit*) at Hamburg University's Institute of Science, Mathematics, and Technology, includes a most useful discussion of Lermontova's research, translating into modern chemical terminology Lermontova's nineteenth-century terms.

Although Juliia Lermontova is widely accepted as being the first of the late nineteenth-century Russian women to be awarded a doctorate in chemistry, it is perhaps worth noting that she shares this honor with the forgotten Lidiia Sesemann, from Viborg, Finland, a semiautonomous grand duchy within Russia from 1808 until 1918. Sesemann received a doctorate in chemistry from Zurich University on 15 May 1874, five months before Juliia Lermontova graduated from Göttingen University (24 October 1874).[31]

One of the three daughters of General Vsevolod Lermontov and his wife, Elisaveta Andréevna Kossikovskaia, Juliia was born in St. Petersburg on 21 December 1846 (Julian calendar).[32] General Lermontov, director of the Moscow Corps of Cadets School, and head of the corps, was a second cousin of the distinguished Russian poet Mikhail Lermontov (1814–1841), a connection of which Juliia was very proud.[33] She grew up in the Lermontov House, the family's home in the Lefortova district of Moscow; her father also owned a country estate south of Moscow, in the Tula district.

In poor health throughout much of her childhood, she did not attend any public school but was given her early instruction at home with her sisters, Olga and Sofia. Both parents were cultured, well-educated people, and the general had a large library that Juliia made good use of, reading widely. Her interest in science, especially chemistry, began early, perhaps stimulated by the technical publications to which she had access; she even succeeded in carrying out various chemical experiments. Language instruction was not neglected, and in addition to the science and mathematics instructors engaged by the Lermontovs she had a number of governesses from Western countries.

Her first career choice was medicine, and for a time she studied histology with Aleksandr Babuchin of Moscow University's medical faculty. But she quickly discovered that the sight of suffering and death caused her acute distress, and she turned to her second subject of interest—agriculture. In 1868, along with two young women friends, she applied for permission to audit courses at the Petrovskaia Academy of Farming and Forestry in Moscow, a recently founded institution already known for its outstanding teaching of the chemical sciences. Although supported by several of the academy's faculty members, the applications of the three women failed to win official approval.

Seeing little alternative if she wanted advanced scientific study, Juliia decided to go to one of the Western universities, a course of action successfully carried out by Russian Nadezhda Suslova (see chapter 1), whose doctorate in medicine from Zurich University in 1867 quickly became a powerful source of inspiration for many young Russian women. Juliia's parents, while encouraging her educational ambitions, remained reluctant to let her go abroad alone. But through her first cousin on her mother's side, Anna Evreinova, who also wanted to study in western Europe, she made contact with another aspiring student, Sofia Kovalevskaia. In 1868 Sofia had entered into a fictitious marriage with a young paleontology student, Vladimir Kovalevskii, in order to free herself from parental control and open her way to advanced study abroad. Sofia Kovalevskaia's older sister, Anna, planned to join the pair, and Juliia's parents were eventually persuaded to let her leave as well. Only Anna Evreinova had to resort to desperate tactics, her father remaining adamantly opposed to her plans for university study abroad.[34]

Juliia Lermontova arrived in Heidelberg at the beginning of the 1869–1870 winter semester, Sofia and Vladimir Kovalevskii having settled there a few months earlier. Heidelberg, one of the oldest of the German universities and widely known in Russia, already had a large number of Russian students. It was well regarded for its strengths in scientific studies and was a relatively liberal institution, but it had not yet admitted any women students. Sofia Kovalevskaia's application for the usual permission to study was rejected, and after lengthy discussion in the academic senate, the university adopted a policy specifying that no general admission be granted to women, although any female applicant was free to ask instructors individually for permission to attend lectures. This meant that Juliia Lermontova, who had also tried and failed to win general permission, had to ask no fewer than eight professors and lecturers (*Dozents*) to let her attend classes and laboratories. Mathematics, physics, mineralogy, meteorology, and climatology as well as chemistry were included in her study plan. Among the famous chemists whose lectures she attended were Robert Bunsen, Gustav Kirchoff, and Herman Helmholz. Laboratory space was by no means assured and her place in Bunsen's up-to-date and well-provisioned laboratory was particularly hard won; overcoming Bunsen's strong prejudice against having a woman, especially a Russian woman, in his domain required the assistance of Sofia Kovalevskaia, with all her persuasive charm.

An outstanding experimental chemist who laid great weight on practical work, Bunsen, one of the leading authorities in analytical chemistry, trained a considerable number of students who went on to distinguished careers, among them Dmitrii Mendeleev. Juliia Lermontova's work in Bunsen's laboratory included an analysis of the platinum group of metals, a study for which she had guidance from Mendeleev, who, although by then established in St. Petersburg, spent parts of the years 1869–1871 in Germany and often visited Heidelberg. Juliia's investigation to determine the exact atomic weight of platinum was of special interest to him in establishing its correct position in the periodic table of the elements he was then working out. At the time the accepted atomic weight of platinum was less than that of its two neighbors, iridium and osmium, which placed it in a position that did not accord with the requirements of Mendeleev's periodic system. Using techniques taught by Bunsen, Juliia Lermontova carried out separations of these three elements designed to produce pure samples of each, as required for accurate atomic-weight determinations.

Overall Juliia found life in Heidelberg very pleasant. The town was small and the two Russian women students were objects of much interest among their fellow students and the townspeople. Long walks with the Kovalevskiis in the beautiful surrounding countryside were special pleasures. But by the summer of 1870 Sofia Kovalevskaia had decided to move to the University of Berlin to study with mathematician Karl Weierstrass, and two semesters later Juliia followed her.

Once again Kovalevskaia had been refused formal permission to attend lectures, and Juliia, despite having excellent recommendations from Heidelberg professors, does not appear to have even submitted a request for such permission. Instead she immediately approached August Wilhelm Hofmann, a well-known chemist with international recognition and a man of generally more liberal outlook than many of his Berlin colleagues.[35] The University of Berlin Chemical Institute that he led was very large, up-to-date, and well equipped; it accommodated a notable number of doctoral students, including many from other countries. Hofmann's own research was extensive, much of it focusing on the chemistry of organic nitrogen compounds, particularly the aliphatic and aromatic amines. He accepted Juliia as a private student, permitting her to attend his lecture courses and carry out research in his laboratory.

The results of her first semester of work under his general supervision were of sufficient interest to warrant prompt publication; presented by Hofmann at the 25 March 1872 meeting of the German Chemical Society and published without delay in the society's *Berichte* (see bibliography), the research concerned the structure of the compound then called diphenin, produced by the reduction of 4,4′-dinitroazobenzene and widely considered to be 4,4′-diaminoazobenzene. Juliia questioned the structural assignment and, expanding on earlier observations by Hofmann, showed that under milder reduction conditions, there could be obtained other products and intermediates for which she suggested structures. She reexamined the structure specified by the eminent French chemist August Laurent (1807–1853) and his colleague Charles Friedrich Gerhardt (1816–1856) for diphenin, and showed that this was not in fact the azo compound, $H_2N\text{-}C_6H_4\text{-}N{=}N\text{-}C_6H_4\text{-}NH_2$, but the hydrazo compound, $H_2N\text{-}C_6H_4\text{-}NH\text{-}NH\text{-}C_6H_4\text{-}NH_2$.

That Lermontova, a young, female chemistry student, should have undertaken to reinvestigate the accepted conclusions of such distinguished elder statesmen of the field as Gerhardt and Laurent is remarkable; it illustrates very clearly the strong influence of Hofmann's teaching—his advice to take up chemical problems, consider hypotheses critically, and then try to back these up in the laboratory. Her method for the preparation of diphenin, 4,4′-diaminohydrazobenzene, was included in Friedrich Beilstein's *Handbuch der Organischen Chemie*, the manual for organic chemists widely used from the time of its first edition (1880–1883) to the present hugely expanded database of organic compounds.[36]

Her major study in Hofmann's laboratory, the subject of her dissertation research, centered on the compound diiodomethane, particularly its role in organic syntheses. She studied the reactions of this compound with both aliphatic and aromatic amines, and with potassium sulfocyanide, the latter a step toward the preparation of the corresponding disulfonic acid. The work was extended to include observations on members of the dihaloethane series.[37]

There being no possibility of the conservative, Prussian University of Berlin conferring a degree on a woman, Lermontova submitted her dissertation, "Zur Geschichte der Methyleneverbindungen" (A contribution to the chemistry of methylene compounds), to the relatively more liberal University of Göttingen on 25 July 1874. In this she was following the path taken by Kovalevskaia, who had submitted her mathematical dissertation and application for consideration for a degree to Göttingen University a week before. A few days later Lermontova requested that the degree be conferred *in absentia*, that is, without the usual oral examination. Here again she was following Kovalevskaia's example. But although Kovalevskaia did avoid an oral examination, thanks to the considerable influence brought to bear on her behalf by Karl Weierstrass and the general inclination of the Göttingen University administration, Lermontova did not. Hofmann, an overworked man who had seven other doctoral students graduating in 1874, supported her request, but he was less vigorous and persistent in stressing Lermontova's case for the degree to be conferred *in absentia* than was Weierstrass for Kovalevskaia. Furthermore, Hofmann's first Göttingen contact, Friedrich Wöhler, was not in an ideal position to help, having by then officially retired. Hans Hübner, who had succeeded Wöhler as full professor of chemistry, initially classed Lermontova's dissertation as *recht gut* rather than giving it the top designation of excellent, but he was willing to grant the degree in absentia. The final decision lay with a full faculty committee, which included the dean, philosopher Hermann Lotze. The committee struggled hard over the basic question of granting degrees to women, a change of policy opposed by Lotze, who expressed his worries about setting precedents and opening the university to a potential rush of women candidates. Such an outcome would, he feared, pose a serious dilemma for the administration, to the point of jeopardizing the university's reputation.[38] As a whole the committee did not find enough evidence to warrant classing the dissertation as excellent at that stage, and it declined to grant the degree without the oral examination.

Lermontova and Kovalevskaia went back to Russia in August 1874, Lermontova going to her parents' home near Moscow where she spent a few weeks preparing for her oral examination. Her two and a half years in Berlin had severely taxed her energy; as well as spending long hours in the laboratory, she had willingly shouldered all the responsibility for the functioning of the small, dingy apartment she shared with Sofia Kovalevskaia, who was absorbed in her mathematics to the exclusion of all else; money was scarce, their food was poor, and their only social contacts were with Karl Weierstrass and his sisters.

The oral examination at Göttingen on 22 October 1874 marked the culmination of this very demanding period in Juliia Lermontova's life. She was examined thoroughly and at length by Hans Hübner in organic chemistry, quite gently by Friedrich Wöhler in inorganic chemistry, and quickly by mathematician Johann Listing in physics. She gave a good account of herself and the degree was granted *magna cum laude*.[39] Wöhler marked the occasion by presenting her with a small memento, a cut stone of the mineral in which the element titanium had recently been identified. She stayed in Göttingen for several weeks, living with Listing's family, until the dissertation was printed and she had somewhat recovered from examination stress. A brief visit to the Weierstrass family in Berlin followed before she returned to Russia in November 1874.

During a short stay in St. Petersburg with the Kovalevskiis en route to Moscow she met several of the most prominent men in the chemical community, including Russia's leading organic chemist, Aleksandr Butlerov, a strong supporter of higher education for women, who suggested that she join his research group. But she went on to Moscow and, since her mother died two months later, remained there with her family for some time.

On Mendeleev's recommendation she was elected a member of the Russian Chemical Society, becoming the second woman, after Anna Volkova (see above), to join the organization. She obtained a place in the Moscow University chemistry laboratory of Vladimir Markovnikov, a former student of Butlerov, and continued the work she had begun in Berlin on halo-substituted aliphatic compounds. This included the elaboration of an efficient method for the preparation of 1,3-dibromopropane, then called trimethylenebromide (see bibliography), an important step in the synthesis of glutaric acid (pyrowine acid), worked out by Lermontova and Markovnikov. Much of this work was published under Markovnikov's name in 1876.[40] About then she became ill with typhus, which developed into meningitis and brought on delirium, causing her family and Sofia Kovalevskaia, who was helping to care for her, to fear for her life. She recovered, although a threat of tuberculosis remained.

During her slow recuperation she made up her mind to go back to St. Petersburg and join Butlerov's group. Beginning in 1876 her studies formed part of Butlerov's extensive program of validating by synthesis his theories of structure in organic compounds. After an interruption of this work in 1877, when her father's death required her return to Moscow for a brief period, she carried out a study of catalyst-assisted alkylation of unsaturated compounds, such as the preparation of 2,4,4-trimethylpentene-2 from isobutylene and isobutyliodide. Further studies formed part of Butlerov's pioneering work on the polymerization of olefins. Lermontova's last research in Butlerov's laboratory concerned the conversion of olefins to acetylenes, such as the preparation of dimethylacetylene from 2-bromobutene-2. Much of the work was incorporated into Butlerov's reports.

Between 1877 and 1879 Juliia Lermontova prepared French-language reports of chemical research in Russia for the *Bulletin* of the Société Chimique de Paris (see bibliography). In addition, she translated into French a chapter from Mendeleev's 1877 book on the rapidly growing and increasingly important oil industry, a field in which Mendeleev was the leading Russian advisor on technological innovations.[41]

Juliia Lermontova's two years in the friendly atmosphere of Butlerov's laboratory passed very pleasantly. Butlerov was a kindly man who took the time to offer his students good practical advice; Lermontova especially admired his clear, logical thinking and much enjoyed his lectures in organic chemistry, which she was allowed to attend as a special student. She lived with the Kovalevskiis, and in contrast to the hermit-like existence she and Sofia Kovalevskaia had led in Berlin, their life in St. Petersburg was bright and lively. They often attended large social gatherings and entertained guests in their own relatively spacious home. But family obligations drew Juliia Lermontova back to Moscow in 1879 for what she expected would be a short period—but which turned out to be for the rest of her life.

She kept in touch with Butlerov about the prospect of a position as instructor in the laboratory that he had organized for the St. Petersburg Higher Courses for Women, but because her real interest was research, this position, as she later wrote in her memoirs, was less than wholly attractive.[42] Before she could teach at gymnasium or university level in Russia she needed to pass the Russian *magistr* examination, her German doctorate not being officially recognized. There was no precedent for a woman to take this step, and Sofia Kovalevskaia's recent applications to present herself for the examination had been refused by both Moscow and St. Petersburg Universities. Lermontova appears to have had hopes and plans in this direction after her return to Moscow in 1879, but no record of any application from her has been found; she must certainly have considered that her prospects were slim.

Markovnikov again made available a place for her in his Moscow University laboratory, and from about 1880 she joined in his major program of systematic studies of the composition of Caucasian petroleum deposits, work that went on until his death in 1904. Despite problems such as inadequate laboratory space and equipment, and lack of funding for chemicals, these studies, which included investigations of the properties of the compounds isolated, were very productive. Markovnikov discovered a whole new class of cyclic paraffins that he called naphthenes, studied their oxidation, and demonstrated their use as convenient raw materials for the preparation of many important chemicals. International recognition came to him in 1900, when he received a gold medal at the First International Petroleum Congress in Paris.[43] Lermontova's part in this research program concerned the development of an efficient method for the separation of aromatic hydrocarbons from the petroleum mixture by catalyst-assisted high-temperature distillation, in particular the production of the tricyclic compound anthracene from oil residues; this investigation, she hoped, would serve as the basis for a profitable commercial venture. Both she and, to a much greater extent, the Kovalevskiis had themselves caught the "oil fever" and had invested in the Ragozin Oil Company.[44] By then, however, the management of her family lands and other demands for her attention were reducing her time for research. She joined the Russian Technical Society in 1881, becoming its first woman member, and discussed the results of her high-temperature cracking work at a meeting of the society's Moscow section on 19 February 1883.[45] But when the method she had developed in the laboratory failed to translate into an efficient refinery process her enthusiasm for applied chemistry investigations lagged considerably, especially since her real interests lay in basic research of the kind she had taken part in during her short interval in Butlerov's laboratory.

About 1882 she made the decision to give up scientific work. She and her sister Sofia Lermontova would appear to have been involved briefly with a Moscow girls' school, but the institution was not very successful and Juliia Lermontova's connection with it stopped after her sister died from typhus in May 1885. From then on agricultural projects on her estate occupied her fully; she studied specialized agronomy publications and informed herself about methods for improving soil fertility and increasing land productivity. In 1889 she visited the Paris World Exhibition, where she found the Agricultural Pavilion, with its exhibitions on agriculture and food products, of great interest. This visit prepared the way for her own considerable success in cheese making; she supplied French-style cheeses for the Moscow market and also tried to establish outlets in the Ukraine.

After Sofia Kovalevskaia's decision about 1880 to return to full-time work in mathematics, Juliia Lermontova had assumed the major responsibility for the upbringing of the Kovalevskiis' daughter, Fufa (Sofia). This child, Lermontova's goddaughter, had held an important place in her life ever since the girl's birth in 1878, when Lermontova was making her home with the Kovalevskiis in St. Petersburg. From 1883, when Kovalevskaia began her teaching career as the first woman professor at Stockholm University, Fufa lived entirely with Juliia Lermontova, except for some relatively short interruptions.[46] In 1885 Kovalevskaia traveled to Russia and attempted to persuade Lermontova to move to Stockholm and live with her. But Lermontova resisted her old friend's blandishments, remained on her estate, Semienkova-by-Tula, outside Moscow, and continued her agricultural work.

Although ill with bilateral pneumonia in the spring of 1889, she had sufficiently recovered by the autumn to visit Kovalevskaia in Stockholm.

From about 1890 she led a very secluded life, staying almost all the time on her estate. The death of Sofia Kovalevskaia from pneumonia in 1891 left only Fufa Kovalevskaia from her earlier circle of close friends; her relationship with the child was close and affectionate, Fufa, by then Lermontova's formally adopted daughter, calling her "Mama Juliia." After the Russian Revolution of 1917, the new administration tried to expel landowner Lermontova from her estate, but the Soviet Commissar of Enlightenment, Anatoly Lunacharski, who was in charge of education and culture, granted her permission to stay on in her house. Two years later, on 16 December 1919, shortly before her seventy-third birthday, she died of a stroke. Her remaining possessions were left to Fufa Kovalevskaia.

Postscript. Juliia Lermontova, the first woman to be granted a doctoral degree in chemistry in accordance with standard examination procedures at a German university, spent only fourteen years in scientific work, giving it up by the age of thirty-five. Her success at Göttingen University in 1874 unquestionably makes her deserving of the wide recognition she is now receiving. However, it should also be borne in mind that her personal success was not quickly followed by notable numbers of women entering German universities; Lermontova may be regarded rather as a (long-forgotten) harbinger, a forerunner, of the women who were acquiring training in the physical and mathematical sciences and contributing to research in these areas by the 1890s. Most of the later women graduated from Swiss or French universities or studied and worked without degrees in St. Petersburg and Moscow. The opening of the German universities to women came only gradually, some only after the turn of the century, following the overall advance of education for girls and women during the last years of the nineteenth century; the few women who were granted degrees at German universities significantly earlier were either officially designated "special cases," such as Kovalevskaia, or, like Lermontova, isolated cases.

Juliia Lermontova's story provides a remarkably clear illustration of the close interaction between German and Russian chemists at this important period of rapid development of both theoretical and practical aspects of the field. As Roussanova puts it, "Das Phänomen der Wechselwirkung zwischen russischen und deutschen chemischen Schulen läuft als roter Faden durch ihre Arbeit."[47] She studied and worked with several of the leading chemists in major research schools of both countries: Bunsen and Hofmann in Germany, Butlerov and Markovnikov in Russia. Her synthetic work and structure studies of organic molecules with Butlerov and Markovnikov were especially noteworthy, as was also the pioneering research with Butlerov on the polymerization of olefins; and even though it did not lead directly to commercial success, her contribution to the development of high-temperature fractionation methods for the isolation of aromatic compounds from petroleum was not insignificant.

A talented and capable chemist, Lermontova worked under conditions few women would consider adequate today; although she must clearly be considered fortunate in being accepted into some of the best research laboratories of the time, she never went on to hold any official position. She was essentially a private student, and in some respects an independent research worker, living on her own, somewhat strained, financial resources. The added complications of poor health, and the family demands on her remarkably kind and generous nature, further reduced her research output. But despite these handicaps Juliia Lermontova ranks high in the history of women in chemistry.

Vera Evstaf'evna Bogdanovskaia[48] (1867–1896), later Bogdanovskaia-Popova, the middle child of three in the family of a well-known St. Petersburg professor of surgery, Evstaf I. Bogdanovskii, was born in St. Petersburg on 17 September 1867. The Bogdanovskiis were comfortably off; Vera, along with her older brother and younger sister, was educated at home until she was eleven years old, when she was sent to the Smolny Institute, the boarding school for girls founded in 1764 by Catherine the Great for the "Upbringing of Noble Girls"; in the nineteenth century the Smolny was still one of the country's elite educational institutions, its curriculum expanded to cover a wide range of subjects from the sciences and mathematics to home economics. By May 1883, when not yet sixteen, Vera had finished her course at the institute, winning a first-class certificate, and in autumn of the same year she started studies in the physical-mathematical department of the St. Petersburg Higher Courses for Women. After completing the program there in 1887 she stayed on in St. Petersburg for two years, gaining further experience through work in chemical laboratories of the Academy of Sciences and the Military Medical Academy as well as at the Higher Courses for Women.

Having decided to study for a doctoral degree, Vera Evstaf'evna moved to Geneva in October 1889, joining the small colony of Russian students there and enrolling in the university's science division. Over the course of five semesters she attended lectures in several areas of interest and found herself especially attracted to the new field of stereochemistry. She carried out her doctoral research under the direction of the widely respected chemist Carl Graebe; although she had initially approached Graebe with her own idea of attempting a synthesis of the phosphorus analogue of hydrogen cyanide (prussic acid), Graebe, considering that to be a difficult and perhaps hazardous undertaking, suggested instead that she examine the course of a reduction reaction in an aromatic ketone, a study that would carry forward investigations in an area in which he had a long and continuing interest.[49] Bogdanovskaia's dissertation research, "Über Dibenzylketon und Dibenzylcarbinol," was published in the German Chemical Society's *Berichte* in 1892 (see bibliography). Accepted as a candidate for the degree of *docteur ès sciences physiques*, she passed the examination and received her University of Geneva degree in 1892.

During her last semester in Geneva (1891–1892), Vera Bogdanovskaia gave tutorials in theoretical and analytical chemistry for students who needed a fuller treatment than that presented in the regular lectures; these were much-appreciated help sessions, given at the request of several of her fellow students, particularly women students of the Russian colony. In addition she presented a series of lectures on stereochemistry, her particular interest throughout her student years.

This teaching experience perhaps served as an introduction to the instructional work she undertook after returning to St. Petersburg in the autumn of 1892. There she was immediately asked to assist Mikhail Lvov, then in charge of teaching inorganic chemistry at the Higher Courses for Women; once again she made special efforts to help beginning students understand basic ideas in chemistry, this time with the blessing of the Council of Classes. At the same time she most likely carried out some work in the laboratory of Gavril Gustavson at the University of St. Petersburg, and being a well-educated young woman and a pleasant conversationalist, she became acquainted with the major figures in chemical research in St. Petersburg at the time, including Butlerov, Mendeleev, and Aleksei Favorskii, the latter then a junior lecturer carrying out his research in Butlerov's laboratory.

However, her situation held little prospect for independent research and advancement in her field, a problem she hoped to resolve by the then not uncommon procedure taken by ambitious and single-minded young Russian women looking for independence and freedom of action; she married a sympathetically minded much older man, Ia. Popov, possibly in a nihilist-style (platonic) arrangement. General Popov, director of the Izhevsk arms factory, a major supplier of guns for the Russian army,[50] was someone who could give her the basic means to follow her ambitions.

Giving up her modest position at the Higher Courses for Women, Vera Bogdanovskaia left St. Petersburg in October 1895 and traveled almost one thousand miles east to the city of Izhevsk, in the western Urals. Once in possession of a laboratory of her own, she ignored the warning of the experienced Graebe and went back to her old idea of preparing the phosphorus analogue of hydrogen cyanide. Using inadequate equipment and without what would now be considered minimal precautions she proceeded with her experiment.[51] As described by General Popov, Vera Evstaf'evna's fatal accident took place at about ten on the morning of 25 April 1896.[52] A glass tube that she was holding contained a mixture of toxic chemicals that had generated considerable pressure. The tube exploded; glass splinters cut her hand, and the chemicals that then entered her bloodstream caused death within five hours.

Her tragic end when still in her twenty-ninth year brought a warm and generous tribute to her from her St. Petersburg colleague Gavril Gustavson. The memorial to her that he published shortly after her death speaks well for the considerable regard in which she was held by her fellow chemists.[53] Her portrait was hung in the laboratory of the Higher Courses for Women in St. Petersburg, and General Popov gave 15,000 rubles to help needy women students.

In addition to her chemical reports, Bogdanovskaia produced several literary works. Her earliest attempts are lost, but by age eighteen she was publishing short stories and essays in *Review of Painting* and in *Rebus*, Russia's first periodical devoted to the then popular subject of spiritualism.[54] After a visit in the summer of 1886 to the village of her forebears in southern Russia, she became interested in beekeeping, a major occupation in the area. Her subsequent study of both the practical and theoretical aspects of this subject resulted in further essays, accepted by the Ministry of Education Library. Her elementary textbook in chemistry for beginning students, published posthumously in

1897, was also approved by the Ministry of Education.[55] An admirably energetic and capable young woman, Vera Bogdanovskaia, in addition to these independent undertakings, is believed to have taken part in the preparations for the posthumous publication of Butlerov's introductory text of organic chemistry, a volume of over seven hundred pages brought out in 1887, a year after its author's death. This work included a supplement by Butlerov's students in German translation as well as in Russian.[56] Being fluent in both languages, Bogdanovskaia brought out a translation from the German of Pet. Jos. Murzel's *Chemie*, making available to Russians a useful contribution on agricultural chemistry from a German specialist.[57]

EVDOKIA ALEKSANDROVNA FOMINA,[58] Mme FOMINA-ZUKOVSKAIA (1860–1894), was born on 4 March 1860 in the town of Lukh, Kostroma *gubernia*, about three hundred miles northeast of Moscow. Her father, a middle-grade civil servant (*kollezhskii sekretar'*), died when she was four, leaving the family without any means of support. They lived in poverty from then on, but Evdokia managed to attend the Kostroma Women Teachers' Seminary and had a further two years at the Samara Girls' Gymnasium. An outstanding student, she gained qualification as a private governess in mathematical subjects.

Intent on acquiring further education, she went when she was about twenty to Geneva, where there was already a small colony of Russian students. She had few financial resources and lived as best she could on her earnings from private tuition, but nevertheless entered fully into her studies, attending lectures in the University of Geneva science faculty and carrying out research under the direction of Carl Graebe. Her stay in Geneva came just a year or two before that of her compatriot, chemistry student Vera Bogdanovskaia-Popova, who followed a similar educational program (see above).

Evdokia Fomina's research on condensation reactions of the naturally occurring pigment euxanthone extended Graebe's long series of investigations of aromatic ketones and also tied in with his interests in colored compounds. She received her degree of *docteur ès sciences physiques* from the University of Geneva in 1889[59] and reported her work in two papers in Swiss and German journals in 1889 and 1890 (see bibliography).

Although offered an assistantship in organic chemistry at the University of Geneva, she went back to Russia hoping to obtain a teaching position there in her own field. But all she could find was work as mathematics instructor for lower-level classes in a private girls' gymnasium in Moscow.

Thanks to help from Vladimir Markovnikov, professor of chemistry at Moscow University and a supporter of higher education for women, she was able to continue work in chemical research. Markovnikov was then carrying out his lengthy and wide-ranging investigations of the Caucasus petroleum deposits; as his private assistant, Evdokia Aleksandrovna became one of the group of several chemists who collaborated with him throughout this period, another being Juliia Lermontova (see above). In his memoirs Markovnikov thanked Evdokia Aleksandrovna for her efforts, referring to her as a pleasant and skillful worker, with a lively and enthusiastic interest in the research being carried out in his laboratory.[60]

Her short report on a particular method for purifying wastewater discharge from a bleaching and dying works reflected Markovnikov's interests in the practical, applied side of chemical research. Analyses by which she demonstrated that the method was ineffective were published posthumously in a joint paper with Aleksandr Jakovin in 1895, and abstracted in a Paris journal in 1896 (see bibliography). For a brief period before her early death she collaborated with Nikolai Dmitrievich Zelinskii on a study of possible synthetic routes to the sulphur analogue of the aromatic ketone indigo, the dye compound, work not unrelated to her doctoral research with Graebe.

She was a member of the Russian Chemical Society and also took part in the activities of the Moscow Society of Amateur Naturalists, helping with the chemical work of its Analytic Commission. A capable research chemist, well regarded by her colleagues, she died in 1894 at the age of thirty-four, her health having suffered severely from her years of living in poverty and her constant hard work. Evdokia Fomina-Zukovskaia remains one of the better-remembered Russian women chemists of that notably sizable pre-1900 generation. One can only speculate about what she might have achieved had she elected to stay in Switzerland or found a more favorable position in Russia.

ZENAIDA VIKTOROVNA KIKINA[61] (fl.1890s–1910) was a student and coworker of Mikhail Ivanovich Konovalov at Moscow University and the Moscow Institute of Agriculture. Konovalov, whose research focused particularly on the nitration of hydrocarbons, was a popularizer of science and a strong advocate of higher education for women. For a year or two in the early 1890s Kikina joined in the nitration studies that were reported in Konovalov's 1896 paper (see bibliography). She then went on to doctoral study at the University of Geneva, where she worked under the guidance of Friedrich Kehrmann. As an assistant of Carl Graebe (see also Bogdanovskaia-Popova and Fomina-Zukovskaia above), Kehrmann had become especially interested in the chemistry of colored compounds, an area in which he went on to an outstanding career. Zenaida Kikina's research concerned nitro and amino derivatives of a yellow compound, the aromatic ketone flavin, work published in French, German, and Russian journals (see bibliography). She was awarded her degree from the University of Geneva in 1902.[62] That year Kikina also coauthored a paper with Konovalov on the introduction of nitro groups into saturated hydrocarbons and in particular into terpene hydrocarbons derived from essential oils of plants.[63] The work, published in 1902, three years after Konovalov had moved to the Kiev Polytechnic Institute, appears to have been her last technical publication. Kikina did, however, contribute a section to the long, multiauthored commemoration of Konovalov in the Russian Physical-Chemical Society's journal following his death in 1906, at the age of forty-eight; her article spoke of the hardworking and dedicated Konovalov from the point of view of his student and coauthor over several years, rather than dwelling on his impressive achievements as scientist and organizer.[64] Her absence from chemical research records after 1903 may be related to Konovalov's early death.

Mikhail Konovalov had at least two other women students and coworkers at the Moscow Institute of Agriculture—ANNA JEBENKO and SOF'IA KOTZINA. Both collaborated in his extensive investigations of the nitration of saturated hydrocarbons and are duly acknowledged in his 1899 paper (see bibliography). No indication of additional technical work by Jebenko or Kotzina has been found.

Notes on the nine women who are known to have studied chemistry in St. Petersburg during the 1880s and 1890s follow, and a sketch of one other, Minna Samelson, a Lithuanian who graduated from the University of Bern, is also included below.

Mme RUDINSKAIA attended lectures at the Higher Courses for Women in the early 1880s and carried out a short research project in collaboration with Ivan Bogomoletz. Her series of experiments involving ammonium parabamate and its conversion to oxaluramide showed that the course of the reaction is influenced not only by temperature but also by pressure. The work made a modest contribution to the extensive research on compounds involved in processes of animal metabolism, such as urine production, then being conducted by some of the leading organic chemists of the time. Reported by Bogomoletz in a joint paper to the Russian Physical-Chemical Society in 1885, it was also abstracted in the London Chemical Society's *Journal* and in the Paris Chemical Society's *Bulletin*, "Correspondance russe" section, in 1886 (see bibliography).

Mme Rudinskaia does not appear to have published any other reports of work in chemistry, at least up to 1914.

ALEKSANDRA NIKOLAEVNA RODZIANKO, Mme, most likely studied chemistry at the Higher Courses for Women during the late 1880s. Her research included work relating to soil fertility using soil samples taken mainly from the Poltava region (now part of the Ukraine). She investigated the absorption of minerals by the mix of organic acids (known as humic acid) produced in the soil by the decay of vegetable matter, a process of considerable practical significance since it regulates mineral bioavailability.

Rodzianko's work was done in the chemical laboratory of the Medical-Surgical Academy, a facility headed by chemist and composer Aleksandr Borodin; reported to the chemical section of the Russian Physical-Chemical Society, it was of sufficient wider interest to be included as an abstract in the London Chemical Society's *Journal* (see bibliography).

A second report of similar work of practical importance is V. A. BALANDINA's note on her quantitative chemical analysis of the waters of Lake Plotbishche in the Siberian district of Yeniseiski.[65] Minerals she found to be present in the highest concentrations were salts of calcium (55.85 milligrams per liter), followed by magnesium, iron, and aluminum. The work was carried out in the laboratory of the Higher Courses for Women and reported to the Russian Physical-Chemical Society in 1900; although omitted from the Royal Society's *Catalogue*, it has been added to the bibliography.

Three students of Gavril Gustavson at the Higher Courses for Women, Mme N. V. BOGOSLAVSKAIA, Mlle H. BULATOVA (later Mme MARKOVA), and Mlle O. POPPER, all made modest contributions to the program of structure investigations in the group of organic molecules known as the alkanes (paraffins).

Gustavson, himself a student of Aleksandr Butlerov, Russia's foremost contributor to structural theory in organic chemistry, joined vigorously in Butlerov's work that led to key concepts such as isomerism and tautomerism. By the mid 1890s, when Bogoslavskaia, Bulatova, and Popper were active in the field, Butlerov's influence was waning, but his students, and their students, to a large extent carried on his program of confirming structural theory by the practical work of synthesizing theoretically possible isomers.[66]

Bogoslavskaia's research using aluminum trichloride as a condensation catalyst was presented by Gustavson at the Ninth Congress of Naturalists and Physicians in Moscow in January 1894 and published in Russian and French journals (see bibliography). Bulatova and Popper both investigated rearrangements in alkanes using methods such as fractional distillation from zinc dust in alcohol. Their results were published as joint reports with Gustavson (see bibliography). No later work by any of these three students has been found in chemical indexes up to 1910.

Both Pelagia Lazareva and Mme L. E. Kaufmann (née Soloveichik)[67] were students at the Higher Courses for Women, Lazareva in the late 1880s, Kaufmann a few years later. Lazareva, working in the laboratory of the Academy of Sciences, carried out research on halo-substituted aromatic hydrocarbons, most likely under the general direction of Aleksandr Butlerov; two papers reporting her findings were published in 1890. Kaufmann, carrying out studies in the same area, assisted in Butlerov's investigations of the preparation of bromo-substituted aromatic hydrocarbons; her results were presented at a meeting of the Russian Physical-Chemical Society in 1900 and published in Russian and French journals (see bibliography). No further technical publications by either Lazareva or Kaufmann were found up to 1907.

Ol'ga Aleksandrovna Davidova, of St. Petersburg, studied chemistry at the Higher Courses for Women in the early 1880s. As an advanced student she carried out two short investigations of organic-reaction pathways under the guidance of Aleksandr Butlerov; these were abstracted in the German Chemical Society's *Berichte* as well as being reported in the Russian Physical-Chemical Society's journal (see bibliography). Proficient in several Western languages, Davidova was responsible for the preparation of the "Correspondance russe" section of the Paris Chemical Society's *Bulletin* between 1882 and 1885 (see bibliography),[68] work that suggests that by then she already had a fairly broad knowledge of chemical research in Russia. No additional publications by Davidova herself were found in the chemical literature, at least up to 1906.

Mme N. P. Rikhter-Rzhevskaia carried out her work in chemistry in St. Petersburg in the late 1890s and for a few years after the turn of the century. She probably studied for a time at the Higher Courses for Women. Her investigation of the hydration speed of acetic anhydride was reported in a short talk to the Russian Physical-Chemical Society in 1900 and abstracted in the Paris Chemical Society's *Bulletin* the following year (see bibliography). A somewhat longer report of her study of bromine vapor pressure above hydrobromic acid solution appeared in 1903,[69] but no further chemical contributions by her were found up to 1914.

Minna Samelson[70] was born on 11 February 1874 in the garrison town and trading center of Kovno (now Kaunas), Lithuania, when that country was part of imperial Russia. No information about her family background has been found. She received her early education in Kovno, and in the summer semester of 1897, when she was twenty-three, she enrolled in the philosophical faculty of the University of Bern to study chemistry.

Her doctoral research was carried out under the direction of Carl Ludwig Friedheim, professor of inorganic chemistry, who had special interests in the analytical and technical branches of the field. She prepared and studied several complex manganomolybdenite salts of the alkali metals, compounds that were of considerable interest around the end of the nineteenth century. Her forty-two-page report appeared in the *Zeitschrift für Anorganischen Chemie* in 1900 (see bibliography), and was also printed in pamphlet form.[71] She received her doctoral degree the same year but does not appear to have published any further research in chemistry under the name Minna Samelson in the period up to 1910.

Footnote: The name of A. M. Shcherbina, who worked with L. A. Chugaev in the chemical laboratory of the Moscow Bacteriological Institute, is listed in the Royal Society's *Catalogue*. A short report by Shcherbina and Chugaev on the action of ethyl oxalate on the cyclic ketone, menthone, was published in the Russian Physical-Chemical Society's journal in 1899. However, the name Shcherbina appears to refer to a man, and so the reference is not included in the bibliography of the present survey.

NOTES

1. "Ziber-Shumova, Nadezhda Olimpievna," *Mir Slovarei: Istoricheskii Slovar'-poisk* [World dictionary: Historical dictionary], http://mirslovarei.com/content_his/ZIBER-SHUMOVA-NADEZHDA-OLIMPIEVNA-17677.html (Aug. 2014); "Nikolai Ivanovich Ziber," *Russian Biographical Dictionary*, http://www.peoples.ru/science/economy/nikolaj_ziber/ (Aug. 2014); Mary R. S. Creese, "Early Women Chemists in Russia: Anna Volkova, Iuliia Lermontova, and Nadezhda Ziber-Shumova," *Bulletin for the History of Chemistry* 21 (1998): 19–24; Bachmann and Bradenahl, "Medizinstudium von Frauen in Bern"; archives, Bern University http://apps.uniarchiv.unibe.ch/syscomm/images/mata/3182_3194.gif (Aug. 2014); J. M. Meijer, *Knowledge and Revolution: The Russian Colony in Zurich, 1870–1873* (Assen, Netherlands: Van Gorcum, 1955), 177. Sources differ on Ziber-Shumova's year of birth; Bern archives gives 1854, *Mir Slovarei* 1856.

2. Nadezhda Ziber's husband, economist Nikolai Ziber, is now recognized as being one of the first to introduce the economic theories of Karl Marx to Russia. While always maintaining close ties with liberal circles in Russia, he spent some years in Switzerland before moving, in 1881, to London, where he came to know both Marx and Friedrich Engels. The author of a number of influential works on Marxist economic theory, Ziber died in Yalta on 28 April 1888, at the relatively young age of forty-four.

3. Nadina Sieber, "Beiträge zur Kenntnis der Schimmelpilze," dissertation, medical faculty, Bern University, 1880.

4. See Louis F. Fieser and Mary Fieser, *Reagents for Organic Synthesis* (New York: Wiley, 1967), 1291.

5. This line of research culminated in the total synthesis of hemin by Hans Fischer (1881–1945) of the Technical Institute in Munich. Fischer was awarded the 1931 Nobel Prize in Chemistry for the achievement.

6. In addition to those listed in the bibliography, one other publication on the field work in the Caucasus in which Ziber-Shumova took part is Marcellus von Nencki, "Die Immunisation gegen die Rinderpest nach den im Institut für experimentelle Medicin in St. Petersburg und auf des Station 'Iknewi' im Gouvernement Tiflis gesammelten Erfahrungen," *Archives International de Pharmacodynamie* 5, nos. 5, 6 (1899): 475–508 (also published in booklet form, 1899).

7. See Oscar Loev, "Marceli Nencki: Die Arbeiten Nencki's und seine Schüler," *Jahresberichte* über *die Fortschritte der Tier-Chemie, oder der Physiologischen, Pathologischen und Immuno-Chemie und der Pharmakologie* 31 (1902): ix–xxxvii, on xiv.

8. See bibliography, and also Marcellus Nencki and Nadina Ziber-Shumova, "Beiträge zur Kenntnis des Magensaftes und der chemischen Zusammensetzung der Enzyme," *Zeitschrift für Physiologische Chemie* 32, Heft 4 (1901): 291–319; *Archives des Sciences Biologiques* (St. Petersburg) 9 (1901): 47–73; *Gazeta lekarska* (Warsaw) 2, no. 21 (1901): 422–28, 455–59, 482–86, 500–505.

9. See bibliography, and also Nadine Ziber-Shumova, "Über die Entgiftung der Toxine durch die Superoxide, sowie tierische und pflanzliche Oxydasen," *Zeitschrift für Physiologische Chemie* 32 (1901): 573–91; Nadine Ziber-Shumova, "Sur les enzymes oxydants," *Gazeta lekarska* (Warsaw) 22 (1902): 271–79; Nadine Ziber-Shumova and Ekaterina Shumova-Simanowskii, "Die Wirkung des Erepsins und Darmsaftes auf Toxine und Abrin," *Zeitschrift für Physiologische Chemie* 36 (1902): 244–56; Nadine Ziber-Shumova and Simon K. Dzierzgowski, "Beitrag zum Studium der Einwirkung der Verdauungsfermente auf das Abrin und über das Schicksal des letzteren im Magenkanal," *Chemische Zentralblatt* 1 (1902): 886; and *Archives des Sciences Biologiques* (St. Petersburg) 8 (1902): 461–82.

10. Nadine Sieber-Shumoff, "M. v. Nenckis Untersuchungen über den Blutfarbstoff und dessen Beziehungen zum Blattfarbstoff," *Münchener Medizinische Wochenschrift* 45 (1902): 1873–76; N.-O. Sieber, "Note sur les travaux scientifiques du prof. M. v. Nencki," *Archives des Sciences Biologiques* 11 (1904): 167–95.

11. See Simon Dzierzgowski, W. Dzierzgowski, and Nadine Ziber-Shumova, "Die Wirkung von Nickel-Salzen auf den tierischen Organismus," *Biochemische Zeitschrift* 2 (1902): 190–218; Nadine Sieber and S. Metalnikow, "Ueber Ernährung und Verdauung der Bienenmotte (*Galleria mellonella*)," Pfluger's *Archiv für gesammte Physiologie* 102, no. 8 (1904): 269–86; Nadine Ziber-Shumova, "Ueber die bakterien feindlichen Stoffe des Blutfibrins," *Centralblatt für Bakteriologie und Parasitenkunde* (Jena), Abteilung 1, 38 (1905): 571–84; Nadine Ziber-Shumova, "Zur Frage nach dem glykolytischen Prinzip des Blutfibrins," *Zeitschrift für Physiologische Chemie* 44 (1905): 560–79; Nadine Ziber-Shumova and Ekaterina Shumova-Simanovskii, "Das Verhalten des Lecithins zu fettspaltenden Fermenten," *Zeitschrift für Physiologische Chemie* 49 (1906): 50–63; Nadine Sieber, "Die Untersuchung von Prof. Emil Fischer seiner Schüler 'über die Synthese der Polypeptiden,'" *Münchener Medizinische Wochenschrift* 53 (1906): 706–8; Nadine Ziber-Shumova, "Die Fettspaltung durch Lungengewebe," *Zeitschrift für Physiologische Chemie* 55 (1908): 177–206; Nadine Sieber, "Die Lipoide der Lunge," *Zeitschrift für Physiologische Chemie* 62, no. 2–3 (1909): 250–53; Nadine Sieber and W. Dzierzgowski, "Die Zusammensetzung der Lunge," *Zeitschrift für Physiologische Chemie* 62, no. 2–3 (1909): 254–58; Nadine Sieber and W. Dzierzgowski,

"Die Purine der Lunge," *Zeitschrift für Physiologische Chemie* 62, no. 2–3 (1909): 259–62; Nadine Sieber and W. Dzierzgowski, "Die Enzyme der Lunge," *Zeitschrift für Physiologische Chemie* 62, no. 2–3 (1909): 263–70; Nadine Sieber, "Zur Frage der Einwirkung von Alkohol auf den Gehalt der tierischen Organe an Phosphatiden," *Biochemische Zeitschrift* 23 (1910): 304–23; Nadine Sieber, "Über die Beziehung der Infektion zu Enzymen," *Biochemische Zeitschrift* 32 (1911): 108–13; Nadine O. Sieber, "Die Hydrolyse der Tuberkelbacillen," *Centralblatt für Bakteriologie und Parasitenkunde, Erste Abteilung* 66 (1913): 554–56; Nadina Ziber-Shumova, "Wasserstoffhyperoxyd als hydrolysierendes Prinzip," *Zeitschrift für Physiologische Chemie* 81 (1913): 185–99; Nadine Ziber-Shumova, [Action of ultra-violet rays on enzymes], *Russki Vrach* 12 (1913): 597–99.

12. Ziber-Shumova's date of death is uncertain; a few sources give 1916 or 1917, rather than 1914.

13. Rodolfo Alessandro Nicolaus, "Bakunin, Marussia," in *Dizionario Biografico degli Italiani* (Rome, 1988), 34:223–24; R. A. Nicolaus, "Maria Bakunin," *La Chimica e l'Industria* 42 (1960): 677–78; R. A. Nicolaus, "Maria Bakunin: Commemorazione," *Rendiconto dell' Accademia delle Scienze Fisiche e Matematiche* 28, serie 4 (1961): 15–21; Pasqualina Mongillo, "Maria Bakunin: La 'Signora' nella storia della chimica," http://www.cdlstoria.unina.it/storiche/Relazione_Mongillo.pdf (Aug. 2014); P. Mongillo, *Marussia Bakunin, una donna nella storia della chimica* (Sovaria Mannelli: Rubbettino, 2008); Marco Ciardi and Miriam Focaccia, "Maria Bakunin (1873–1960)," in *European Women in Chemistry*, ed. Jan Apotheker and Livia Simon Sarkadi (Weinheim: Wiley-VCH, 2011), 51–54.

14. Mark Leier, *Bakunin: The Creative Passion* (New York: St. Martin's Press, 2006), especially 288, 293.

15. See also the section on Sof'ia Bakunina in chapter 1.

16. See J. H. van't Hoff, "Chemistry in Space," in *Dix Années dans l'Histoire d'une Theorie*, trans. James E. Marsh (Oxford: Clarendon Press, 1891).

17. Sources do not specify which of the two Erlenmeyer chemists entered the competition—the father, Richard August Carl Emil (1825–1909), or the son, Friedrich Gustav Carl Emil, sometimes known as Emil Jr. (1864–1921). Dates suggest the son as the most likely, but it is worth noting that the senior Emil Erlenmeyer, who had retired from his professorship at the Munich Polytechnic School in 1883, also worked on isomerism in the cinnamic acids. The area is still one of interest; recent work includes the exploration of factors influencing isomeric distribution in preparation procedures, coupled with theoretical calculations of energy differences between isomers. See István Pálinkó, Béla Török, Gyula Tasi, and Tamás Körtvélyesi, "Experimental and Computational Tools for Mechanistic Study: A Modified Perkin Condensation Leading to Alpha-Phenylcinnamic Acid Isomers," *Electronic Conference on Trends in Organic Chemistry, ECOTC-1* (1995), online at http://www.ch.ic.ac.uk/ectoc/papers/45/ECTCIN.html (Aug. 2014).

18. A.-J.-J. Vandevelde, "Recherches sur l'acide phénoxyacétique.—Acide phénoxycinnamique. Deuxième communication," *Bulletin de l'Académie Royale de Belgique* 33, serie 3 (1897): 221–29.

19. A sample of Bakunin's publications includes Maria Bakunin, "Sulla formazione degli indoni in rapporto colle stereoisomerie e su di un nuovo metodo per la preparazione di indoni, anidridi ed eteri," *Gazzetta chimica Italiana* (hereafter *Gazz.*) 30, parte 2 (1900): 340–64; Maria Bakunin, "Sulle condensazioni in presenza dei metalli e loro cloruli," *Rendiconto dell' Accademia della Scienze Fisiche e Matematiche,* Naples (hereafter *Rend.*) 10, serie 3 (1904): 17–22; Maria Bakunin and Gaetano Altieri, "Sintesi del benzil-β-naftolo e suoi derivati," *Rend.* 10, serie 3 (1904): 33–37; Maria Bakunin and Michele Barberio, "Sintesi del benzil-α-naftolo e suoi derivati," *Rend.* 10, serie 3 (1904): 23–32; Maria Bakunin and E. Lanis, "Relazioni fotochimiche dei nitrofenilindoni. Nota 1," *Gazz.* 41, parte 2 (1911): 155–84; Maria Bakunin, "Gli indoni ed i loro prodotti di trasformazione al sole. Loro comportamento con l'onozo. Nota IV," *Rend.* 17, serie 3 (1911): 379–86; Maria Bakunin and T. Angrisani, "Sulla costituzioni del fenilortonitroindone e sui prodotti di scissione," *Rend.* 20, serie 3 (1914): 184–91; Maria Bakunin and E. Lanis, "Reazioni fotochimiche dei nitrofenilindoni. Memoria 1," *Atti della Reale Accademia della Scienze Fisiche e Matematiche* 15, no. 3, serie 2 (1914): 1–22; Maria Bakunin and F. Giordani, "Reazioni fotochimiche dei fenilnitroindoni. Produtti di trasformatione dei fenilortonitroindone. Nota V." *Rend.* 22, serie 3 (1916): 22–30; Maria Bakunin and G. Fisceman, "Meccanismo della reazione del Perkin," *Gazz.* 46 (1916): 73–103; Maria Bakunin and F. Giordani, "Soluzioni colloidali e sintesi organiche. Meccanismo della sintesi dei diacetati di nitrobenzilidene. Nota II," *Rend.* 23, serie 3 (1917): 39–50; Maria Bakunin, "Sul meccanismo della sintesi di Perkin-Oglialoro," *Rend.* 2, serie 4 (1932): 20–22; Maria Bakunin and D. Peccerillo, "Sul meccanismo della sintesi Perkin-Oglialoro," *Gazz.* 65 (1935): 1145–61.

20. Maria Bakunin and G. Dragotti, "Contributo alla conoscenza dei pigmenti melaninici," *Rend.* 10, serie 3 (1904): 222–40.

21. Maria Bakunin and Vincenzo Petitti, "Recerche tossicologiche della morfina," *Gazz.* 34, parte 1 (1904): 1–38.

22. Maria Bakunin, "L'industria degli scisti ittiolici dell' Italia Meridionale," *Atti del Reale Istituto d'Incoraggiamento di Napoli* 9, serie 6 (1911): 215–37; Maria Bakunin and F. Giordani, "Gli scisti bituminosi di

Castroreale e Barcellona (Sicilia)," *Annali di Chimica Applicata* 15 (1925): 265–72; Maria Bakunin, "Sulla utilizzazione delle rocce bituminose con speciale riguardo alla Sicilia," *Atti del XI Convegno Nazionale di Chimica puro ed applicata*, Palermo, I, II, III (1927): 468–513.

23. Maria Bakunin, "La sorgente di Valle di Pompei," *Atti della Accademia Pontaniana* 37, no. 12 (1907): 1–6; Maria Bakunin and F. Arena, "La sorgente minerali di Valle di Pompei. Relazione dell' analisi chimica e batteriologica," *Napoli, Atti della Reale Accademia delle Scienze Fisiche e Matematiche* 14, no. 3, serie 2 (1910): 1–21.

24. Among the several memorials she contributed to the Naples *Rendiconto* one, coauthored with Agostino Oglialoro-Todaro, was to Cannizzaro: "In memoria di Stanislao Cannizzaro," *Rend.* 16, serie 3 (1910): 135–42.

25. Faber Renato Fabbris, "Biography of Renato Caccioppoli," *MacTutor History of Mathematics Archive*, School of Mathematics and Statistics, University of St. Andrews, http://www-history.mcs.st-andrews.ac.uk/Biographies/Caccioppoli.html (Aug. 2014).

26. Maria Bakunin's many publications not already noted include the following: "Sulla sintesi degli acidi non saturi e sui loro prodotti di disidratazione," *Napoli, Atti della Reale Accademia delle Scienze Fisiche e Matematiche* 11, serie 2, Memoria 2a (1902): 1–9; "Sulla eterificazione di acidi con fenoli," *Gazz.* 32, parte 1 (1902): 178–85, and *Napoli, Atti della Reale Accademia delle Scienze Fisiche e Matematiche* 11, serie 2, Memoria 4a (1902): 1–6; "Azione del cloruro di benzile sugli amidofenoli," *Rend.* 11, serie 3 (1905): 361–75, and *Gazz.* 36, parte 2 (1905): 211–27; Maria Bakunin and V. Majone, "Ricerche tossicologiche sulla stricnina," *Rend.* 11, serie 3 (1905): 333–60, and *Gazz.* 36, parte 2 (1906): 227–52; Maria Bakunin and L. Parlati, "Studio sui prodotti di disidrazione dell' acido fenilortocinnamico e dei produtti che accompagnano quest'acido nella sintesi del Perkin," *Gazz.* 36, parte 2 (1906): 264–82; Maria Bakunin and L. Parlati, "Contributo alla conoscenza delle isomerie spaziali negli acidi fenilnitro-cinnamici (anidridi, indoni, cloruri, fenilidrazi, idrazoni, ossime)," *Rend.* 12, serie 3 (1906): 503–15; Maria Bakunin and E. Lanis, "Reazioni fotochimiche dei nitrofenilindoni. Nota I," *Gazz.* 41, parte 2 (1911): 155–84; "Sull' azione dei raggi ultravioletti sugli stereoisomeri della serie cinnamica. Nota II," *Rend.* 17, serie 3 (1911): 372–75; Maria Bakunin and E. Lanis, "Reazioni fotochimiche dei nitrofenilindoni, Memoria I," *Napoli, Atti della Reale Accademia delle Scienze Fisiche e Matematiche* 15, serie 2 (Memoria) (1914): 1–22.

27. Iusuf S. Musabekov, *Iuliia Vsevolodovna Lermontova, 1844–1919* (Moscow: Nauka, 1967), especially 8–9 in chapter 1, "Pervie zhenshchiny - khimiki" [The first women chemists]; *Bol'shaia sovetskaia entsiklopediia*, in English as *Great Soviet Encyclopedia*, ed. A. M. Prokhorov, 3rd ed. (Moscow: Izd-vo Sovetskaia entsiklopediia, 1970–1981; New York: Macmillan, 1973–1982), 5:574 (English), 5:308–9 (Russian); Creese, "Early Women Chemists in Russia," 19–24.

28. In 1878, when the Russian Chemical Society joined the Russian Physical Society (founded in 1872), it became part of the joint Russian Physical-Chemical Society.

29. Marina Loskutova, "Public Science as a Network: The Congresses of Russian Naturalists and Physicians in the 1860s –1910s," *Baltic Journal of European Studies* (formerly *Proceedings of the Institute for European Studies*) 1 (o.s. 9), no. 1 (2011): 153–69, on 159.

30. Musabekov, *Iuliia Vsevolodovna Lermontova*; Ann Hibner Koblitz, *A Convergence of Lives: Sofia Kovalevskaia: Scientist, Writer, Revolutionary* (Boston: Birkhäuser, 1983); Hibner Koblitz, "Science, Women, and the Russian Intelligentsia: The Generation of the 1860s," *Isis* 79 (1988): 208–26; Cordula Tollmien, "Zwei erste Promotionen: Die Mathematikerin Sofja Kowalewskja und die Chemikerin Julia Lermontowa," in *"Aller Männerkultur zum Trotz": Frauen in Mathematik und Naturwissenschaften*, ed. Renate Tobies (Frankfurt/Main: Campus Verlag, 1997), 83–129; Tollmien, "Julia Wsewolodowna Lermontowa," http://www.cordula-tollmien.de/julialebensdaten.html (Aug. 2014); Tollmien, *Fürstin der Wissenschaft: Die Lebensgeschichte der Sofia Kowalewskaja* (Weinheim and Basel: Beltz Verlag, 1995); Elena Roussanova, *Julia Lermontowa (1846–1919): Die erste promovierte Chemikerin* (Norderstedt: Books on Demand, 2003). Shorter accounts of Juliia Lermontova's life and work include Marianne Offerein, "Julia Lermontova (1846–1919)," in *European Women in Chemistry*, ed. Apotheker and Sarkadi, 27–30; Anne C. Hughes, "Lermontova, Iuliia Vsevolodovna (1847–1919)," in *The Biographical Dictionary of Women in Science: Pioneering Lives from Ancient Times to the Mid-20th Century*, ed. Marilyn Ogilvie and Joy Harvey (New York and London: Routledge, 2000), 2:775–76; Creese, "Early Women Chemists in Russia."

31. Mary R. S. Creese, *Ladies in the Laboratory II: West European Women in Science, 1800–1900* (Lanham, Md.: Scarecrow Press, 2004), 22–23; matriculation register, Zurich University, http://www.matrikel.uzh.ch/active/static/27227.htm (Aug. 2014). Sesemann's doctorate appears to have been awarded without any particular attention being given to the event.

32. By the Gregorian calendar this is 2 January 1847.

33. Mikhail Lermontov claimed descent from a Scottish adventurer, Captain George Learmont, who entered Russian service early in the seventeenth century; George Learmont was thought to be descended from the Scottish poet Thomas the Rhymer (Thomas Learmonth). Mikhail Lermontov's claim of Scottish descent has never been

proved or disproved, but his belief in a Scottish connection was expressed in two of his poems, "The Coffin of Ossian" (1830) and "A Wish" (1831). In the latter he wrote of "a field of my forefathers blooms, / Where their ashes rest / In an empty castle on the foggy slopes." See Ian G. Anderson, *Scotsmen in the Service of the Czars* (Edinburgh: Pentland Press, 1990), 162–63.

34. Anna Evreinova fled on foot at the end of November 1870, and in darkness and mist crossed the border into German territory. In 1873 she received a law degree from Leipzig University, the first awarded to a woman in Europe.

35. August Wilhelm von Hofmann (1818–1892), a student of Justus von Liebig at the University of Giessen, was a *Privatdozent* (lecturer) at the University of Bonn in 1845, when the Royal College of Chemistry in London was set up as a school for teaching practical chemistry. Largely through the influence of Queen Victoria's consort, Albert, Prince of Saxe-Coburg-Gotha, Hofmann was appointed the college's first director, a post he held for twenty years. In 1864 he returned to Bonn, and the following year became professor of chemistry and director of the laboratory at the University of Berlin. London's Royal College of Chemistry became the first constituent college of Imperial College, London, and eventually became the chemistry department of Imperial College. Hofmann's twenty-year stay in London may well have given him a more open-minded approach than that of his fellow University of Berlin faculty members when it came to the question of the admission of women to university education. Tollmien, "Zwei erste Promotionen," 93. However, scientific education for girls and women was at a very early stage in London also during Hofmann's years there. Mary R. S. Creese, *Ladies in the Laboratory? American and British Women in Science, 1800–1900* (Lanham, Md.: Scarecrow Press, 1998), 265–72.

36. F. Beilstein, *Handbuch der Organischen Chemie*, II Abteilung. Specieller Theil. Aromatische Reihe (Hamburg and Leipzig: Leopold Voss, 1883), 982.

37. Lermontova's use of ball-and-stick depictions to clearly and simply illustrate the structures of her compounds followed Hofmann's development of this innovative technique. He is thought to have first demonstrated the art at a talk he gave to the Royal Institution just before leaving London. See A. W. Hofmann "On the Combining Power of Atoms," *Proceedings of the Royal Institution of Great Britain* 4 (1862–1866): 401–30.

38. But no flood of women applicants appeared. The next woman to request consideration for a degree to be conferred by Göttingen University was British mathematician Grace Chisholm Young, who took her oral examination and received her degree *magna cum laude* in the spring of 1894. Not until 1900 was a second doctorate in chemistry awarded by a German university to a woman, Clara Immerwahr, who graduated from Breslau University that year. See Creese, *Ladies in the Laboratory?*, 192–94, and Creese, *Ladies in the Laboratory II*, 142–45. Lermontova lapsed into obscurity, her story buried in Göttingen University's files collection, and was rediscovered only in recent years. See also Tollmien, *Fürstin der Wissenschaft*, 87.

39. Julie von Lermontoff, "Zur Kenntniss der Methyleneverbindungen," thesis, philosophical faculty, University of Göttingen, 1874 (Göttingen: Druck der Universitäts-Buchdruckerei von E. A. Huth, 1874).

40. Vladimir Markovnikov, "Ueber die Isomeren Pyroweinsäuren," Liebig's *Annalen der Chemie* 182 (1876): 324–46.

41. Dmitrii Mendeleev, *Neftianaia promyshlennost' v Sievero-amerikanskom shtatie Pensil'vanii i na Kavkazie* [The oil industry in the North American state of Pennsylvania and in the Caucasus] (St. Petersburg: Tip. tovarishchestva "Obshchestvennaia Pol'za," 1877). For Lermontova's French-language translation of one chapter, see Dmitrij Mendeleeff, "L'origine du pétrole," *Revue Scientifique de la France et de l'Étranger* 18 (1877): 409–16. See also Francis Michael Stackenwalt, "Dmitrii Ivanovich Mendeleev and the Emergence of the Russian Petroleum Industry, 1863–1877," *Ambix* 45, part 2 (1998): 67–84.

42. Iuliia Lermontova, *Vospominaniia o Sofya Kovalevskoi* [Memories of Sofia Kovalevskaia], 1891 (Archives, Academy of Sciences, St. Petersburg), printed in *Sofya Kovalevskaia: Vospominaniia i Pis'ma* [Sofia Kovalevskaia: Memories and letters], ed. Solomon IA. Straikh (Moscow: Izd-vo Akademii nauk SSR, 1961). See also Roussanova, *Julia Lermontowa*, 90.

43. "First International Petroleum Congress, Paris, 17–28 August 1900," *Petroleum Review*, Supplement, issued in parts (25 August 1900, 1, 8, 15, and 22 September 1900); "Oil of Russia," http://www.oilru.com/or/24/413 (Aug. 2014). Markovnikov was one of several Russian oil industry men who received awards at this congress. Russia was then the world's major oil-producing country, its output amounting to 51.1 percent of the total; the United States came second with 42.6 percent.

44. See also Hibner Koblitz, *Convergence of Lives*, 146–47, 150–52, 168–73.

45. Juliia Lermontova, *Teknik* 20 (1883): 10–13. (*Teknik* was a Moscow news sheet.)

46. Fufa Kovalevskaia's father, Vladimir, committed suicide in April 1883 (Hibner Koblitz, *Convergence of Lives*, 171–72). After Sofia Kovalevskaia's death in Stockholm in February 1891, Fufa, who had for a time been living with her mother, completed her primary schooling in Sweden while living with friends and then returned to Juliia Lermontova in Moscow. She attended the Higher Courses for Women in Moscow, qualified as a doctor,

and worked in the Red Cross in the Soviet Union and abroad. After retiring she became a medical librarian and also translated works from Swedish into Russian. She lived until she was seventy-four, dying about 1952 in the Soviet Union. Hibner Koblitz, *Convergence of Lives*, 236 note 72.

47. Roussanova, *Julia Lermontowa*, 114: "The phenomenon of interaction between Russian and German schools of chemistry runs like a red thread throughout her work."

48. Eleanor S. Elder and Sue-Dee Lazzerini, "The Deadly Outcome of Chance—Vera Estaf'evna [*sic*] Bogdanovskaia," *Journal of Chemical Education* 56, no. 4 (1979): 251–52; Obituary, *Nature* 56, no. 1441 (1897): 132; "Bogdanovskaja, Vera Evstafjevna," in *Lexicon der Frau*, ed. Gustav Keckeis (Zurich: Encyclios-Verlag AG, 1953), 1:467; Anne C. Hughes, "Bogdanovskaia, Vera Evstaf'evna (1867–1896)," in *Biographical Dictionary of Women in Science*, ed. Ogilvie and Harvey, 1:153; Ia. Popov, [Introduction II], in Vera Evstaf'evna Bogdanovskaia, *Ocherki, stati, otryvki* [Essays, articles and extracts], *Pamiati Very Evstafevny Bogstanovskoi, 1867–1896* [In memory of Vera Evstaf'evna Bogdanovskaia, 1867–1896] (St. Petersburg: Ia. Popov, tip. V. Kirshbauma, 1898), 1:iii–ix. This biographical note by General Popov is the source most heavily relied on for the details given here.

49. In 1869 Carl Graebe, working in Adolf von Baeyer's laboratory in collaboration with Carl Liebermann, synthesized alizarin (an aromatic ketone) starting from anthraquinone. Synthetic alizarin was first placed on the market in 1871, only two years later, and within a short time it supplanted the natural product. Graebe and Liebermann's achievement was later recognized as marking the start of Germany's very important dye industry, which supplied a huge international market that included the manufacturers of military uniforms. See Louis F. Fieser and Mary Fieser, *Organic Chemistry*, 2nd ed. (London: George G. Harrap, 1953), 908–10.

50. Founded as a munitions factory in 1807 by Tsar Aleksandr I, the Izhevsk factory's output of military hardware for the Russian army by the 1890s included the new, breech-loading, bolt-action, magazine-fed Mosin-Nagant rifle used by the military for the next one hundred years. The city of Izhevsk is still an important industrial center, noted specially for its military-hardware industry.

51. The phosphorus analogue of hydrogen cyanide was first prepared electrolytically, by Du Pont chemist T. E. Gier, sixty-five years after Vera Bogdanovskaia's death. A very reactive, colorless gas, it can be stored only below its triple point of −124±2°C. See T. E. Gier, "HCP, a Unique Phosphorus Compound," *Journal of the American Chemical Society* 83 (1961): 1769–70.

52. The dates of death given in Russian and Western accounts differ because of calendar differences; by the nineteenth century the Julian calendar, used in Russia, was twelve days behind the Gregorian calendar used in western countries. See Nautical Almanac Offices of the United Kingdom and United States, *Explanatory Supplement to the Astronomical Ephemeris and the American Ephemeris and Nautical Almanac* (London: H. M. Stationary Office, 1961).

53. G. Gustavson, [A few words about Vera Evstaf'evna Bogdanovskaia], *Journal of the Russian Physical-Chemical Society* 29 (1897): 147–51.

54. See Julia Mannherz, "Mysterious Knocks, Flying Potatoes and Rebellious Servants: Spiritualism and Social Conflict in Late Imperial Russia," paper presented at the Fifth International Postgraduate Conference on Central and Eastern Europe, School of Slavonic and East European Studies, University College, London, 6–8 November 2003, available from UCL Discovery, http://eprints.ucl.ac.uk/16345/1/16345.pdf (Aug. 2014). In late nineteenth-century Russia, as in western Europe, spiritualist beliefs and practices appealed to diverse groups, which included some of the country's leading scientists and intellectuals. Chemist Aleksandr Butlerov was among those taking part in spiritualist activities in St. Petersburg. See also the sketch of Eleanor Sidgwick, assistant to Cambridge physicist Lord Rayleigh, in Creese, *Ladies in the Laboratory?*, 219–20.

55. No mention of this work has been found in the usual records. The *Lexikon der Frau* sketch (see note 48) gives its title as *Nachalnye ychebnik khimii* [Beginning textbook in chemistry].

56. A. M. Butlerov, *Vvedenie k polnomu izucheniiu organicheskoi khimii* [Introduction to a complete study of organic chemistry] (St. Petersburg: tip. V. Demakova, 1887), 719 pp.

57. P. J. Murzel, *Chemie* (Berlin: P. Parey), Unterrichtbücher; Landwirtschaftliche, 2nd ed., 1894, Bd. 10.

58. "E. A. Fomina-Zhukovskaia," *Journal of the Russian Physical-Chemical Society* 26 (Chem.), (1894): 432–34; Musabekov, *Iuliia Vsevolodovna Lermontova*, 24–25; Anne C. Hughes, "Fomina-Zhukovskaia, Evdokiia Aleksandrovna (1860–1894)," in *Biographical Dictionary of Women in Science*, ed. Ogilvie and Harvey, 1:457–58.

59. E. Phomina, "Recherches sur quelques combinaisons du groupe de l'euxanthone," dissertation, faculty of science, University of Geneva, 1889, listed in the catalogue of the Bibliothèque Nationale de France, http://catalogue.bnf.fr/ark:/12148/cb311024697 (Aug. 2014).

60. Vladimir Vasil'evich Markovnikov, *Izbrannye trudy* [Selected works], ed. Al'fred Feliksovich Platé and Georgii Vladimirovich Bykov, Classics of Science Series (Moscow: Akademii nauk SSSR, 1955). See chapter entitled "Khimiia alitsiklicheskikh soedinenii" [Chemistry of alicyclic compounds], 576, for Markovnikov's mention of Fomina's work and pleasant personality.

61. Information collected from technical papers and chemical indexes.

62. Zénaïda Kikine, "Sur quelques dérivés de la flavinduline," dissertation, faculty of science, University of Geneva, 1902–1903 (Geneva: Ch. Zoellner, 1903); there the authorship is given, in error, as L. Kikine in the published version.

63. Mikhail Konovalov and Zenaida Kikina, [Action nitrifique de l'acide nitrique sur les carbures saturés et sur leur dérivés: Nitration du dihydrocamphène et du chlorhydrate de pinène], *Journal of the Russian Physical-Chemical Society* 34 (1902): 935–44.

64. Z[enaida] V. Kikina, "Vospominaniia o M. I. Konovalovie ego ychenitsy" [Memories of M. I. Konovalov by his student], *Journal of the Russian Physical-Chemical Society* 42 (1910): 52–57.

65. See Musabekov, *Iuliia Vsevolodovna Lermontova,* 15 note 8.

66. See Alexander Vucinich, *Science in Russian Culture, 1861–1917* (Stanford, Calif.: Stanford University Press, 1970), 141–44, 382–85.

67. For a brief reference to Kaufmann, see Musabekov, *Iuliia Vsevolodovna Lermontova*, 14–15.

68. In this work Davidova followed a former student of Butlerov, Aleksandr Aleksandrovich Krakau, later a lecturer at the Higher Courses for Women and from 1891 professor of chemistry at the St. Petersburg Electrotechnical Institute.

69. N. P. Rikhter-Rzhevskaia, [Sur la tension de vapeur du brome en solution dans l'acide bromhydrique], *Journal of the Russian Physical-Chemical Society* 35 (1903): 441–49.

70. Archives, Bern University, http://apps.uniarchiv.unibe.ch/syscomm/images/mata/8221_8232.gif (Aug. 2014); information in technical paper.

71. Minna Samelson, "Über Permanganmolybdate," inaugural dissertation, philosophical faculty, University of Bern, 1900 (Leipzig: Druck Metzger & Wittig, 1900). Samelson dedicated the work to her brother.

Chapter Three

Life Sciences

Botanists, Zoologists, and Entomologists

BOTANISTS

Only five late nineteenth-century Russian women published papers in the botanical sciences that were listed in the Royal Society's *Catalogue*, a relatively low number that stands out in contrast to the large groups of women in North America and Britain who were active in botany at that time.[1]

Of the Russian contributors to the plant sciences listed in the bibliography—Fedchenko, Karsakova, Malinesco, Sokolova, and the neurologist-physiologist Schipilova—the most widely known is Fedchenko. The indefatigable assistant and companion of her husband, Aleksei Fedchenko, throughout his short life, she continued and greatly extended his work after his death. Among her major contributions to botany was the editing of the extensive records of the Fedchenko journey through Turkestan in 1868–1871, an expedition in which she took part.

Ol'ga Aleksandrovna Fedchenko,[2] née Armfeld (1845–1921), the third of nine children of Aleksandr Osipovich Armfeld (1806–1868) and his wife, Anna Vasilievna Dmitrovskaia (1821–1888), was born in Moscow on 18 November 1845.[3] Of German-Russian ancestry, Aleksandr Armfeld was professor *ordinarius* of legal medicine and health at Moscow University, and from 1838 also held the post of inspector of classes at the Nikolaevski Institute of the Moscow Foundling Hospital, a secondary school for young ladies and also a teacher-training institution.

The large and close-knit Armfeld family, well known and highly respected in Moscow society, provided a stimulating atmosphere for growing children, thanks in large part to Anna Vasilievna, daughter of a state councillor and landowner, who had been educated at the Ekaterinski Institute for girls. Anna Vasilievna's drawing room became one of the centers of the city's literary and artistic life; frequent guests included playwright and novelist Nikolai Gogol, actor Mikhail Shchepkin, writer Sergei Aksakov and his writer sons, and the poet Mikhail Lermontov; Lev Tolstoy was another of her friends.

Ol'ga and her sisters received their early instruction at home from governesses, and then she and her younger sister Natalia attended the Nikolaevski Institute, Olga entering in September 1857, shortly before her twelfth birthday. Concentrating on western European languages, music, drawing, and painting, she studied there for seven years and graduated in 1864 with the diploma of *candidat*, the lowest-level degree in Russia's higher-education system. She had further art training from landscape artist Aleksei Savrasov, an instructor at the Moscow School of Painting, Sculpture, and Architecture.

Ol'ga Aleksandrovna's interests in natural history began early and she had excellent opportunities to develop in this direction during summers, when the Armfeld family moved to Aleksandr Armfeld's estate in Mozhaisk, a country district about seventy miles west of Moscow in the valley of the Moskva River.[4] She made small collections of minerals, shells, insects, and birds' eggs, some of which were later incorporated into the holdings of the Moscow University Zoological Museum. Her first herbarium, made up of the plants of the region that she assembled in 1861 and 1862, was used by Nikolai Kaufman in his famous *Moskovskaia flora*—a creditable contribution for a girl of sixteen.[5] In 1864, just after her graduation from the Nikolaevski Institute, she became a founding member of the Society of Enthusiasts of Natural Science, Anthropology and Ethnography (hereafter Moscow

Society for Natural Science), a group that started at Moscow University at that time. Ten years later this society elected her an honorary fellow.

For several years, beginning about 1861, she occupied herself teaching languages and drawing, visiting the zoological museum, and collecting plants. Her knowledge of various branches of natural history expanded, and she made good use of her art skills and language training. The tasks she undertook included helping with collections and assisting in research at the museum, translating biology texts from French, German, and English into Russian, communicating with foreign scientists, and preparing natural history illustrations.

Among the naturalists she met in the new society was Aleksei Pavlovich Fedchenko (1844–1873), a recent graduate of Moscow University. They worked together for some time in the field, and she assisted him in such matters as translating letters to foreign scientists and recording anthropological data. Ol'ga Armfeld married Fedchenko in 1867, a year before the Moscow Society for Natural Science recommended him to Turkestan governor-general Konstantin Petrovich von Kaufman (1818–1882) for the scientific exploration of Turkestan. That territory, part of which was still undergoing conquest by imperial Russia, was then under the control of the military; outside the oasis cities it was a primitive, somewhat dangerous place for travellers.[6] The projected expedition was an ambitious undertaking to explore territory as far south as the northern ranges of the Pamirs. Ol'ga Fedchenko was an official, but unpaid, member of the expedition, her major responsibility being botanical work. To prepare herself and extend her natural history background, she visited several museums in Russia and accompanied Aleksei Pavlovich on trips to study collections in Sweden, Austria, and Italy, all places where she carried out a substantial amount of collecting as well.

The Fedchenkos started their exploring work in what was then the Zaravshanskii region (later Samarkand Province) of eastern Uzbekistan, territory only recently conquered by the Russians. Travelling on horseback through unexplored country, they made good progress despite difficulties at times with local inhabitants. After investigations around the cities of Tashkent and Samarkand, they went out into the Kyzyl Kum desert region, the vast expanse of land between the Amu Darya and Syr Darya rivers that stretches across more than one present-day republic and is now well recognized as an important wildlife region (see fig. 4). Further exploration took them to the Zarafshan Range of the northern Pamirs; the

Figure 4. Region of southern Russia explored by botanist Olga Fedchenko.

region around Fergana in the fertile Fergana Valley, along which went the northern Silk Road to China; and to the most northerly of the Pamir ranges, the Trans-Alay Range that forms the border between Kyrgyzstan and Tajikistan, a world of spectacular scenery with peaks above twenty thousand feet, immense glaciers, lakes, and broad, flat upland valleys.

The confidence of the Moscow Society for Natural Science in Ol'ga Fedchenko's botanical abilities was well placed; she is said to have collected 1,527 plants. In addition, being the expedition's only artist, she painted landscapes typical of the regions she was moving through, drew maps, collected insects, and had charge of such matters as the accounts. When they returned to Moscow and began the task of analyzing the massive volume of records and collected materials (botanical, entomological, zoological, and geographical), Ol'ga labeled collections, made necessary contacts with Russian and foreign specialists, and translated English-language articles on the geography of Turkestan into Russian.

In the autumn of 1872, after a brief stay in Moscow, the Fedchenkos went again to western Europe. Their son, Boris, was born that year, but Ol'ga continued her translating, assisted her husband in his ever-increasing correspondence with foreign scientists, and in addition took lessons in botany in Leipzig. In 1873 Aleksei Fedchenko died in a violent storm on the Col du Géant, Mont Blanc, where he had gone to investigate alpine glaciers in order to compare them with those in the mountains of Turkestan.[7] Ol'ga arranged for his burial at Chamonix and returned with her infant son to Moscow. There she was asked by the Moscow Society for Natural Science to continue the editing of the Turkestan expedition reports, the first parts of which had already appeared in serial form. Thanks to her persistence, the reports were successfully completed and published in twenty-four issues, the last of which, written by Ol'ga herself, was a detailed, systematic treatment of the expedition's botanical findings.[8]

Her first major publication, brought out in 1873, was her English-to-Russian translation of Sir Henry Yule's essay on the upper regions of the Amu Darya, or Oxus, River, a work that won the medal of the Russian Geographical Society.[9] The society elected Fedchenko a "cooperating member" in 1877, in recognition of her work on the Turkestan expedition.[10] Beginning about 1877 she was secretary of the Moscow Society for Natural Science while continuing the work of classifying and ordering collections in Moscow Botanic Gardens and translating publications on central Asia by foreign authors.[11]

These many undertakings during the 1870s and early 1880s earned her only a modest income on which to support herself and bring up her young son. Added to her anxieties throughout these years was the succession of distressing, even tragic, events befalling members of the Armfeld family. Her younger sister Natalia, who in 1872 returned to Moscow after studies in Heidelberg, joined the student movement to spread revolutionary ideas; arrested in 1879 and exiled to Siberia, she died there eight years later.[12] The youngest of the family, Nikolai, also active in student revolutionary circles, died in 1880, and Anna Armfeld in 1887. Ol'ga Fedchenko's only scientist sibling, her older brother Aleksandr, ran into difficulties in the early 1870s when his promising career at Novorossisk University (now the University of Odessa) came to an abrupt end and he returned to Moscow.[13]

Only after a delay of more than a decade was Ol'ga Fedchenko free to resume her own research, and only in the 1890s did she again take part in botanical expeditions. One of the first of these was an 1891–1892 trip to the southwestern Urals (later the Republic of Bashkortostan); her investigations there, over terrain that varied from steppe land to high mountains, resulted in the first complete flora of the territory. In 1893 she collected plants in the Crimea, and the following year in the Caucasus. On these trips she regularly had the company of her son, Boris, then a student at Moscow University, whose early interest in botany was already well developed.

From 1897 until the end of her life, more than two decades later, she devoted her energies to the study of the flora of Turkestan, with a special emphasis on the Pamirs, that alluring "Roof of the World." After a brief trip to the region in 1897, she spent the following two years studying collections in the important west European botanic gardens, including those in Berlin, Paris, and London, where she was well known in the botanical communities. Then, in 1901, she fulfilled a long-held ambition. That year, at the age of fifty-six, she set out on a second extended trip to Turkestan, accompanied by Boris Fedchenko, himself a specialist in Turkestan flora by that time.

Still an isolated colonial outpost, Turkestan had remained under military administration, but it was somewhat easier to approach than at the time of the first Fedchenko expedition; a link of the Trans-Caspian Railway now went as far as Tashkent. Botanical exploration had not progressed beyond the early stages. The fifty-two-day second Fedchenko expedition, which targeted some of the more accessible parts of this botanically diverse country, went along the Pamirs as far as the Afghan border, a journey that required considerable stamina. Ol'ga Aleksandrovna spent long days on horseback, took little rest, and subsisted on tea and dry bread. Her reports were published under the titles *Flora Pamira*, with four supplements (1903–1905), and *Definition of Pamir Plants* (1910); the former became a classic.[14]

She made two more journeys to the mountains of Turkestan, in 1910 and 1915, and brought back several hundred plants, a number of them new to science. Overall she covered the whole range of plants native to the region. Her *Conspectus of Turkestan Flora*, the most important of her works on Turkestan, was published simultaneously in Russian and German early in 1902 and in several later editions, some jointly authored with her son.[15] She was widely recognized as a leading specialist on Turkestan flora. Her monograph on the genus *Eremurus*, a thorough and comprehensive study of this genus, typical of the plants of Turkestan, was considered especially outstanding; its publication in German made it widely accessible to Western readers. The dwarf species *Eremurus olgae*, as well as several other Turkestan plants, was named after her.[16] Also of great interest is her study of the genus *Allium*.[17]

From 1900 St. Petersburg was her usual place of residence, and her place of work the Imperial Botanic Garden, where Boris Fedchenko was a staff member from 1900 and chief curator of the Turkestan herbarium from 1905.[18] In 1895 she left St. Petersburg temporarily to establish a large botanic garden at Ol'gino, in the Mozhaisk district west of Moscow that she had known in her childhood. Here, in this last of Russia's great private gardens, she cultivated primarily plants of Turkestan, started from seeds and specimens she had brought back from her travels, and further enriched by exchanges with botanists and botanic gardens in western Europe, including London's Kew Gardens. Many of the plants were critical forms and some groups were worthy of monographic treatment. The garden and its plants did not survive the upheavals that followed the Russian Revolution.

After her move to St. Petersburg in 1900 she carried out major research in plant classification and systematics, working at the Imperial Botanic Garden; she was formally elected an honorary member of the institution in 1912 and was always known there simply as Ol'ga. Overall she brought out about seventy papers and monographs, including a number of very valuable reports.[19] All her writings were notable for their clarity, unpretentiousness, and methodical precision in their presentation of facts; her statements were always carefully verified, and were considered safe to build on. Beginning with her translation of Henry Yule's essay on the upper Amu Darya in 1873, her published writings spanned a period of forty-eight years, ending with her "Novye materialy k poznaniiu roda *Eremurus*" (New material toward knowledge of the genus *Eremurus*), brought out in 1921, a month before her death. Her career in botanical exploration and plant systematics, as measured from the start of her first Turkestan expedition, lasted more than half a century. It has been said that "scientific work was her life."[20]

Widely known abroad as well as in Russia, she held memberships in a number of scientific societies, including l'Académie Intérnationale de Géographie Botanique (Le Mans). In the spring of 1912 she journeyed to the Royal International Horticultural Exhibition in Chelsea, London, and about this time also visited herbaria and botanic gardens in Berlin and Geneva, as well as in London. Her last trip to western Europe was in 1914, when she collected about one hundred plants in the region of Naples. She died in her seventy-sixth year, on the night of 28–29 April 1921, in her apartment at Pesochnaia Street, Petrograd, where she had been living with her son and his family.

Although she was not one of the young women of the 1860s who took part in the effort to open university education to women, Ol'ga Fedchenko, largely self-educated, played a pioneering role in another aspect of the advancement of women; she most effectively demonstrated the capabilities of women in difficult and demanding botanical fieldwork, and in addition produced outstanding studies in plant classification and systematics. Her entry into botanical exploration via marriage to a trained naturalist was not unconventional, but the outstanding work she did independently after her husband's early death testifies to her own considerable drive and abilities. Her election as corresponding member of the Russian Academy of Sciences in 1906 attests to her acceptance by her male colleagues and their recognition of her contributions to science.[21] Ol'ga Fedchenko was one of a small, select group—the leading eastern European women scientists of her generation—capable of recognizing a clear path before her and endowed with the vitality to follow where it led.

Both Natal'ia Karsakova and Ekaterina Sokolova, whose careers began in the 1890s and lasted well into the twentieth century, were largely laboratory workers concerned with questions in plant morphology and systematics.

Born in Moscow, NATAL'IA VLADIMIROVNA KARSAKOVA[22] (1863–1941) studied natural history at the Sorbonne, probably starting in the late 1880s. Among her teachers was French botanist Philippe Van Tieghem, known for his work on fungus development.

Over the course of two years, 1891–1892, Karsakova carried out research at the Roscoff Biological Station in Brittany, under the guidance of the notable phycologist Edouard Bornet. Her first investigation, building on recent work by English algologist Thomas Hughes Buffham along the coasts of Devon and Cornwall, concerned the development of the motile reproductive cells, or zoospores, in the genus *Myriotrichia*, a small brown annual alga, two species of which were common along the Brittany coasts;[23] Karsakova's report appeared in 1892 (see bibliography). Four years later she brought out a second paper that reported her structure and classification studies on algae from the Canary Islands collected in 1895–1896 by the British-French algologist Anna Vickers.[24] Among the new species described by Karsakova in her 1896 paper was the small red alga of the Ceramiaceae family, *Vickersia baccata* (J. Agardh) Karsakoff (see bibliography). *Vickersia* is now recognized as a distinct genus. In 1902 she published a French translation of *Fiziologiia rastenii*, by the Russian botanist and biochemist Vladimir Palladin (1859–1922).[25]

The date of Karsakova's return to Moscow has not been established. However, at least by postrevolutionary times she held a research position in botany at Moscow University, carrying out systematic studies on algae from the Barents Sea and the waters around Novaya Zemlya, Arctic waters being one of the regions of considerable interest to Russian biologists from the later decades of the nineteenth century. In 1925 she published a seventeen-page report, with B. K. Flerov, on the algae of the southeastern region of the Barents Sea;[26] a second report by Flerov and Karsakova, in 1932, described 115 species of algae they found in waters of the littoral zone and in material thrown ashore around Novaya Zemlya. Many of these, which included twenty-six Chlorophyceae, forty Rhodophyceae, forty-seven Phaeophyceae, and one member of the Cyanophyceae, were new for the region.[27]

Her scientific studies continued steadily through the 1930s and were quite likely included in the publications of the Moscow Oceanographic Institute, but these have not been searched. Natal'ia Karsakova died when she was in her late seventies in Moscow, in 1941, a time in the Second World War when German armies had penetrated deep into Russian territory.

EKATERINA (CATHERINE) MIKHAILOVNA SOKOLOVA[28] (fl. 1890s) was a student of plant embryologist Ivan Nikolaevich Gorozhankin (1848–1904), founder of the Russian school of plant morphology, professor at Moscow University, and director of the university's botanic garden.

Sokolova's fifty-one-page paper, published in 1891, reported her painstaking and scrupulously detailed morphological investigation of the fertilization process in gymnosperms. Her work was carried out under the supervision of Gorozhankin, who had taken up the relatively neglected field of gymnosperm studies as one of his special areas of interest. Sokolova investigated the intricate process of the union of the male spore with the female reproductive cell. She published at least one other long paper during the 1890s, a 110-page report on the growth of root hairs and rhizoids—the filaments that serve as anchors, or roots, in fungi (see bibliography).

OLGA MALINESCO[29] (b. 1864) studied algae under the direction of Robert Chodat at the University of Geneva. After receiving her Certificate of University Studies from Geneva University in 1892, she spent a summer semester at Bern University.

Mme Olga was born in Bukovina, a border region on the northern slopes of the Carpathian Mountains, at that time a province of the Austro-Hungarian Empire; later it united with Bulgaria and then became part of Soviet Russia.

She coauthored two short reports with Chodat, both published in 1893, one on structure and the other on subdivision processes in algae of the class Cyanophyceae, the blue-green algae (see bibliography). Her work involved an examination of granules found within the cell protoplasm of these unicellular algae; opinion was divided on how the granules were formed and what their function was, some botanists considering them to be produced by division of a primitive central nucleus, others seeing them simply as pockets of reserve materials. Chodat and Malinesco's studies led them to conclude that the granules were not formed by subdivision of a central nucleus. The Cyanophyceae are now classed as cyanobacteria, photosynthetic organisms common in fresh and salt water. Malinesco's second joint study examined the division process in the small green alga *Dactylococcus infusionum*. The variety of forms produced had led to much confusion among botanists; Chodat and Malinesco offered the relatively more simple explanation of phase changes.

ZOOLOGISTS AND NATURALISTS

Among seven zoologists and naturalists whose work is sketched here, three—Sof'ia Pereiaslavtseva, Mar'ia Rossiiskaia, and Ekaterina (Catherine) Wagner—carried out research at the Sevastopol Marine Biological Station. Pereiaslavtseva was director for ten years of this new research laboratory, which was dedicated to the investigation of the flora and fauna of the Black Sea; Rossiîskaia and Wagner both carried out periods of research there. The other four contributors discussed in this section were a Moscow-based naturalist; two scientific illustrators and coworkers of Ilya Mechnikov, a Russian-Italian working at the University of Pavia; and a Moscow-based science writer and translator.

Sof'ia Mikhailovna Pereiaslavtseva[30] (1849–1903), the daughter of Mikhail Pereiaslavtsev, a colonel in the Russian army, was born on 29 September 1849 in Voronezh, a major city in southwest Russia. She received her early education in Kursk, about 130 miles west of Voronezh, where, except for a year in St. Petersburg, she lived until 1870.[31] Then followed two years in Kharkov, the home of a long-established university and several technical and scientific institutes. Her earliest published work included a study of the Lepidoptera of Voronezh Province; she also carried out taxonomic investigations of the fauna of Kharkov district.[32]

She enrolled at Zurich University in the winter semester of 1872 and specialized in zoology, particularly embryology; her doctoral degree was awarded in 1876.[33] Like a great many of the Russian women students in Zurich at that time, she took part in the political activities that centered around the exiled Russian revolutionary intellectuals Mikhail Bakunin and Peter Lavrov. A follower of socialist Lavrov, who tended more toward reform than revolution, she was appointed librarian of the Lavrist library in January 1874, this smaller library having some time previously split off from the original Bakunist library. For some reason her management did not satisfy the subscribers, but she remained librarian until she graduated in 1876 and left Zurich.[34]

In 1878 Sof'ia Pereiaslavtseva was elected director of the recently established Sevastopol Marine Biological Station that belonged to the Novorossisk Society of Naturalists, a group attached to Novorossisk University in Odessa; it was the first such station set up to advance marine exploration of the flora and fauna of the southern Crimea waters.[35] Pereiaslavtseva's directorship was probably the first held by a woman in a scientific institution in Russia. Also the station's chief zoologist, she remained director until 1888, when the post was taken by the eminent embryologist Aleksandr Onufrievich Kovalevskii (1840–1901), a faculty member at Novorossisk University and secretary of the Novorossisk Society of Naturalists.[36] Pereiaslavtseva stayed on as the station's chief zoologist until 1891, when a serious quarrel with Kovalevskii caused her to resign and leave Sevastopol.

During her years at the biological station her research centered on the embryonic development of Amphipods (small crustaceans) and Turbellaria (flat worms). She brought out many journal articles and one or two comprehensive, beautifully illustrated monographs on Black Sea fauna (see bibliography). These included her especially valuable monograph on Turbellaria, the first in that field.[37] In 1889 she was elected chairperson of the zoological section of the Eighth Congress of Russian Naturalists and Physicians, a notable honor.

Having no alternative employment and no income to replace her station director's salary, she lived for several months with relatives in St. Petersburg and tried to earn money by translating foreign-language articles.[38] Her scientific colleagues were not oblivious to her difficulties. In 1892 the Moscow Imperial Society of Naturalists awarded her a travel grant, which made possible a year's stay at the famous Naples Zoological Station, a facility used by many marine biologists. Pereiaslavtseva explored the waters of the Gulf of Naples and the French coast. The following year more help came from the scientific community; the Ninth Congress of Russian Naturalists and Physicians raised enough money to let her travel to Paris, where, for over a year, she worked in the Muséum national d'histoire naturelle. With the cooperation of algologist Philippe van Tieghem and entomologist Louis Eugène Bouvier (known for his many studies on Crustacea), she investigated the development of several species of the Pedipalp order of Arachnids, starting at the egg stage (see bibliography). How long she was able to eke out her funds in Paris is unclear, but in 1903 she went to Odessa hoping to continue her zoological research. Despite some financial help from the St. Petersburg Academy of Science her situation became increasingly difficult. By then seriously ill, she died in Odessa on 1 December 1903, when she was fifty-four; some of her friends said her death was from starvation.

A thorough and conscientious investigator, Periaslavtseva demanded much of herself. Her 1884 paper on the development of Rotifera (see bibliography) illustrates her research methods; instead of selecting one or another phase of development as being the most important, the procedure commonly used in most previous work, she placed several of these microscopic sweet-water worms under the object glass of a microscope and waited until one of them laid an egg. She then observed continuously for the next thirty to thirty-five hours and, with only two- to three-hour interruptions, for subsequent phases. As the *Nature* reviewer of the 1884 article remarked, the method had its "inconveniences," namely it prevented sleep for two nights; but it allowed all stages of development to be figured and, since she studied ten different species, a type designated for the development of the egg.

One of Russia's outstanding women scientists of the time, she was well regarded by fellow zoologists, western European as well as Russian. Perhaps her failure to retain her position in the scientific research community could be taken as an illustration of how difficult life could be in those days for a female scientist lacking an independent source of income or a close connection with an influential male scientist who was able to further her work.

Zoologist Mariia Aleksandrovna Rossiîskaia,[39] later Rossiîskaia-Kozhevnikova, was born on 1 January 1861 in the small town of Iarmalinuak, Kingdom of Poland, part of the Russian region of Poland after the country's three-way partition in 1795. Her father, the son of a much-travelled sailor who wrote of his voyages, grew up in St. Petersburg and made his career in the Russian army; he later commanded the Galitski regiment, one of the regular regiments of the Russian army. Her mother, the daughter of a landowning family, was educated at the Kharkovsk gymnasium for girls.

Mariia entered the Kharkovsk gymnasium in 1869, at the age of eight, and five years later transferred to the Orlovsk gymnasium for girls, where in 1876 she completed the course, finishing as the top student and gold medalist. From 1877 until 1879, following a year of teacher training at the gymnasium, she taught beginning classes in Russian language at the gymnasium and arithmetic at a private girls' school. Then aged eighteen, she began studies at the Higher Courses for Women in St. Petersburg, taking classes mainly in the mathematics-physics department but also attending lectures in the natural sciences. Thanks to private tuition in mathematics during her high school years she did well and graduated in 1883. From 1884 until 1887 she held a position as an assistant to zoologist Nikolai Petrovich Vagner at the Higher Courses for Women while also studying in St. Petersburg University's Department of Zoology with Konstantin Sergeevich Merezhkovskii, who taught her advanced techniques.

Rossiîskaia's first research was carried out during the summer of 1886 at the Sevastopol Zoological Station where, under the direction of station director Sof'ia Pereiaslavtseva, she began her study of amphipod embryonic development; the area was one in which both she and Pereiaslavtseva made important contributions over the next decade (see bibliography). Her first paper, reporting the initial part of her investigation of the development of the small crustacean *Orchestia littorea*, was read at a meeting of the St. Petersburg Naturalists' Society in December 1886; she was elected a member of the society soon after, on the recommendation of Nikolai Vagner, Philip Vassil'evich Ovsyannikov, and Ivan Sechenov. Rossiîskaia's 1888 paper, published in the *Bulletin* of the Moscow Naturalists' Society, was especially noteworthy, and has been widely cited over the years.[40]

About 1890 Mariia Rossiîskaia married Moscow University zoologist and bee specialist Grigori Kozhevnikov, and over the next several years published under the name Rossiîskaia-Kozhevnikova. She collected around the Sevastopol station throughout much of the 1890s and brought out at least three reports (see bibliography), but how long she continued scientific work has not been determined. In 1912 she published a short biographical work commemorating the life of the young Russian poet Semen Iakovlevich Nadson (1862–1887), who died of tuberculosis at the age of twenty-five.[41]

CATHERINE WAGNER (Mme)[42] lived in St. Petersburg during the last decades of the nineteenth century. Interested in marine fauna, and well prepared, she acquired laboratory experience and skills, perhaps in the Museum of Invertebrate Zoology at St. Petersburg University, a study museum founded early in the nineteenth century. By the mid-1880s, under the directorship of Nikolai Vagner, collections in that museum had expanded to include material from the Black Sea, and work in comparative embryology was being encouraged.

In the summer of 1887, Catherine Wagner had an opportunity to visit the Sevastopol Marine Biological Station. Sof'ia Pereiaslavtseva, the station director, suggested that she undertake a study of the embryological stages of development of the amphipod *Melita palmata*; a member of the *Melita* genus of small crustaceans, *Melita palmata* was expected to have a pattern of early development similar to that already observed by Pereiaslavtseva in the closely related genera *Grammarus* and *Caprella* (see bibliography). This guess remained to be verified. Wagner's microscopic work, from the stage of endodermal, mesodermal, and ectodermal cell-layer differentiation to the first stages of development of the organs to which these cell layers gave rise, indicated that *Melita palmata*'s development did indeed follow the same routes as those Pereiaslavtseva had reported in 1888–1889 for *Grammarus* and *Caprella*. Catherine Wagner published her observations in an eight-page paper in the *Bulletin* of the Moscow Naturalists' Society in 1892 (see bibliography).

MARIIA DMITRIEVNA BALASHEVA,[43] a Moscow-based naturalist, carried out a long series of extensive field studies of environmental influences on mollusk populations around the Moscow region, a land of mixed forest and agriculture; her observations continued over at least a decade, from the early 1880s until the mid-1890s. She kept in touch with several of the prominent figures in natural history, particularly Aleksandr Kovalevskii, a man with wide ecological interests in addition to his major embryological studies. Kovalevskii, a faculty member at Odessa University, gave her much encouragement in her fieldwork, as did Nikolai Zograf (1851–1919), another of the outstanding naturalists of the time and a strong advocate of careful fieldwork. The favorable reception Balasheva received at the Seventh Congress of Naturalists and Physicians in Odessa in 1883, when she presented her first report, would surely have bolstered her enthusiasm for the demanding program of work she had set herself.

Two fairly substantial papers, published in 1887 and 1888 in the *Mémoirs* of the Odessa New-Russia Society of Naturalists (see bibliography), reported her detailed survey of populations of the gastropod class of mollusks, with particular emphasis on the freshwater snails of the genera *Planorbis* and *Limnaeus*. Of worldwide distribution, *Planorbis* is of special interest in being the only shelled gastropod that holds its shell upside down. Balasheva's studies involved visiting ponds and lakes in the Moscow region, measuring their areas and depths, taking into account environmental features, and counting snail numbers. Two further papers, published in the Moscow Naturalists' Society *Bulletin* in the early 1890s, reported her extension of these studies to include the influence of plants on freshwater mollusk populations (see bibliography).

OL'GA NIKOLAEVNA MECHNIKOVA, née BELOKOPYTOVA[44] (1858–1944), one of eight children of Nikolai Belokopytov, was born in Odessa in 1858. The family was well off, thanks to Belokopytov's extensive land holdings in Kiev and Kherson provinces. A man of the old Russian patrician school, his ideas about girls' education were out of tune with the changing times; on the other hand, the sympathies of his much younger wife were all with the younger generation. In her youth Mme Belokopytova had painted and played the cello, interests she passed on to Ol'ga, but early marriage and a large family had required her to give up her artistic aspirations.

Growing up in Odessa, Ol'ga Nikolaevna, along with her siblings, attended the nearby lycée. At this time her family lived in an apartment above rooms occupied by Ilya Mechnikov (Elie Metchnikoff), who held the position of *Dozent* at the new University of Odessa. As Mechnikov, then a widower in his late twenties, got to know the Belokopytov children, he found that Ol'ga was interested in natural science and he offered to give her lessons in zoology as well as help in preparing for the lycée final examinations. Before these examinations were taken, however, seventeen-year-old Ol'ga Nikolaevna and Ilya Mechnikov married; Ol'ga's father had disagreed with Mechnikov's modern ideas about education, and so, rather than postponing marriage until after Ol'ga had completed her lycée education, Mechnikov cut through the differences with her father by asking for her hand. The ceremony took place in February 1875 and the resulting partnership was satisfactory; he and Olga worked well together for the rest of Mechnikov's life.

After completing her lycée education, Ol'ga Mechnikova continued biology studies under her husband's direction and attended his lectures at the University of Odessa; within a short time she had progressed sufficiently to be able to carry out some research in his laboratory. Her investigation, part of Mechnikov's wide-ranging study in evolutionary biology, focused on the pelvic and shoulder girdles in cartilaginous fishes; these structures, which support the pectoral and pelvic fin muscles in fishes, are of special evolutionary significance as the forerunners of the shoulder and hip bones of land vertebrates. Ol'ga's work was carried out on *Scaphirhynchus hermanni* Kessler (1877), or dwarf sturgeon, a species of cartilaginous fish of the Aralo-Caspian region with a long fossil record that shows little morphological change; her reports appeared in German journals in 1880, after a short note in the *Mémoires* of the New Russian Society of Naturalists (see bibliography).

Mechnikov taught at the University of Odessa until 1882, by which time he felt that the institution had been so damaged by government intervention and political turmoil, especially after the 1881 assassination of Tsar Aleksandr II, that he could no longer carry out his work. In 1879, during a stay at the Naples Zoological Station, he had continued research on evolutionary changes by studying the very early development of the larvae of two very simple and primitive groups that showed evolutionary innovation, namely members of the Echinodermata phylum (widely recognized forms of which are the star fishes and sea urchins), and species of what was then called the Coelenterata or Cnidaria phylum (jellyfishes, sea anemones, corals, etc.), whose position in the evolutionary tree has long been debated. Olga contracted typhoid fever during this visit and throughout much of the following year was seriously ill. By 1881 Mechnikov also was ill and depressed, but nevertheless was beginning to develop his theory of phagocytosis, that is, the engulfing of invading pathogens and other particulate matter by mobile host cells; the work eventually led him to his concept of host defense strategies and immunity.[45]

Following the deaths of both of Ol'ga's parents in 1881, the Mechnikovs had charge of the large family and the extensive Belokopytov land holdings in Kiev and Kherson provinces. Turning over the management of these estates to his nephews, Mechnikov, now financially independent, resigned from his university post. In 1882 he and Ol'ga went to Messina, Sicily, whose surrounding waters were then rich in marine life; Ol'ga's two sisters and three young brothers accompanied them.

After three years in Messina, where work on the phagocytosis theory continued, the Mechnikov family went back to Odessa, Mechnikov having been appointed director of a new Bacteriological Institute, founded by the municipality of Odessa and the *zemstvo* of Kherson Province. Disagreements between the institute and the local medical community, as well as disharmony among laboratory personnel, seriously hampered scientific productivity, however, and Mechnikov resigned after about two years.

In 1888 the Mechnikovs moved to Paris, Ilya Mechnikov having accepted an invitation from Louis Pasteur to join his institute. A few years later, when the new Pasteur Institute was opened, Mechnikov had his own large laboratory, and finding the general atmosphere congenial and Pasteur in sympathy with his research ideas, he stayed there for the rest of his life. Studies on infectious diseases, closely connected to his overarching theories of immunity, occupied much of his attention at this time. Ol'ga's lengthy, single-authored 1891 report on anthrax vaccination (see bibliography) was part of these investigations; she also published a note in 1901 on the effect of microbes on tadpole development, work that demonstrated the necessity of having present the microbes of the natural environment for building immunity and allowing normal development; tadpoles kept in sterile conditions from the egg stage on failed to grow and develop normally and lived significantly fewer days than those in a control group.[46]

Throughout these years Ol'ga's artistic interests remained strong and were encouraged by Mechnikov. When in 1898 they bought a small house in Sèvres, on the western outskirts of Paris, she built her own studio and began to make the acquaintance of a number of Paris artists, including sculptor Auguste Rodin and one of the sons of Lev Tolstoy. Work in the Mechnikov laboratory continued, with daily train journeys to the city from Sèvres after 1905, when they gave up their Paris apartment. Studies on immunity, artificial immunity, and immunity from infectious diseases were all areas in which Ol'ga assisted. Mechnikov was also, over a long period, marshaling strong defenses for his phagocyte theory against equally strong attacks from most of the German biologists, especially microbiologist and pathologist Paul von Baumgarten of Tübingen University and bacteriologist Robert Koch, director of the Prussian Institute for Infectious Diseases. In 1908 Mechnikov shared the Nobel Prize in Physiology or Medicine with German research physician Paul Ehrlich; Ehrlich was honored for his development of the antisyphilitic arsenical drugs, and Mechnikov for his work on immunity. This shared Franco-German award is said to have been made in large part to call a truce in a divisive war.[47]

In 1911 Ol'ga accompanied her husband and other members of the Pasteur Institute on a bacteriological expedition through the Kirghiz and Kalmuck country just north of the Caspian Sea. The Kalmucks, then a primitive and nomadic people of the open steppes, were particularly susceptible to tuberculosis and plague, characteristics that made them of special interest for Mechnikov in his immunity studies.

With the start of the First World War, scientific work in the Pasteur Institute stopped. By then Mechnikov's health was declining and he died in July 1916. As he had requested, Ol'ga wrote an account of his life and work, a biography that was very well received when it appeared in 1920 and was quickly translated into a number of other languages. She also published an epilogue to this work (essentially a summary of Mechnikov's major ideas) in a paper in the *Annales* of the Pasteur Institute in 1919.[48] Most of Mechnikov's papers she sent to Moscow, where a Mechnikov Museum was established in 1926. Ol'ga Mechnikov stayed on in France, living through the upheavals of the Second World War and the German occupation. She died in Paris in 1944, when she was about eighty-six.

Ludmilla Vassilievna Mechnikova (Lussia),[49] née Fedorova, (1845?–1873), zoologist and bacteriologist, Ilya Mechnikov's first wife, was born in Orenburg, but spent much of her childhood in the Siberian trade center of Kyakhta, on the Russian-Mongolian border. Later the Fedorovich family moved to Moscow. In January 1869, although already seriously ill with tuberculosis, Ludmilla married Ilya Mechnikov.

Shortly after their marriage the Mechnikovs, hoping that a milder climate would benefit Ludmilla's health, spent some time in northwest Italy, in the coastal city of Spezia. Here Ilya Mechnikov carried out studies concerning the genealogy of the lower marine animals, work that established a link between the Echinodermata and worms. A talented artist, Ludmilla helped to prepare the illustrations for his subsequent reports. The following year, at San Remo on the Mediterranean coast near the French-Italian border, she was well enough to take part in Mechnikov's research on the very early stages of the development of Siphonophora and Medusae, members of the ancient group that includes such animals as jellyfish and sea anemones. The lengthy paper reporting this work was jointly authored (see bibliography).

Ludmilla Mechnikova died in Madeira on 20 April 1873, after several months of intense suffering. She was about twenty-eight.

Very little information has been found about the late nineteenth- and early twentieth-century naturalist Tatiana Vedenissova;[50] her name strongly suggests that she was Russian, or at least of Russian descent, but because Italy was her residence and place of work, she might well have been placed with the Italian group for the purposes of this survey. She was most likely a doctoral student in zoology and natural history at the University of Pavia in the early 1890s; by at least 1904 her home was in Milan, some twenty-five miles north of Pavia.

About 1893 naturalist and zoologist Pietro Pavesi (1844–1917), director of the Museum of Natural History in the University of Pavia zoology department, suggested that she undertake the determination of a collection of recently acquired East African Crustaceae; she appears to have had a particular interest in this area of natural history.

The collection had been brought back to Pavia by L. Bricchetti-Robecchi, an engineer journeying through the region then known as Italian Somaliland, an Italian colony from 1855 until 1944, and now part of Somalia. In 1890–1891 Bricchetti-Robecchi collected during a long trip, the first leg of which took him over more than five hundred miles north along Somaliland's Indian Ocean coast, from Obbia to Alula on the Gulf of Aden; he also went into the interior to Sinadogo, on the edge of higher country, and the dry region around the Shebelle River and its tributaries, several of which are seasonal rather than permanent.

Tatiana Vedenissova identified, named, and, where required, provided detailed descriptions of over twenty species. The greater part of the collection consisted of marine forms, such as prawns and shrimps from the Indian Ocean coast and the waters around the islands at the southern end of the Red Sea; many of these were well-known forms, widely distributed. On the other hand, the terrestrial and freshwater specimens turned out to include a number of rare forms, among which were three new species—two isopods, *Armadillidium nitidum* of the genus *Armadillidium*, relatives of the common sowbug, *Peryscyphis nigro-maculatus*, found both inland and at a coastal location, and *Apus somalicus*, found inland at Sinadogo and on the coast at Obbia.

Her sixteen-page report in the *Bollettino* of the Italian Entomological Society in 1894 (see bibliography) was published as a contribution from the Zoological Laboratory of the University of Pavia.

CATH. (CATHERINA?) KRANTZ was a linguist who, in the early 1890s, translated scientific articles for the French Société nationale d'acclimatation, and also supplied the society with reports of scientific investigations carried out in other countries that had not yet appeared in print.[51] Her accounts included syntheses into a connected whole of contributions to a particular area of investigation made by a number of authors over several years.

Of the four reports by Krantz published in the society's *Bulletin* (see bibliography), two are outlined here by way of illustrating her broad scientific competence and linguistic dexterity.

Her contribution concerning two species of tortoise found along the western and southern sides of the Caspian Sea described in detail a study carried out by M. Solomon, a member of the Russian Imperial Society for the Acclimatization of Animals and Plants. Eight representatives of these two species, *Testudo graeca* (the Greek tortoise or spur-thighed tortoise) and *Cistudo lutaria* (now *Emys orbicularis*), were brought back to Moscow by Solomon in 1887. Most appear to have been accommodated in the Moscow Zoological Gardens; Solomon, the main observer, later gave one of the females a home in his house and grounds, with free access to a much-favored warm place near his cooking stove.

Food preferences, feeding habits, and extreme sensitivity to colder weather were painstakingly described, but perhaps the most important and interesting parts of the report were the observations made on the female's mating activities, her laborious and elaborate nest preparation, and the egg-laying, which took place in mid-June. Unfortunately the attempt at propagation in captivity in the Moscow climate failed. When the nest was opened two and a half months later the eggs were found to have rotted, and examination of the contents suggested that they had never been fertilized. Krantz gave no indication of the existence of any Russian-language published report of this experiment; she appears to have either personally observed the tortoise's activities or received a detailed account from Solomon.

Another of her French-language contributions, on a subject far removed from the little tortoises of the warm, dry Caspian region, was a discussion of the findings of several investigators, mostly Swedish, on the herring populations then still found in the shallow coastal waters of the western Baltic, particularly the areas around Denmark and in the Kattegat, the sea channel between Denmark and the west coast of Sweden. Included in her article were natural history studies by F. Henicke on local varieties of herring found at particular locations around the Swedish west coast; the question of whether these populations all differed from the herring of the ocean was one of much interest at the time. Other work discussed was the demonstration by Swedish zoologist and fisheries specialist Axel Ljungman of a sixty-year cycle of periods of abundance and scarcity. To complete her account Krantz outlined the contribution of Swedish zoologist August Malm of the Göteborg Natural History Museum. Malm, one of the country's fisheries experts, addressed the problems then facing Swedish herring fishermen, whose small-scale operations, even in years of abundant catches, were insufficient to allow them to compete in continental salt-herring markets with the well-established Norwegian and Scottish producers. He diagnosed the difficulties as being poor and uneven product quality, resulting from inadequate cleaning and packing procedures, lack of uniform salting with good-quality salt, and the tendency to use low-quality barrels—all of which had led to 50 percent of the product being unfit for sale. The remedy was to move the whole process to a well-regulated, centralized plant with high standards; Scottish herring experts were to be asked to advise on techniques. It might be noted that one decade after this account by Krantz was published, the Royal Swedish Fisheries Research Center was established at Lysekil on Sweden's west coast.

Of the remaining two articles by Krantz listed in the bibliography, the first discussed the distribution of freshwater fishes in Russia, and the factors bringing about the spreading of species from one river and lake system to another, including the role played by the rock formations through which the rivers flowed. Particular note was taken of efforts to control inroads of the predatory Brochet, the northern pike (*Exox lucius*). Krantz collected her information from the works of three Russian scientists: biologist Leonid Sabaneev (1844–1898), known for his studies of freshwater fish biology; naturalist Nikolai Severtsov (1827–1885), who carried out several wide-ranging investigations; and geologist Sergei Nikitin (1851–1909), who was then investigating the movements of fish stocks in the pond and lake system on his grounds near Moscow.[52] The other article reported a successful experiment carried out at Potsdam by a German stockbreeder, Herr Winter, on the cross-breeding of the Wapati, or North American elk, with what was referred to as the common deer, or European cerf.

ENTOMOLOGISTS

Few Russian women published work on entomological topics in major science journals before 1901. Discussed here are the contributions of three—Elizavet Freîmut and Olga Tikhomirova of Moscow, and Olga Poletaeva of St. Petersburg. There were undoubtedly a number of others who were interested in insects and contributed collections and observational notes to museums and local natural history society holdings; the career of one such contributor, botanist and explorer Olga Fedchenko, who made substantial insect collections for Moscow University's natural history museum, is sketched above.

Elizavet Karlovna Freîmut published one report in 1870 in the *Bulletin* of the Moscow Society for Natural Science. She discussed fairly extensive studies of the Tenthredinidae family, the sawflies, describing several new species, including, in particular, *Pompholyx dimorpha*, which she found to be wingless (see bibliography).

Olga Tikhomirova (Mme) carried out a number of studies on insect development during the 1890s. Her research was done in fairly close collaboration with Aleksandr Andreevich Tikhomirov of Moscow University's zoology department.[53] One might wonder if she was Tikhomirov's wife, but no definitive information on the point has been found. Tikhomirov's major work concerned questions in insect embryology and physiology.

Olga Tikhomirova published a short account of *Pulex serraticeps*, a species of the flea family, in the *Bulletin* of the Moscow Society for Natural Science in 1890; the work involved microscopic examination of her preparations of cell-level structures. Two years later she spoke at the International Congress of Zoology in Moscow on her research of very early development immediately after fertilization in a species of lacewings, *Chrysopa perla*. She investigated steps in the formation of the mesoderm, one of the germ layers in the developing embryo that give rise to specific organs, such as the heart or the muscular system. Her work on the biology of the pseudoscorpions of the Moscow region appeared in 1894 (see bibliography).

She carried out one further pre-1901 study, a lengthy investigation of a possible food source for *Bombyx mori*, the silkworm. The climate of the Moscow region being unsuited to the growing of mulberry trees, an alternative leafy plant was needed. Tikhomirova undertook the work at the request of the Committee on Agriculture and of Aleksandr Tikhomirov, then director of the Moscow Imperial Society of Agriculture, one of whose special interests was the early development of the silkworm. The series of silkworm feeding trials she conducted in 1893 demonstrated that, with suitable preparation, the leaves of *Scorzonera hispanica* (black salsify) were accepted as food by the worms from the newly hatched stage on. However, she was unable to obtain enough leaves for adequate nutrition and so silk yields suffered. Her report was published in the major French agricultural journal in 1896 (see bibliography) and again in German translation as a ten-page booklet in 1926.[54]

Ol'ga Gustavovna Poletaeva[55] (b. 1854), a member of the Russian Entomological Society (St. Petersburg), was born in Novgorod. She received her early education at a St. Petersburg boarding school, and later (1879–1881) studied at the newly opened Higher Courses for Women, where she concentrated on science subjects. Interested in insect anatomy, she carried out an investigation of the respiratory organs of odonate larvae while she was still a student; her major, sixteen-page report appeared in the Russian Entomological Society's *Horae* in 1880 (see bibliography). About the same time she brought out her notable contribution to knowledge of the Odonata of the St. Petersburg region. A detailed study, it included work on systematization with reference to a variety of previously published sources on European Odonata. She added a few more species to her St. Petersburg region list in 1885 (see bibliography).

In her second anatomical study, published in 1886, Poletaeva examined insect hearts and the accessory structures. Using adult forms of the *Bombus* genus (the bumble bees), and larvae of the *Cimbex* genus of the sawfly group, she carried out a painstaking investigation of the construction of the organs in both genera, which allowed her to draw her conclusions about the functions of the several chambers she observed; some of her interpretations differed from those of earlier workers, and she did not hold back from offering alternative explanations. Her report on the smoothing out of the wings of flying insects during their change from pupa to adult stage, which appeared in the Russian Entomological Society's *Trudy* (see bibliography), was considered of sufficient importance to be offered as an offprint booklet.

NOTES

1. For a discussion of nineteenth-century American and British women botanists, see Mary R. S. Creese, *Ladies in the Laboratory? American and British Women in Science, 1800–1900* (Lanham, Md.: Scarecrow Press, 1998), 2–63.

2. Anatolii Petrovich Bogdanov, *Materialy dlia istorii nauchnoi i prikladnoi deiatel'nosti v Rossii po zoologii i soprikasaiushchimsia s neiu otrasliam znaniia, preimushchestvenno za poslednee tridtsatipiatiletie, 1850–1887g.* [Material for the history of science and applied fields in Russia, in zoology and related new branches of knowledge, mainly in the latest period, 1850–1887], in *Izvestiia Imperatorskago Obshchestva Liubitelei Estestvoznaniia, Antropologii i Etnografii* [Proceedings of the Imperial Society of Natural History, Anthropology, and Ethnography] (Moscow, 1888), vol. 1, sheet 26; Sergei IUlevich Lipschitz, "Taxonomy, Floristics and Phytogeography," in L. P. Breslavetz, B. L. Issatschenko, N. A. Komarnitzky, S. IU. Lipschitz, and M. P. Maximov, *Ocherki po Istorii Russkoi Botaniki* [Essays on the history of Russian botany] (Moscow: Société des Naturalistes de Moscou, 1947), 1:32, 33, 53; V. L. Komerov, "Obituary: Ol'ga Aleksandrovna Fedchenko," *Izvestiia Rossiiskoi Akademii Nauk* 15, s. 6, no. 1 (1921): 247–48; Audrey le Lièvre, "Fedchenkoana," *The Garden* 112, no. 1 (1987): 65–68 (port.); Olga Yu Elina, "Private Botanical Gardens in Russia: Between Noble Culture and Scientific Professionalization (1760s–1919)," in *The Global and the Local: The History of Science and Cultural Integration of Europe,* ed. M. Kokowski, Proceedings of the Second ICESHS, Cracow, Poland (6–9 September 2006): 579–85; O. A. Valkova, *Ol'ga Aleksandrovna Fedchenko* (Moscow: Nauka, 2006), and Valkova, "The Conquest of Science: Women and Science in Russia, 1860–1940," *Orisis* 23 (2008): 135–65, especially 145, 149, 158; Anne C. Hughes, "Fedchenko, Ol'ga Aleksandrovna (1845–1921)," in *The Biographical Dictionary of Women in Science: Pioneering Lives from Ancient Times to the Mid-20th Century*, ed. Marilyn Ogilvie and Joy Harvey (New York and London: Routledge, 2000), 1:438–39. I thank especially Ol'ga Fedchenko's biographer, Olga Valkova, of the Institute of the History of Natural Sciences and Technology, Moscow, for valuable help with Fedchenko.

3. There would appear to have been nine children in the family, but Olga Valkova's extensive research found only six, of whom the second, Nadezhda (b. 1844), died at the age of fourteen. Possibly the missing three died in infancy. See Valkova, *Ol'ga Aleksandrovna Fedchenko*, 17–18.

4. The picturesque and agriculturally rich region of Mozhaisk, one of the most important orthodox centers in Muscovy, has many famous names associated with it, including those of writers Nikolai Gogol (1809–1852), Lev Tolstoy (1828–1910), and Boris Pasternak (1890–1960); it was the location of the Battle of Borodino in the Napoleonic Wars and was a frontline position in the German onslaught in 1941.

5. Nikolai N. Kaufman, *Moskovskaia flora, ili, opisanie vysshikh rastenii i botaniko-geograficheskii obzor Moskovskoi gubernii* [Moscow flora, or, accounts of the higher plants and a survey of the plant geography of Moscow province] (Moscow: izd. knigoprodavtsa A. I. Glazunova, 1866), 708 pp.

6. The region known as Russian Turkestan encompassed a wide stretch of oasis country south of the Kazakh steppes. Control of important subregions had passed from one group to another through the many centuries of the area's complex history, but at the time of the Russian colonial expansion in the 1860s the Uzbek Turks were prominent. The Silk Road to China traversed the region, passing through its famous ancient cities, such as Tashkent and Samarkand, focal points for cultural diffusion.

7. The Fedchenko Glacier in the Pamirs of northern Tajikistan is named after Aleksei Fedchenko. Considered the longest glacier in the world outside polar regions, it was discovered in 1878; full exploration was carried out in 1928 by a German-Soviet team led by Willi Rickmers.

8. A. P. Fedchenko, E. Regel, A. Bunge, and O. A. Fedchenko, *Puteshestvie v Turkestan* [Travels in Turkestan], 2 vols. (St. Petersburg: Knizhnyi sklad M. Stasiulevichna . . ., 1876–1902), repr. Moscow, 1950, in 1 vol., 466 pp.

9. Henry Yule, "The Geography and History of the Upper Waters of the Oxus," in John Wood, *A Journey to the Source of the River Oxus* (London: John Murray, 1872; repr. Karachi, 1976); Genri Iulia, *Ocherk geographii i istorii verkhov'ev Amu-Dar'i*, trans. O. A. Fedchenko, with additions and footnotes by A. P. Fedchenko, N. V. Khanykov, and H. Yule (St. Petersburg: tip. V. Bezobrazova, 1873). Henry Yule (1820–1889), from Inveresk, near Edinburgh, served in the Bengal engineers (Army of the East India Company) for about twenty-two years. After retiring in 1862 he began a second career exploring the historical geography of central Asia, a subject on which, by 1875, he had become a leading authority. President of the Hakluyt Society from 1877 to 1889, and of the

Asiatic Society from 1885 to 1887, he was vice president of the Royal Geographical Society from 1887 to 1889; an accomplished scholar and a fine linguist, he is best remembered for his authorship, with Arthur Coke Burnell, of a glossary of Anglo-Indian words and phrases: *Hobson-Jobson: Being a glossary of Anglo-Indian Colloquial Words and Phrases, and of Kindred Terms* . . . (London: John Murray, 1886); see Felix Driver, "Yule, Sir Henry (1820–1889)," in *Oxford Dictionary of National Biography* (Oxford: Oxford University Press, 2004), 60:976–78.

10. The Russian Geographical Society, the first Russian learned association to accept contributions from women to its publications, elected one other woman "cooperating member" in 1877, the folk customs specialist A. Ia. Efimenko, for her pioneering studies among northern peoples. See also Alexander Vucinich, *Science in Russian Culture, 1861–1917* (Stanford, Calif.: Stanford University Press, 1970), 86.

11. Ol'ga Fedchenko also assisted foreign authors by translating Russian articles into English. Thanks to her rough translation for Henry Yule of the article "O Karakorume i o drugikh razvalinakh bliz Orkhona" [On Karakhorum and other ruins near the Orkhon], published in *Isviestiia Imperatorskago Russkago Geograficheskago Obshchestva* 9, no. 10 (1873): 355–60, Yule was able to bring out a short English-language account of the 1873 visit of the Russian traveler and explorer Innokentii V. Paderin to the region around Karakhorum, the medieval capital of the Mongol Empire; Paderin was then Russian consul in Urga (now Ulaanbaatar). See Henry Yule, "Visit of Mr. F. [*sic*] Paderin to the site of Karakorum," *Geographical* Magazine 1 (1 July 1874): 137–39.

12. Natalia Armfeld (1850–1887) spent two or three years in Heidelberg shortly after her father's death in 1868. She lived in the small students' commune set up by the Korvin-Krukovskii sisters, Sofia Kovalevskaia and Aniuta Korvin-Krukovskaia, along with Julia Lermontova and one or two others (see the section on Sofia Kovalevskaia in chapter 4). Natalia studied music, astronomy, and mathematics, receiving tuition in the latter subject from Kovalevskaia. But, like Aniuta Korvin-Krukovskaia, rather than continuing studies, she chose an active role in the revolutionary movement. Back in Moscow by 1872, she joined the student campaign to spread propaganda among workers in the region and for several years she played a major role in this underground work in which her younger brother Nikolai Armfeld (1858–1880) was also involved. Natalia was arrested in Kiev and exiled to Siberia in 1879. She married one of the alleged student leaders, A. I. Komov, who was also condemned and sent to Siberia. Natalia died when she was about thirty-seven, along with her new-born infant, in Kara, eastern Siberia, a labor camp on the Manchurian border, one of the most feared of the tsarist prisons. See Valkova, *Olga Aleksandrovna Fedchenko*, 132–33, and later entries for Armfeld, Natalia.

13. A graduate in the physical sciences at Moscow University, Aleksandr Armfeld (1842–1897) was especially interested in cattle breeding and livestock farming. About 1869 he became head of the university farm at the new Novorossisk Imperial University. In 1872 he was appointed *Privatdozent* in agronomy and agricultural science, and also editor of the *Zapiski Imperatorskago Obshchestva Sel'skogo Khoziaistva Iuzhnoi Rossii* [Reports of the Imperial Society of Rural Economy in southern Russia]. Before that year was out, however, he was accused of mismanagement of the farm and failure to comply with its stated purpose; these accusations may have been false, Aleksandr Armfeld becoming the sacrificial victim caught in the middle of a struggle between factions of professors, but he was dismissed and compelled to leave the university. After returning to Moscow he gradually renewed his connections with Moscow University and other scientific groups in the capital. He worked in the university museum's zoological collections and carried out valuable studies on rare breeds of cattle. Valkova, *Olga Aleksandrovna Fedchenko*, 130–31.

14. Ol'ga Fedchenko, *Flora Pamira* . . . (St. Petersburg: tip. Gerol'd, 1903–1905), and later supplements; Ol'ga A. Fedchenko, *Opredelitel' pamirskikh rastenii* [Definition of Pamir plants] (IUr'ev [Tartu]: tip. K. Mattisena, 1907), and later editions.

15. Ol'ga Aleksandrovna Fedchenko and Boris Alekseevich Fedchenko, *Conspectus Florae Turkestanicae* . . . (IUr'ev [Tartu]: tip. K. Mattisena, 1913). See also O. A. Fedchenko and Boris A. Fedchenko, *Perechen' rastenii Turkestana i Kirgizskogo kraia = Conspectus florae Turkestanicae et Kirgizskogo kraia* [List of plants of Turkestan and the Kirghiz lands], (Leningrad: Glavnyi botanicheskii sad., 1924–).

16. Ol'ga Fedchenko, "*Eremurus*: Kritische Übersicht der Gattung," in *Mémoires de l'Académie Impériale des Sciences de St. Petersburg,* s. 8, Classe physico-mathématique, 23, no. 8 (1909), 210 pp. Reprinted as O. Fedtschenko, *Eremurus: Kritische Übersicht über die Gattung*, ed. J. Cramer and H. K. Swann, Plant Monograph Reprints (Lehre, Niedersachsen: J. Cramer, 1968). A genus of the Asphodelaceae family, *Eremurus* was one of Ol'ga Fedchenko's particular interests since her first journey to Turkestan in 1868–1871, when she collected seven species, of which three were new to science. She increased her count by four species, including one new to science, in her 1897 and 1901 visits. Most of the other specimens important in defining species she was able to examine in herbaria of museums and in private collections; in her own Ol'gino garden she cultivated some thirteen species or hybrids, plants that played a major role in her taxonomic studies.

Dramatically handsome plants also known as desert candles or foxtail lilies, the many species of *Eremurus* are native to the mountainous regions across western Asia, from the Caucasus to Tian Shan. Development of the blooming stage, with the production of flower spikes that can reach ten feet in height in the larger species, often takes several years; blossom color varies from white to a deep copper tone; leaves form tufts of narrow, strap-like blades rising from tubers anchored by thick, lengthy roots in the larger species. *E. robustus* roots may be boiled and eaten, their taste resembling that of asparagus; they are also used as a source for a type of glue. For a useful review of the *Eremurus* monograph see W. Botting Hemsley, "The Genus *Eremurus*," *Gardner's Chronicle* 49, s. 3, no. 11 (1909): 181.

17. Ol'ga Aleksandrovna Fedchenko and Boris Alekseevich Fedchenko, "Rasteniia Turkestana preimushchestvenno *Alaia*, sobrannyiia vo vremia puteshestvii 1901 i 1904 g.g = Plantae Turkestanicae imprimis Alaicae, in itineribus annorum 1901, et 1904 lectae [I]–[II]," in *Trudy Imperatorskago St. Petersburgskago botanicheskago sada* 24, no. 3, and 28, no. 1 (1905–1912). *Allium*, the onion genus, one of the world's largest plant genera, contains several hundred species, including many important vegetables, such as onions, leeks, and garlic. Most species are native to the Northern Hemisphere, mainly Asia; the native Turkestan Onion, sometimes referred to as the Ornamental Onion, is often used in gardens to discourage squirrels.

18. Le Lièvre, "Fedchenkoana," 66.

19. For a list of Ol'ga Fedchenko's publications by year, see Valkova, *Ol'ga Aleksandrovna Fedchenko*, 290–98.

20. Valkova, *Ol'ga Aleksandrovna Fedchenko*, 281 and note 1050.

21. Sofia Kovalevskaia was elected a corresponding member of the Russian Academy of Sciences seventeen years earlier, in 1889, just after receiving the Paris Academy of Science Prix Borodin; Marie Curie's election came in 1907, four years after she was awarded (jointly with Pierre Curie and Henri Becquerel) the Nobel Prize in Physics.

22. Sergei IUlevich Lipschitz and Vladimir Nikolaevich Sukachev, *Russkie botaniki: biografo-bibliograficheskii slovar* [Russian botanists: biographical-bibliographical dictionary], 4 vols. (Moscow: izd-vo Moskovskogo obshchestva ispytatelei prirody, 1947–1952), 4:90. I thank Alla Polekutina, Academy of Science, Moscow, for help with Karsakova.

23. For Buffham's paper see T. H. Buffham, "The Plurilocular Zoosporangia of *Asperococcus bullosus* and *Myriotrichia clavaeformis*," *Journal of Botany* 29, no. 347 (Nov. 1891): 321–23.

24. Algologist Anna Vickers, over many years, worked in the Roscoff region; see Creese, *Ladies in the Laboratory?*, 47.

25. Vladimir Ivanovich Palladin and N. Karsakova, *Physiologie des Plantes*, trans. from the 3rd Russian ed., revised and corrected by N. Karsakova (Paris, 1902), 276 pp.

26. B. K. Flerov and N. V. Karsakova, "Vodorosli jugo-vostochnoi chasti Barentsova morei" [Algae of the southeastern region of the Barents Sea] (Moscow: Izdanie Plovuchego Morskogo Nauchnogo Instituta, 1925).

27. B. K. Flerov and N. V. Karsakova, "Spisok vodoroslei Novoi Zemli," *Gosudarstvennyi Okeanograficheskii Institut. Trudy (USSR)* 2, no. 1 (1932): 46–73; English trans. by V. Astvazaturov, *List of the Algae of Novaya Zemlya* (Washington, DC: Naval Oceanographic Office, Ft. Belvoir Defense Technical Information Center, 1974), 52 pp.

28. Antonia Alekseevna Shcherbakova, Nina Alekseevna Bazilevskaia, and Konstantin Fedorovich Kalmykov, *Istoria botaniki v Rossii: Darvinovskii period, 1861–1917g g.* (Novosibirsk: Akademia Nauk CCCP, Sibirskoe Otdelenie, tsentral'nye siberskii botanicheskii sad., 1983), 105.

29. Archives, Bern University, http://apps.uniarchiv.unibe.ch/syscomm/images/mata/6737_6748.gif (Dec. 2013).

30. N. IA. Kuznetsov, obituary for Sof'ia Mikhailovna Pereiaslavtseva, *Revue Russe d'Entomologie* 6 (Dec. 1903): 422; matriculation register, Zurich University, http://www.matrikel.uzh.ch/active/static/26699.htm (July 2014); Valkova, "Conquest of Science," 147, 149, 150; Alexandru S. Bologa, "Development of Marine Biological Institutions around the Black Sea," in *Ocean Sciences bridging the Millennia: A Spectrum of Historical Accounts*, selected papers from the Sixth International Congress on the History of Oceanography, Qingdao, People's Republic of China, 15–20 August 1998, ed. Selim Marcos et al. (Paris: UNESCO, 2004), 209–22; Ann Hibner Koblitz, "Science, Women, and the Russian Intelligentsia: The Generation of the 1860s," *Isis* 79 (1988): 208–26; Alphonse Rebière, *Les Femmes dans la Science: Notes recueilles*, 2nd ed. (Paris: Nony, 1897), 223 (port. opp. 224); J. M. Meijer, *Knowledge and Revolution: The Russian Colony in Zurich, 1870–1873* (Assen, Netherlands: Van Gorcum, 1955), especially 151; Diana Lipscomb, "Women in Systematics," *Annual Review of Ecology and Systematics* 26 (1995): 323–41, on 334.

31. One might speculate that she attended the Kursk school for young noblewomen, an institution with high academic standards established in the eighteenth century by Empress Catherine II, and which, in 1870, became the Mariinskaia Grammar School for Girls. See "Kursk State University: The Oldest but Young at Heart," http://kursksu.ru/documents/ksu.pdf (Sept. 2014).

32. Kuznetsov, "Obituary," wrote about these studies and stated that the Lepidoptera work was published in *Trudy* (Works) of the Kharkov Nature Society (1872). The reference has not been verified.

33. Sof'ia Mikhailovna Pereiaslavtseva, "Vorläufige Mitteilungen über die Nase der Fische," inaugural dissertation, Zurich University, 1876 (Zurich: Zürcher und Furrer, 1876).

34. See Pereiaslavtseva's entry in matriculation register, Zurich University.

35. The station had its origins in the ideas of a group of biologists in Odessa in the late 1860s. A small station was started in Odessa in 1871 and later transferred to Sevastopol, where a permanent facility was established. In 1892 it was integrated into the Russian Academy of Science.

36. Aleksandr Kovalevskii was the brother-in-law of mathematician Sofia Kovalevskaia.

37. Sof'ia Pereiaslavtseva, *Monographie des Turbellariés de la Mer Noire* (Odessa: Novorossiiskoe Obshchestvo Estestvoispytatelei, 1892).

38. Perhaps it is worth restating that paid employment, other than in teaching and medicine, was not common for women in the sciences at that time.

39. A. P. Bogdanov, "Mariia Aleksandrovna Rossiiskaia" in *Materialy dlia istorii nauchnoi i prikladnoi deiatel'nosti* [Material for the history of science and applied fields . . .], vol. 1 (1888), sheet 9, port. on 8; information in technical papers.

40. For examples of later references to Rossiiskaia's work, see Gerhard Scholtz and Carsten Wolff, "Cleavage, Gastrulation, and Germ Disc Formation of the Amphipod *Orchestia cavimana* (Crustacea, Malacostraca, Peracarida)," *Contributions to Zoology* 71, no. 1/3 (2002): 1–23; Matthias Gerberding and G[erhard] Scholtz, "Cell Lineage of the Midline Cells in the Amphipod Crustacean *Orchestia cavimana* (Crustacea, Malacostraca) during Formation and Separation of the Germ Band," *Development Genes and Evolution* 209, no. 2 (Feb. 1999): 91–102.

41. Mar'ia Aleksandrovna Rossiiskaia-Kozhevnikova, *Vospominaniia o S. Ia. Nadsonie: k 25-lietiiusodnia konchiny* (Moscow: tip. T-va I. N. Kushnerev, 1912), 10 pp.

42. Information in technical papers.

43. Information in technical papers.

44. Ol'ga Mechnikova, *Life of Elie Metchnikoff, 1845–1916*, trans. and preface by Sir Edwin Ray Lankaster (London: Constable, 1921), esp. viii, translated from the French *Vie d'Élie Metchnikoff* (Paris: Librairie Hachette, 1920), 272 pp.; obituary for E. Metchnikoff, *Proceedings of the Royal Society of London* 89, B (1917): li–lix; E. Metchnikoff, *Souvenirs: Recueil d'articles autobiographiques*, trans. L. Piatigorski (Moscow: Éditions en langues étrangères, 1959), translated from the Russian, Ilya Mechnikov, *Stranitsy vospominanii: Sbornik avtobiograficheskikh statei*, ed. with commentary by A. Gaissinovitch (1946); Joy Harvey and Marilyn Ogilvie, "Metchnikova, Olga (Belokopytova) (1858–1944)," in *Biographical Dictionary of Women in Science*, ed. Ogilvie and Harvey, 2:887–88.

45. For a full discussion of the phagocytosis theory, see Alfred I. Tauber, "Metchnikoff and the Phagocytosis Theory," *Molecular Cell Biology*, Nature Reviews, 4 (Nov. 2003): 897–901.

46. Olga Metchnikoff, "Note sur l'influence des microbes dans le développement des têtards," Travail du laboratoire de M. Metchnikoff, *Annales de l'Institut Pasteur* 15 (1901): 631–34.

47. Tauber, "Metchnikov and the Phagocytosis Theory," 897.

48. "Epilogue" (dernier chapitre du livre de Mme. Metchnikoff, *Vie d'Élie Metchnikoff*), *Annales de l'Institut Pasteur* 33 (1919): 53–59.

49. Mechnikova, *Life of Elie Metchnikoff*, esp. 69–79; Metchnikoff, *Souvenirs*.

50. *Naturalists' Universal Directory*, comp. Samuel Edson Cassino, 19th ed. (Salem, Mass.: S. E. Cassino, 1905), part 2, 154. This reference gives Vedenissova's address as Via Pontaccio 14, Milano. See also "Istituto di Zoologia," in *Annuario della R. Università di Pavia, Anno Scolastico 1899–1900* (Pavia: Tipografica successori Bizzoni, 1900), 138–42, esp. 142. Included is a listing of the publications from the Zoological Laboratory since its founding in the 1870s; Dr. Tatiana Vedinissova's paper on the East African Crustaceae appears in this list.

51. See also the work of chemist Ol'ga Davidova, who contributed a number of Russian-to-French translations of scientific articles to the "Correspondance russe" section of the *Bulletin* of the Paris Chemical Society during the 1880s. This kind of work had considerable value, since it brought many research reports not otherwise readily accessible to the attention of western European scientists. Cath. Krantz lived and worked in the Moscow region in the 1890s, and on that basis she has been placed in the Russian section of this survey. No information about her place of birth or family background has been found.

52. See, for instance, S. N. Nikitin, "Notes sur les modes de propagation des poissons des eaux douces," *Bulletin de la Société Impériale des Naturalistes de Moscou* 3 (1890): 423–35.

53. See remarks on pages 242, 244, and 246 in Aleksandr Andreevich Tikhomirov, *Développement du ver à soie du mûrier* ("Bombyx mori") *dans l'oeuf* (Lyon: A. Rey, 1892), 170 pp., extract from *Rapport des travaux du "Laboratoire d'études de la soie" pour l'année 1891*. Russian text originally published in 1882.

54. O. Tichomirow, *Die Skorzonera als Futter für die Seidenraupe*, trans. G. Von Trentovius (Ilmenau, Germany: A. Schroeter, 1926).

55. Bogdanov, *Materialy dlia istoria nauchnoi i prikladnoi deiatel'nosti* . . . [Material for the history of science and applied fields . . .], vol. 2 (1889), 1.

Chapter Four

Mathematicians and Astronomers

MATHEMATICIANS

Two Russian women, Sofia Kovalevskaia and Vera Shiff, published papers in mathematics that were listed by the Royal Society in its *Catalogue* of nineteenth-century journal publications; one other early Russian female mathematician, Liubov' Zapol'skaia, is also noted here.

Sofia Kovalevskaia, the first female mathematician to be remembered primarily for achievement in mathematics, is one of the most extensively studied of all early women scientists, holding as she does a remarkably strong attraction for historians. Books and articles examining the many facets of her short, colorful life appear regularly, their numbers rivaling those devoted to Marie Curie, the prime example of a successful early woman scientist. This great wealth of detailed information on Kovalevskaia being readily available, only a short sketch of her life and work is offered below.

Sofia Vasilievna Korvin-Krukovskaia, later Sofia Kovalevskaia[1] (1850–1891), the daughter of General Vasilii Vasilievich Krukovskii (1800–1875) and his wife, Yelisaveta Fedorovna (1820–1879), was born in Moscow on 15 January 1850. Her father, the commandant of the Moscow artillery garrison, came from a line of Cossack generals. In his later years, after persistent effort, he had his claim to descent from Hungarian king Matthias Korvin accepted by the College of Heralds, and duly joined the ranks of the minor nobility, changing his name to Korvin-Krukovskii. His wife (*née* Shubert) was descended from a family of German astronomers. Sofia had an elder sister, Aniuta (1844–1887), and a younger brother, Fedor (1855–1919).

In 1858 General Korvin-Krukovskii retired from the military and moved his family to his estate of Palibino, in the Vitebsk district of Belorussia. A traditional Russian patriarch, he was autocratic and somewhat stern, but highly cultured and well educated; he completely dominated his young and elegant wife, often treating her like a child, although she too was well educated and cultivated. During her earliest years Sofia was brought up by her nanny, one of the household serfs, who took care of her lovingly and was her steady and constant comfort and companion. Nanny gave her freedom to wander in the surrounding woods with her sister, but she also told her tales of terrifying dangers and strange monsters that gave the child nightmares. This had the result of bringing her even closer to Nanny, who reinforced her notion that her parents favored her little brother, the long hoped-for son and heir, and that her mother preferred the prettier Aniuta to her. When Sofia was eight years old, her father, realizing that his daughters were to some extent running wild despite the presence of a French governess in the household, banished Nanny to washerwoman status, dismissed the Frenchwoman, and brought in two new people, a Polish tutor, Iosif Ignatevich Malevich, and an English governess, Margaret Smith. Miss Smith strove to impose some English order on the chaotic Russian feudal household but after five years had to admit defeat in her efforts to convert her wild young Russian charge into an English lady; Sofia rebelled against Miss Smith's regime and regularly found ways around her strict regulations. Her innate exuberance, curiosity, and vitality remained intact, and her childhood was by no means dreary; life on a Russian estate changed with the seasons and, especially for a child, was exciting and full of adventure.

Sofia's interest in mathematics was first awakened by her conversations with an uncle who had no formal knowledge of the subject but who liked to philosophize about such topics as asymptotes to curves and squaring the circle. Her first systematic instruction came from Iosif Malevich. She was soon so absorbed in algebra and geometry that she took to reading an algebra book in secret to avoid Miss Smith's restrictions. Perhaps Sofia's well-known account of her first encounter with the formulae and symbols of differential and integral calculus is worth repeating here. In her *Russian Childhood* reminiscences she tells how the walls in the children's rooms at Palibino came to be papered with lithographed copies of notes of calculus lectures by Maxili Ostrogradskii, one of Russia's leading mathematicians. Sofia, then about eleven years old, spent hours on end before the mysterious walls trying to decipher at least individual phrases and discover the proper order of these hieroglyphics. Eventually the formulae and even the wording became imprinted on her brain. Five years later, when she had her first calculus lessons, her tutor was amazed at the ease with which she grasped the concepts of limits and derivatives. The memorized lecture notes now began to make sense.[2]

In 1864, when Sofia was fourteen, an important event in her mathematical progress occurred. On one of his visits to Palibino a neighboring landowner, Nikolai Tyrtov, professor of physics at the St. Petersburg Naval Academy, brought a copy of his recently published textbook on elementary physics for Sofia's father.[3] She tried reading it and, coming upon trigonometric functions in the chapter on optics, needed to understand what was meant by a sine function. She then went on to devise for herself the concept of chord and used it to understand the text. Tyrtov was so impressed that he persuaded her father to let her study advanced mathematics; two years later, in the summer of 1866, during a visit to St. Petersburg with her mother and sister, Sofia started studies with Aleksandr Strannoliubskii. A former student of Tyrtov, Strannoliubskii was a good mathematician, an excellent tutor, and a supporter of higher education for women. Also of great importance that summer was the meeting of Aniuta, an aspiring writer, with Fyodor Dostoyevsky, who captivated Sofia's interest as well. Dostoyevsky remained a lifelong friend and influential figure for both sisters.

University education in tsarist Russia in the late 1860s not being open to women, Sofia, Aniuta, and Aniuta's friend Zhanna Evreinova, who wanted to study law, decided they must get their advanced education abroad, as several Russian women had already done. Such a course of action was not easy to follow. Both General Korvin-Krukovskii and Zhanna's father had refused permission for their daughters to leave the country to study at university level. Faced with these decrees, the young women decided that one of them must contract a nihilist-style marriage with a willing young man, a tactic used by a number of Russian women at the time to gain independence from parental control by acquiring the status of a married woman. General Korvin-Krukovskii was eventually persuaded to give his consent to Sofia's marriage to Vladimir Kovalevskii, a young student of paleontology who had already studied in Germany.

For a short time after their marriage in September 1868, Vladimir and Sofia remained in St. Petersburg, Sofia studying mathematics intensively with Strannoliubskii and attending lectures on one or two other subjects as well, thanks to special permissions from willing professors. But the situation was far from satisfactory, and in April 1869 the Kovalevskiis moved to Heidelberg.

Although Sofia had expected to enroll as a regular student at Heidelberg University, she found that all she could do was request special permission from individual professors to attend their lectures. She began her studies in the summer semester of 1869, taking a very full load of eighteen to twenty lecture hours a week. At the end of the semester she and Vladimir visited London and Sofia met the novelist and essayist George Eliot (Mary Ann Evans, 1819–1880), a social radical and a supporter of higher education for women; Sofia remained a close friend of Eliot for the rest of the Englishwoman's life, and her writings significantly influenced Sofia's own literary ambitions.

In the late autumn of 1869 Sofia and Vladimir were joined by Sofia's friend Juliia Lermontova (see chapter 2), who wanted to study chemistry. Juliia fitted in well, causing little disruption in the Kovalevskii living arrangements, but the subsequent arrival of both Aniuta and Zhanna Evreinova, and the resulting crowding of the small apartment, severely strained relationships. Very soon the group broke up. Sofia, impressed by the glowing accounts from her Heidelberg mathematics professors of the teaching of Berlin mathematician Karl Weierstrass, made up her mind to study with Weierstrass. Enrolling as a regular student at Berlin University was not then possible for a woman, but Sofia, having distinguished herself in a trial test Weierstrass gave her when she first visited him, was accepted as his private student. It might be noted that he also satisfied himself, by judicious enquiries of his Heidelberg colleagues Leo Konigsberger and Paul DuBois Reymond, about Sofia's character and possible associations with Russian student revolutionary groups.

She began studies in Berlin in the autumn of 1870 and quite soon became Weierstrass's favorite pupil. He repeated his lectures to her and discussed his research. Indeed he considered her one of the brightest and most hardworking of his students. All of her nine published mathematical papers developed themes he had discussed with her, and he acknowledged that her remarks over the long period of their contact had inspired some of his own work.[4]

During the spring and summer of 1871 these studies were interrupted for several months when Sofia, worried about her sister Aniuta, went to Paris to assist her in a difficult situation. Aniuta had become deeply involved in the Paris Commune, a short-lived bourgeois revolutionary group that had broken away from the relatively conservative French Republican government. Sofia and Vladimir, who accompanied her, lived in the city for more than a month, Sofia helping to nurse the casualties in the commune's fight against the French Republican army. Further dangers for Aniuta followed after Paris fell to the army, but by the autumn Sofia was back in Berlin, and Aniuta—with her husband, Victor Jaclard, one of the leaders of the Commune—had escaped to Zurich.[5] Vladimir moved from Heidelberg University to Jena, where he began his doctoral dissertation work on the evolution of ungulates—a brilliant piece of research, which initially stirred up considerable controversy but some years later led to Kovalevskii's recognition as one of the founders of Russian paleontology.

Sofia then began a period of intense study broken by intervals of helping Aniuta or spending time with Vladimir. She was joined in Berlin by Juliia Lermontova, perhaps fortunately for Sofia, who left the entire functioning of their small, dingy apartment to Juliia, while she spent all her time on mathematics. Over the four years she worked under Weierstrass's guidance she wrote three papers, all of which Weierstrass considered to have more than sufficient mathematical content to meet doctoral degree requirements. These were "On the Theory of Partial Differential Equations," work later known as the Cauchy-Kovalevski theorem; "On the Reduction of a Certain Class of Abelian Integrals of the Third Rank to Elliptic Integrals"; and "Supplementary Remarks and Observations on Laplace's Research on the Form of Saturn's Rings" (see bibliography). All three were submitted to Göttingen University in support of her request for consideration for a degree, there being no possibility of a woman graduating from Berlin University. Thanks to Weierstrass's considerable influence in mathematical circles, he was able to persuade the Göttingen administration and the faculty examiners to grant Sofia a doctorate *in absentia*, as a "special case," without the normal oral examination. She received the degree in July 1874, when she was twenty-four.[6] Her friend Juliia Lermontova was awarded her doctorate the following autumn. Juliia followed the normal route, passing the usual oral examination, thereby becoming the first woman to satisfy the university's regular doctoral degree requirements (see the section on Lermontova in chapter 2). This means that Juliia Lermontova, and not Sofia Kovalevskaia, took the first step along the way that later women would follow, although by no means immediately; the change depended on the slow rearrangement over the next two decades of academic (and public) opinion toward the goals long sought by the women's movement.[7]

Sofia and Vladimir returned to Russia in August 1874 and settled in St. Petersburg, where they were soon joined by Juliia Lermontova. All three were warmly welcomed by St. Petersburg's scientists and society in general, but Sofia was soon disillusioned in her hopes for a teaching position, the Russian system of higher education still being closed to women. A teaching post in the upper grades of a gymnasium or in a university was open only to holders of a Russian *magistr* degree, a qualification for which, by law, women were ineligible.[8] Of the three friends who came back to Russia in 1874 with high German academic credentials, only the chemist, Juliia Lermontova, was able to find a position reasonably in keeping with her qualifications.

Sofia was active in various capacities on the governing committee of the St. Petersburg Higher Courses for Women, which opened in 1878 and provided courses taught by University of St. Petersburg faculty; she herself was barred from teaching there. She became a celebrity in high society and in literary circles, publishing short articles in the radical newspaper *Novoe vremia* (New times), which Vladimir Kovalevskii edited for a few years. Sofia's articles were chiefly theater reviews and surveys of technical and scientific developments. Weierstrass sent long letters to her that included discussions of some of his best research, but she rarely replied.

Vladimir, despite his outstanding work in paleontology, remained unable to find an academic position. Hoping to finance his own and Sofia's scientific projects, and with Sofia's cooperation, he engaged in business speculations on a grand scale. In 1880, when he was finally appointed to a lectureship at Moscow University, his finances collapsed, leaving him in serious trouble. The family moved to Moscow, and Vladimir took on the additional job of managing director of a Caucasus oil firm.

Matters in the Kovalevskii family, which now included a two-year-old daughter, Sofia Vladimirovna (Fufa), did not improve. Sofia concluded that Vladimir was inextricably entangled in financial problems he was unable to resolve; further, his difficulties were severely affecting his mental state. She returned to her mathematical studies, despite her retreat of several years, and with encouragement from the eminent mathematician Pafnutii Chebyshev, she prepared a talk for the mathematical section of the Sixth Congress of Russian Naturalists and Physicians in St. Petersburg in 1879. Her presentation, the six-year-old but still unpublished results of her work on abelian integrals, never duplicated in the intervening years, was well received. In the audience was Swedish mathematician Gösta Mittag-Leffler, a former student of Weierstrass; in 1881 Mittag-Leffler was appointed professor of mathematics at the new University of Stockholm, which had been founded as a liberal alternative to the venerable, church-dominated Uppsala University. Much impressed by Sofia and her work, he soon began his considerable efforts to secure a teaching post for her in Sweden.

She paid a brief visit to Berlin in the autumn to renew her acquaintance with Weierstrass, and in March 1881, having decided to leave Vladimir rather than sink with him into a morass of legal and financial difficulties, she left Moscow for Berlin, taking her two-and-a-half-year-old daughter with her. Placing the child in the care of a young Russian nursemaid, she worked intensely for two years in Berlin and Paris on a new problem Weierstrass had suggested, namely, the solution of certain differential equations to describe the propagation of light in doubly refracting crystals. Soon after she had started on this she became interested in another problem, the question of how the rotation of a rigid body around a fixed point is represented mathematically. This greatly reduced the amount of time she spent on the light-propagation work, and it was only in August 1883 that she presented the latter at the Seventh Congress of Russian Naturalists and Physicians in Odessa.

Sofia had begun discussing with Mittag-Leffler the possibility of a *Dozent* position at Stockholm University early in 1881, but Weierstrass, from whom Mittag-Leffler had requested a recommendation for Sofia, was of the opinion that she should not attempt to secure a teaching post before she had a publication out, a condition not met until she presented her 1883–1884 paper (see bibliography).

In the spring of 1883 her family situation changed. Vladimir, finally overwhelmed by his many difficulties, committed suicide. Sofia did what she could to sort out some of Vladimir's legal and financial legacy, and arranged to place five-year-old Fufa with her old friend Juliia Lermontova, now settled on the Lermontov family estate outside Moscow. Juliia had known the child since her birth and was Fufa's much-loved godmother, "Mama Juliia."

Sofia Kovalevskaia's appointment at Stockholm University in 1883 as professor of mathematics, with a probationary period as an unpaid *Privatdozent*, caused immense controversy in Sweden, not least at Uppsala University, although there was also opposition in the Stockholm community. Sofia always felt grateful to Sweden for the career opportunity she had been given, but compared to St. Petersburg, Berlin, and Paris, Stockholm was a small town, and as time went on she found its prevailing liberal-bourgeois social atmosphere somewhat irksome. The fact that she was not bringing up her daughter herself was especially unacceptable to many Stockholm families, far from unheard-of though the practice may have been among the Russian noble classes. More than anything else she felt psychologically isolated; she missed the automatic understanding that connected her with her Russian friends, and most of all she missed the lively political discussions. But she quickly learned Swedish and made many friends, especially among the city's liberal-minded women. She and Mittag-Leffler's sister, Anna Carlotta, who had already published several works on social themes, collaborated closely for a time on further literary projects.

With much backing from Mittag-Leffler and a number of other Stockholm professors, her mathematical career moved slowly ahead. She began lecturing in February 1884 and was well liked by her students. In July of that year she was appointed professor extraordinary in advanced analysis (the equivalent of assistant professor in the United States) for a period of five years, at the modest salary of 4,000 Swedish kronor a year. In the late spring of 1885, when Hjalmar Holmgren, professor of theoretical mechanics at the university and Stockholm technical school, became ill, it was proposed that Sofia, the most suitable person available, take over his lectures temporarily. When Holmgren died the following autumn, Sofia had hopes that the university would appoint her to fill the position permanently; but even Mittag-Leffler and his allies were unable to win over the forces who opposed such an appointment, not because Sofia was female but to a large extent because of her widely known radical politics and her nihilist background.

In 1885 she published a second paper on the propagation of light through crystals; this work contained a nontrivial error, discovered by the young Italian mathematician Vito Volterra, who did not contact Mittag-Leffler or Weierstrass about his discovery until after Sofia's death. A number of other reports followed, including two on the research she had carried out as possible dissertation material, one of them on the reduction of abelian integrals to simpler elliptic integrals, and the other on the shape of Saturn's rings.

An undertaking Sofia embarked on as soon as she arrived in Stockholm was the encouragement of Russian and western European mathematicians to contribute papers to *Acta Mathematica*; the first Scandinavian mathematical journal, *Actà Mathematica* had been founded by Mittag-Leffler in 1882. Sofia had many friends in several countries, among mathematicians the closest being Karl Weierstrass in Berlin, Pafnutii Chebyshev in St. Petersburg,[9] Charles Hermite in Paris, and Leopold Kronecker in Berlin, contacts that brought her remarkable success in procuring contributions to *Acta Mathematica*. In addition, she had such an unerring judgment on the value of the papers submitted that she was officially made a coeditor in 1884, an appointment that made her the first woman on the editorial staff of a scientific journal. Her work on *Acta Mathematica* meant that she effectively functioned as an intermediary between Russian mathematicians and those in western Europe, to some extent reducing the gap between the two schools.

Her greatest mathematical triumph came in 1888, when the Paris Academy of Sciences awarded her its prestigious Prix Bordin. That year the academy's choice of subject for the Bordin prize was the classical problem of the motion of a rigid body about a fixed point. A topic that had challenged men such as Leonhard Euler, Siméon Denis Poisson, and Joseph Louis Lagrange, each of whom had solved it for particular cases, it had been set previously for the Bordin competition without any prize being awarded. Of the fifteen submissions the academy received in 1888, Sofia Kovalevskaia's so impressed the judges that the prize money was increased from the originally designated 3,000 to 5,000 francs.[10] The prize was presented at a special session of the academy a week after the winner was announced. Sofia became the toast of Paris; she was delighted with her success and very glad to have the money.

More honors came to her in 1889. Although the cautious and conservative Swedish Academy of Sciences resisted all the efforts of Mittag-Leffler and his friends to elect Sofia a member, they awarded her a prize for extensions to her work on the rotation of a solid body around a fixed point; and in November of the same year the Russian Academy of Sciences elected her a corresponding member. She was the first woman to be so honored; her election had required an amendment to the academy's charter to allow women to become members, a change brought about in large part by Pafnutii Chebyshev.

Sofia Kovalevskaia's last mathematical paper was a theoretical note presenting a simplified proof of a theorem of Heinrich Bruns in potential theory using her own early work in partial differential equations (see bibliography).

Her intellectual life was by no means confined to mathematics, and from the 1880s on she published a notable number of essays, novellas, and memoirs, as well as poems and plays; many more were unfinished when she died. In all these writings her radical political opinions are discernible, a quality that made them popular with the liberal and radical intelligentsia in Russia. She usually presented her political messages with subtlety to ensure passage past the tsarist censors, although one or two of her works were banned from publication in Russia. With the exception of *Vospominaniia detsva* (Memories of childhood), published in two parts in *Vestnik Evropy* (Messenger of Europe) in 1888–1889, her work is not considered to have exceptional literary value. *Vospominaniia detsva*, however, a well-crafted and heavily fictionalized account of her childhood days, offers a lively portrait of provincial Russian family life among the lower aristocracy in the late 1850s and early 1860s. As well as the wealth of domestic detail she presents, Sofia describes the strong influence on her of her nursemaid, and also clearly traces the political development of her older sister and of herself, despite the strict conservatism of her father and the shy meekness of her mother. Her major novel, *A Nihilist Girl*, published in Swedish in 1892 under the title *Vera Borontsoff*, was the story of a young woman who sacrifices her own happiness to marry a condemned revolutionary, since a married man had a better chance of having his sentence reduced. Perhaps of particular interest to her English-speaking readers is her memoir of George Eliot, *Vospominaniia o Dzhorzhe Elliote*, published in the Russian periodical *Russkaia mysl'* in 1886. Sofia had kept in contact with Eliot after meeting her during her visit to London in 1869.[11]

In 1888 the Russian sociologist Maxim Maksimovich Kovalevskii, a distant cousin of Sofia's late husband, was invited to Stockholm to deliver a series of lectures; Kovalevskii had recently been dismissed from his chair of state law at Moscow University on suspicion of holding radical views. Homesick for Russia, Sofia spent so much time with him that she was in danger of not completing the work intended for her entry for the Paris Academy of Sciences competition; but all went well, and Kovalevskii accompanied her to Paris for the prizegiving and the festivities that followed.

Relations between Sofia and Maxim Kovalevskii became close; he asked her to marry him but she declined. For her, marriage would have meant giving up her Stockholm professorship, and this she did not want to do. In February 1891, after spending the winter vacation at Kovalevskii's villa in the south of France, Sofia developed a cold; despite this she stopped in Paris and Berlin on her way home to talk with mathematical colleagues. By the time she reached Stockholm she was seriously ill with pneumonia. She died at her home on 10 February 1891, a month after her forty-first birthday. Her daughter, Fufa, who had lived with her in Stockholm during what turned out to be her last few years, was formally adopted by Juliia Lermontova.

Sofia Kovalevskaia was accepted and respected as a colleague by many of the great and near-great mathematicians of late nineteenth-century Europe. None of her works were groundbreaking, she never founded a new school of analysis or mechanics, and she died suddenly at the height of her mathematical and literary success. But her name is remembered in the history of mathematics as well as in the history of women in science; the Cauchy-Kovalevski theorem, an important extension of Cauchy's earlier work, is basic in the theory of partial differential equations and continues to appear in textbooks on the subject. Furthermore, her research on the revolution of a solid body about a fixed point, a brilliant application of abstract mathematics to a problem in mechanics, inspired a great deal of work in this area by later mathematicians.

Vera Iosifovna Shiff[12] (Frau, d. 1919), the other Russian woman mathematician whose pre-1901 work is listed in the Royal Society's *Catalogue*, is said to have been of French background and to have grown up in France,[13] but so far little information about her early years is readily available.

She attended the St. Petersburg Higher Courses for Women, graduating in 1882, and continued studies in mathematics. Over the next decade she developed a strong interest in the teaching of mathematics, and in particular she emphasized the value of teaching students the importance of practice in problem solving. Her first mathematical publication, her nine-page paper on an alternative method to that recently published by Modest Postnikov for investigating fourth-order curves,[14] appeared in the *Communications* of the Kharkov Mathematical Society in 1893 (see bibliography). The following year her *Methods for Solutions of Problems in Elementary Geometry* was published in St. Petersburg.[15] In 1901 she received a doctoral degree in mathematics from Göttingen University, perhaps after a period of advanced study there in the 1890s. The same year she became a junior instructor at the St. Petersburg Higher Courses for Women, where she taught mathematics courses until at least 1913.

Vera Shiff wrote and edited a large number of books on mathematics, especially works relating to instruction.[16] These appear to have been much in demand among students in the physical sciences; physicist Paul Epstein, in his reminiscences of his undergraduate days at Moscow University, described the collection of mathematical problems by Shiff that he had used as "very good," in contrast to others, by famous scientists, that he found "excruciatingly dull." Indeed, he implied that the critical problem-solving skills he developed sprang to a significant extent from the help he derived from Shiff's books.[17]

Like several others of that early generation of women scientists, Vera Shiff did not survive the period of civil disruption and the hardships of two harsh winters that followed the First World War. She died in 1919.

One other female mathematician, LIUBOV' NIKOLAEVNA ZAPOL'SKAIA[18] (1871–1943) is noted briefly in this nineteenth-century survey, although none of her publications is listed in the Royal Society's *Catalogue*. Zapolskaia, after receiving a doctorate from Göttingen University, went on to become the first woman to be awarded a Russian *magistr* degree in abstract mathematics, three decades after Sofia Kovalevskaia was denied any chance of acquiring this valuable qualification.

The daughter of Nikolai Nikitovich Zapolskii and his wife, Sophia Vasil'evna, Liubov' Zapol'skaia was born on 7 August 1871 in the village of Bigil'dino-Marmots, Ryazan province, about one hundred miles southeast of Moscow. Her father, the son of a missionary, was born in Siberia and had theological training, initially at Irkutsk Theological Seminary and later at St. Petersburg Theological Academy. He earned his living as a school teacher, for a time in the city of Ryazan; later, after the family moved to St. Petersburg sometime in the 1870s, he became principal teacher in a military academy. Often in poor health, he died in the 1880s, leaving the family ill provided for. Possibly with some material help from a cousin, Aleksandra Losev, Liubov' was able to complete her high school education at one of the schools for girls run by the Department of Empress Maria Institutions, schools that included science and mathematics in their curricula. She graduated in 1887, with a gold medal, and immediately went on to a three-year course of teacher training; this she completed in 1890, winning another gold medal. Then aged nineteen, she decided to continue her studies, and that year enrolled in the physical-mathematical division of the St. Petersburg Higher Courses for Women, where she attended lectures on a wide variety of subjects, among them physics, chemistry, astronomy, and mechanics, as well as several branches of mathematics, all taught by highly respected St. Petersburg University lecturers. By the time she had completed these studies, in 1894, she had made up her mind to go further into the field of higher algebra.

In 1895 she went to Göttingen, where, with official permission from the university and with her expenses paid by her Losev relative, she enrolled in the philosophical faculty and began studies under the guidance of two famous mathematicians, David Hilbert and Felix Klein. She passed her oral examination in 1900 and defended her dissertation, a study in algebraic number theory directed by Hilbert, in 1902.[19]

On her return to Russia the following year she was received a teaching position at the Moscow Higher Courses for Women. She continued her studies on the theory of algebraic fields and successfully defended a dissertation for the Russian *magistr* degree at Moscow University in 1905; the work was published a year later.[20]

Apart from a short return visit to Göttingen University for further mathematical studies in 1910, she spent the rest of her career teaching mathematics at girls' schools and pedagogical institutes attached to regional universities. In 1919 she published one more scientific paper, a fairly lengthy study that appeared in the *Proceedings of the Yaroslavl Pedagogical Institute*.[21] Ill health caused her to retire in 1930, when she was about sixty, and return to Ryazan, where she lived with her cousin Aleksandra Losev, then growing old and in need of help. She remained very interested in the teaching aspects of science and mathematics, became a member of the Ryazan Scientific Society, and gave much of her time to tutoring schoolchildren and engineering students at Ryazan State University. Liubov' Zapol'skaia died on 3 November 1943, at the age of seventy-two.

ASTRONOMERS

Among several Russian women who worked in astronomy during the last decades of the nineteenth and the early years of the twentieth centuries, the publications of three—Mariia Zhilova, Countess Nadezhda Bobrinskaia, and Evgeniia Maximova—have been found listed in the Royal Society's *Catalogue*.[22]

MARIIA VASIL'EVNA ZHILOVA[23] (1870–1934) was born in the town of Rybinsk, in the Yaroslavl region northeast of Moscow. Her father was a merchant with a successful trade in bread in surrounding communities, and the family lived comfortably.

Mariia attended a girls' gymnasium in Rybinsk, where she was an outstanding student. She graduated from the eighth class in 1888, winning a gold medal. A year later she moved to St. Petersburg to continue her education, a course of action permitted by her parents only on the condition that she lived with a married sister who acted as her guardian. She attended lectures in the physical-mathematical division of the Higher Courses for Women from 1889 to 1894, a time when there were still relatively few women taking advanced training in the physical sciences; those that were benefited from close attention from many of the distinguished faculty in a wide range of fields. Among Mariia Zhilova's instructors were Swedish-Russian astronomer Oskar Andreevich Backlund, director of the Pulkova Observatory, as well as physicist Orest Danilovich Khvol'son, geologist and explorer Ivan Vasil'evich Mushketov, and chemist Dmitrii Mendeleev. Backlund awakened her interest in astronomy and encouraged her to develop skills in practical work; Khvol'son, recognizing her exceptional abilities, suggested she take up work in electromagnetism; and Mendeleev's frequent informal talks, which ranged over many areas of science important to Russia, were a lasting influence.

After much thought about career choice, she accepted an offer of employment from Backlund, who then was in the process of enlarging the computer staff at Pulkova Observatory. Backlund went to considerable effort to get staff positions for women workers, but like his contemporary Edward Pickering, director of the Harvard Observatory, he did not succeed.[24] The women computers in Russian observatories were at that time classed as temporary workers paid by the piece, and many left after a few years for more steady employment. Having some financial resources of her own, Zhilova stayed on.

By the time she had worked at the Pulkova Observatory for more than a decade, director Backlund was ready to officially state that the acceptance of scientifically educated women had greatly benefited the observatory, even beyond his expectations. His 1908 report testified that their diligence and scientific understanding had considerably speeded up the processing of observations and that, in addition to their assigned work, they had successfully carried out observations of their own.[25]

During her first years at Pulkova, Zhilova processed photographic plates that recorded positions of planets, star clusters, and double stars, measured star spectra, and carried out calculations, these tasks being set by the observatory astronomers, not all of whom had welcomed women into their workplace. Some of the astronomers had denied women the use of observational instruments for independent work.

One of Zhilova's major contributions to the work of the observatory was her share in compiling tables of Bessel numbers.[26] She also contributed much to the preparation of the Pulkova Star Catalogues, which listed the most precise positions of 558 stars for 1845, 1865, 1885, 1906, and 1930. One of these was Pulkova's contribution to the huge *Carte du Ciel* project, which involved the photographic preparation, by international cooperation, of a complete photographic sky chart, work carried out mainly in the 1890s. Zhilova was active in presenting reports to scientific societies, and even found time to give a number of popular public lectures. Perhaps in recognition of her many contributions to the observatory's work she was included in two Pulkova expeditions (1896 and 1914) to observe solar eclipses.

With encouragement and guidance from Oskar Backlund, she gradually moved on to independent work in celestial mechanics, determinations of the orbits of small planets.[27] In 1905, on Backlund's recommendation, she received the Russian Astronomical Society's prize, and became a fellow of the society.

After the October Revolution, when new regulations for state observatories were established, women computers were eligible for staff status. In 1921 Zhilova became adjunct astronomer. Her work on the motion of minor planets and the determination of their orbits was widely acknowledged in imperial Russia and the Soviet Union as well as abroad as a valuable contribution to celestial mechanics. She became a member of the Association of Astronomers of the Soviet Federal Socialist Republic.

After many years of collaborative work in the computing bureau, Mariia Zhilova would appear to have moved on; she may have been granted a pension, and having settled in the village of Pulkova near the observatory, she continued her very productive small-planet observational work. She also had a small, portable telescope of her own, which she used to observe sun flares.

Long troubled with nerve-related illnesses, she died on 16 May 1934, when she was about sixty-three; burial was in the Pulkova cemetery. Minor planet (1255) Schilowa was named after her; it was discovered by Grigorii Nikolaevich Neujmin, on 8 July 1932, at the Simeis Observatory in the Crimea.

Nadezhda Aleksandrovna Polovtzeva, later Comtesse Bobrinskaia[28] (1865–1920), was born in St. Petersburg on 27 May 1865. A daughter of Aleksandr Aleksandrovich Polovtzov (1832–1909) and his wife, Nadezhda Mikhailovna (1843–1908), she came of noble background, her mother being an illegitimate daughter of Grand Duke Mikhail Pavlovich. Readily available sources give no indication of Nadezhda Aleksandrovna having attended a girls' gymnasium; most likely she had private tuition. In 1883, at the age of eighteen, she married Count Aleksei Aleksandrovich Bobrinskii (1852–1927); the first of her five children, daughter Nadezhda Alekseevna (1884–1896), was born the following year.

Aleksei Aleksandrovich Bobrinskii's family held a prominent place among the Russian nobility; the family traced its line back to Aleksei Bobrinskii (1762–1813), an illegitimate son of Catherine the Great. Bobrinskiis were prominent in imperial Russia's government circles and also played a major role in building railways and developing fledgling Russian industry throughout the nineteenth century; several were active in scientific work as well. Comtesse Nadezhda's husband was a noted historian, archaeologist, author of scholarly works, and chairman of the Imperial Archaeological Commission. He carried out a number of excavations in the region around the Black Sea; particularly notable were his investigations of Scythian mounds near Kerch and Kiev. Also a leading figure in St. Petersburg politics, he was president of the St. Petersburg Zemstvo Assembly and from 1906 headed the Council of United Nobility; from 1912 he served on the State Council of imperial Russia. Bobrinskii fled to France after the October Revolution.

Nadezhda Bobrinskaia worked at the Pulkova Observatory for several years in the 1890s and probably for a time in the early 1900s. Three of her astronomical reports published in the St. Petersburg Academy of Science *Bulletin* are listed in the Royal Society's *Catalogue*, beginning with an orbit determination for comet 1890 VI [1893]. Her study of G.C.4294=M92 (General Catalogue 4294, Messier Catalogue 92) appeared a year later. A cluster of minute stars in the constellation Hercules, G.C.4294 is one of the brighter globular clusters in the Northern Hemisphere; it is now known to have a very low abundance of elements other than hydrogen and helium, signifying that it is one of the oldest clusters in the Milky Way population. In 1897 Comtesse Nadezhda Aleksandrovna brought out a table of daily positions (ephemeris) for the small planet (147) Protogeneia (see bibliography). She published one further paper, a 1901 contribution reporting a small correction in the orbit of the minor planet (300) Geraldina (discovered in 1890), along with a calculation of its future position; the latter was one of the early attempts to reliably establish the periodic perturbations of Geraldina by Jupiter and Saturn, work that continued for many decades.[29]

Comtesse Bobrinskaia's work reflected her lively interest in a broad spectrum of current astronomical research; she was clearly an enthusiastic amateur astronomer. With the outbreak of severe fighting in the Russo-Japanese dispute about Far East territory (1904–1905), she turned her energies to urgently needed work in the Red Cross, service for which she received an award. She appears to have stayed on in Russia throughout the upheavals of the October Revolution and most of the 1918–1921 civil war between Bolshevik and White Russian forces. She died on 20 March 1920, two months before her fifty-fifth birthday, at Fort-Aleksandrovsk (now Fort-Shevchenko), a military base town on the east side of the Caspian Sea. Count Aleksei Aleksandrovich married Raissa Novikova in 1920, in Nice; he died in Grasse in 1927, at the age of seventy-five.

Evgeniia Aleksandrovna Maksimova[30] (fl. 1890s–1900s), was a fellow student of Mariia Zhilova in the physical-mathematical division of the St. Petersburg Higher Courses for Women during the 1890s. When she and Zhilova completed their studies there in 1894, both were invited by Pulkova Observatory director Oskar Backlund, an astronomy instructor at the Higher Courses, to join the Pulkova computing bureau. Maximova started her work there in 1895, and over the next six years, in addition to her assigned computing for observatory astronomers, she studied the motion of the very large asteroid (209) Dido, one of the small planets of the main belt that circles between Mars and Jupiter. Two papers, published in 1895 and 1896, reported her ephemeris studies for the asteroid for these years; work on orbit determinations for Dido followed in 1900 and 1901.[31]

Shortly after that, finding life too difficult on her scanty earnings, she gave up her observatory work and became a teacher.

NOTES

1. Works about Sofia Kovalevskaia include the following: Ann Hibner Koblitz, *A Convergence of Lives: Sofia Kovalevskaia: Scientist, Writer, Revolutionary* (Boston: Birkhäuser, 1983); Tollmien, *Fürstin der Wissenschaft: Die Lebensgeschichte der Sofia Kowalewskaja* (Weinheim and Basel: Beltz Verlag, 1995); Jan-Erik Bjork, "Sonja Kovalevsky," in *Operator Methods in Ordinary and Partial Differential Equations: S. Kovalevsky Symposium*, University of Stockholm, June 2000, ed. Serfio Albeverio (Basel: Birkhäuser, 2002), reproduced from *Operator Theory: Advances and Applications* 132 (2002): 11–53; Michele Audin, *Remembering Sofia Kovalevskaya* (London: Springer, 2011), trans. from the French, *Souvenirs sur Sofia Kovalevskaya* (n. p.: Calvage et Mounet, 2008); Gary J. Tee, "Sof'ya Vasil'yevna Kovalevskaia," address delivered at the Eleventh New Zealand Mathematics Colloquium, Palmerston North, 21–24 May 1976, *Mathematical Chronicle* 5 (1977): 113–39; Don H. Kennedy, *Little Sparrow: A Portrait of Sophia Kovalevsky* (Athens: Ohio University Press, 1983); O. A. Valkova, "The Conquest of Science: Women and Science in Russia, 1860–1940," *Osiris* 23 (2008): 146–47. See also Sofya Kovalevskaia, *A Russian Childhood*, ed. and trans. Beatrice Stillman (New York and Berlin: Springer, 1978); Alexander Vucinich, *Science in Russian Culture, 1861–1917* (Stanford, Calif.: Stanford University Press, 1970), 343–48.

2. Kovalevskaia, *Russian Childhood*, 25.

3. Nikolai Pisarevskii and Nikolai Tyrtov, *Nachal'nyia osnovaniia fiziki* [Foundations of physics] (St. Petersburg: Éduard Veimar, 1860).

4. Karl Weierstrass and S. V. Kovalevskaia, *Pis'ma Karla Veierstrasse k Sof'e Kovalevskoi, 1871–1891; Briefe von Karl Weierstrass an Sofie Kowalewskaia, 1871–1891*, ed. P. IA. Polubarinova (Kochina Polubarinova), (Moscow: Nauka, 1973). Some additional material appears in *Briefwechsel zwischen Karl Weierstrass und Sofja Kowalewskaja*, comp. and ed. Reinhard Bölling (Berlin: Akademie Verlag, 1993).

5. See Koblitz, *Convergence of Lives*, 104–110, for a fuller account of the activities of Sofia and Aniuta during the period of the Paris Commune of 1871.

6. Sofia Kovalevskaia was the second woman to be awarded a doctorate *honoris causa* by Göttingen University, the first being Dorothea Schlözer eighty-seven years earlier. Although the two cases may be too different for a comparison to be significant, Dorothea's story is worth putting on the record. The daughter of historian and journalist August Ludwig Schlözer, she was born in Göttingen on 10 August 1770. Believing that the intelligence of women was no less than that of men, August Schlözer gave his daughter a broad and thorough education that included ancient and modern languages, mathematics, and the sciences, taught by the best of private tutors. In 1787, after successfully passing a private examination in a range of subjects (such as mathematics, logic, languages, and literature) before a Göttingen University faculty committee, Dorothea was awarded a doctorate, as a "special case"; she was then seventeen. She took part in the preparation of at least one scholarly work, an account of the coinage of imperial Russia in the eighteenth century, brought out by her father in 1791. Her help is duly acknowledged in August Schlözer's *Vorrede*. See August Ludwig von Schlözer, *Münz, Geld- und Bergswerke-Geschichte des Russischen Kaiserthums: Vom J. 1700 bis 1789: Meist aus Urkunden beschreiben* (Göttingen: im Verlag bei Vandenhoek und Ruprecht, 1791). The author's seven-page preface includes a technical note signed by Dorothea Schlözer. In 1792 Dorothea married Matthäus Rodde, a wealthy merchant in Lübeck and also a senator who was granted baronial status in 1803. She died in Avignon, in poor health, on 12 July 1825, a month before her fifty-fifth birthday. For further details see Personen, Stadtarchiv Göttingen, http://www.stadtarchiv.goettingen.de/personen/schloezer_dorothea.htm (Aug. 2014). A short biographical account of Dorothea Schlözer was published in 1839 in Émile Bégin, *Villers, Madame de Rodde et Madame de Staël* (Metz: Verronnais, 1839), but this work has not been examined.

7. See Tollmien, *Fürstin der Wissenschaft*, 88. For Grace Chisholm Young, the first woman to be awarded a doctorate in mathematics by Göttingen University by the regular route, see Mary R. S. Creese, *Ladies in the Laboratory? American and British Women in Science, 1800–1900* (Lanham, Md.: Scarecrow Press,1998), 192–94.

8. This restriction severely limited the prospects of Sofia Kovalevskaia's friend Elizaveta Litvinova as well. Elizaveta Fedorovna Ivaschkina (later Litvinova or Litvinova-Ivaschkina), the daughter of Fedor Ivaschkin, a landowner in the Tula district, was born on 21 November 1847. She completed her early education at a girls' gymnasium in St. Petersburg and had ambitions for further advanced study. In 1866 she married Viktor Litvinov, who declined to allow her to travel to western Europe and enroll at a university but did not block further study in St. Petersburg. For several years she had private lessons from Aleksandr Strannoliubskii, the excellent tutor with whom Sofia Kovalevskaia had also studied. In 1872, after a separation from Viktor Litvinov (by some accounts his death), Litvinova went immediately to Zurich and enrolled at the Zurich Polytechnicum, where she concentrated on mathematics. Her diploma from the University of Zurich School for Teachers was conferred in October 1874, by which date she had already enrolled in Bern University's philosophical faculty. The first female student in that faculty, she graduated with a doctoral degree in 1879. Her inaugural dissertation, "Lösung einer Abbildungsaufgabe," a solution to a problem in the representation of surfaces, was printed as a twenty-eight-page

booklet by the St. Petersburg Academy of Sciences in 1879. Like Kovalevskaia a few years earlier, she was denied the opportunity of presenting herself for the Russian *magistr* examination and so could not teach in a university or in the upper grades of a gymnasium. Litvinova finally accepted a post as a teacher in the lower grades of a girls' high school. About a decade later, however, following repeated requests to the Ministry of Education, she became the first woman in Russia to teach mathematics in the upper grades of a gymnasium, although without the rights and privileges of male teachers of this rank.

Throughout the 1890s she supplemented her meager income by writing. Her biographies of some of the more famous scientists and mathematicians were published in the series *Zhizn' zamechatel'nykh liudei* (Lives of outstanding people), brought out by the liberal St. Petersburg publisher Florentii Pavlenkov. They included a work on Aristotle (1892), one on Laplace and Euler (1892), and others on Struve (1893), Kovalevskaia (1894), and Lobachevskii (1895); another account of Laplace's life and work was published in Petrograd in 1919. However, Litvinova is remembered more for her contributions to the philosophy and practice of teaching mathematics. She published over seventy articles in this area, mainly in the journal *Zhenskoe Obrazovanie* (Women's education) and its successor, *Obrazovanie*, and became known as one of Russia's foremost pedagogues. The revolutionary activist Nadezhda Krupskaia (wife of Lenin), who was Litvinova's student at the A. A. Obolenskaia gymnasium, recalled that Litvinova not only taught the usual mathematics, but showed her students how to think logically and to extract laws and principles from a series of separate cases; overall she served as a constant source of inspiration and encouragement for her female students, several of whom went on to become successful scientists. Litvinova also wrote scientific articles and biographical sketches aimed at a wide general audience, often subtly incorporating egalitarian themes in a way that mostly avoided offending the tsar's censors.

She took part in the women's movement in Europe, contributing a number of articles dealing with Russian affairs to the Paris-based feminist monthly *Bulletin de l'Union universelle des Femmes* throughout the 1890s and early 1900s; she was one of four Russian delegates to the 1897 International Women's Congress in Brussels. Litvinova retired from teaching about 1914, after a career of some thirty-five years. She is thought to have spent the rest of her life in the country with her sister, through the upheavals of the First World War and the revolution. She most likely died about 1919. See matriculation register, Zurich University, http://www.matrikel.uzh.ch/active/static/26313.htm (Aug. 2014); archives, Bern University, http://apps.uniarchiv.unibe.ch/syscomm/images/mata/2933_2944.gif (Aug. 2014). For a full listing of Litvinova's publications and further information, see Ann Hibner Koblitz, "Elizaveta Fedorovna Litvinova (1845–1919?)," in *Women of Mathematics: A Biobibliographic Sourcebook*, ed. Louise Grinstein and Paul Campbell (Westport, Conn.: Greenwood Press, 1987), 129–34.

9. Sofia Kovalevksaia had a steady correspondence with Chebyshev over several years; she translated from Russian to French at least two of his papers published in *Acta Mathematica*. P. Tchebycheff, "Sur la représentation des valeurs limites des intégrals par des résidues intégraux," trans. Sophie Kowalevski, *Acta Mathematica* 9 (1887): 35–56; P. Tchebycheff, "Sur les sommes composées des coefficients des séries à termes positives," letter addressed to Sophie Kowalevski, *Acta Mathematica* 9 (1887): 182–84.

10. Mathematician Gary Tee has succinctly described Kovalevskaia's Prix Bordin research as follows: "Euler had solved the problem of the motion of a rigid body in free fall, or (equivalently) suspended at its centre of mass in a gravitational field. Lagrange had solved . . . [it] in the case where it is suspended at a point such that the body's ellipsoid of inertia is symmetrical about an axis, and the centre of mass also lies on that axis of symmetry. Sof'ya considered all cases in which the coordinates of the rigid body can be expressed as power series of time, and she showed that, apart from the cases of Euler and Lagrange, all such cases consist of a body suspended at a point for which the body's ellipsoid of inertia is an oblate spheroid of revolution with the equatorial radius twice the polar radius, and with the centre of mass lying in the equatorial plane. In a brilliant display of virtuosity she constructed the coordinates explicitly as ultra-elliptic functions of time. (Actually she had overlooked the possibility of the coordinates having an essential singularity at time zero but that minor omission was rectified by A. M. Lyapunov . . .)." Tee, "Sof'ya Vasil'yevna Kovalevskaia," 132.

11. See Hibner Koblitz, *Convergence of Lives*, ch. 14, for a full discussion of Kovalevskaia's literary works.

12. Valkova, "Conquest of Science," 158; Rebière, *Femmes dans la Science*, 158; Alice Epstein, interview with Paul S. Epstein (1883–1966), 1965–1966, Oral History Project, California Institute of Technology Archives, http://resolver.caltech.edu/CaltechOH:OH_Epstein_P (Aug. 2014). Epstein received his undergraduate education and carried out his master's degree work at Moscow University, beginning in 1902. By the early 1920s, following a meeting in Leiden with American physicist Robert Millikan, he had accepted a teaching position at the California Institute of Technology.

13. Epstein, interview with Paul S. Epstein, part 1, "Undergraduate Years in Moscow."

14. Modest V. Postnikov, "Etîudy po teorii krivykh' chetvertago poriadka" [Studies on the theory of curves of the fourth order], *Proceedings of the Physico-Mathematical Section of the Society of Naturalists of the Imperial University of Kazan* 7 (1889): 118–58.

15. Vera Shiff, *Metody rieshenii voprosov elementarnoi geometrii* (St. Petersburg: tip. V. S. Balasheva i K. N. Fontanki, 1894.)

16. These included *Sbornik uprazhnenii i zadach po differentsial'nomu i integral'nomu ischisleniiam* [Problems and exercises in differential and integral calculus] (St. Petersburg: M. O. Vol'f), which might be called a best seller; it was brought out in two parts, each of which ran to several editions between the late 1890s and 1910 (1902, 1905, 1909, with an enlarged edition of part 2 in 1910). Other books were [Introduction to the theory of determinants] (1907), [Problems and exercises in analytic geometry] (1904; 2nd ed., 1908), and [Rectilinear trigonometry] (2nd ed., 1910), all of which were published in St. Petersburg.

17. Epstein, interview with Paul S. Epstein, part 1, "Undergraduate Years in Moscow."

18. Valkova, "Conquest of Science," 154; N. S. Reshetina, "Liubov' Nikolaevna Zapol'skaia," Science Library, Ryazan State University, http://library.rsu.edu.ru/blog/wp-content/uploads/2013/01/zapolskaya.pdf (Aug. 2014).

19. Liubov Zapolskaia, "Über die Theorie der relativ–abelschen-cubischen Zahlkörper," inaugural dissertation, higher philosophical faculty, Göttingen University, 1902.

20. Liubov' Zapol'skaia, *Teoriia algebraicheskikh oblastei ratsional'nosti, obrazuiushchikhsia pri reshenii uravnenii 3-i stepeni* [Theory of algebraic fields expressed efficiently in the form of third-degree equations] (Moscow: University tip., 1905), 370 pp.

21. Liubov' Zapol'skaia, "O svoistvakh nekotorykh chislovykh luchei," *Proceedings of the Yaroslavl Pedagogical Institute, Physical-Mathematical Collection* T.2, Vyp. 4, Yaroslavl, (1919): 19–34.

22. For information about other early Russian women astronomers, see, for instance, Alla G. Masevich and Aleksandra K. Terenteva, "Zhenshchiny-astronomy," *Na rubezhakh poznaniia vselenna. Istoriko-astronomicheskie issledovaniia* 23 (1991): 90–111; Maria N. Neuimina, "Zhenshchiny-astronomy Pulkovskoi Observatorii," *Istoriko-astronomicheskie issledovanniia* 10 (1969): 229–39. See also Joseph S. Tenn, "Oskar Backlund: The Eleventh Bruce Medalist," *Mercury* (November/December 1991): 175, 178–79.

23. Neuimina, "Zhenshchiny-astronomy," 229–34

24. For a note on Pickering, see Creese, *Ladies in the Laboratory?*, 230.

25. From Neuimina, "Zhenshchiny-astronomy," 231, quoting from the director's report for 1908.

26. See M. V. Zhilova in *Tables des quantités Besseliènnes pour les années 1905–1910, 1911–1913, 1914–1915, 1916–1920*, (St. Petersburg: Imprimérie de l'Académie des Sciences, imprimé par ordre de l'Observatoire Central Astronomique [Soviet Union], 1900–1926). Bessel numbers are used by astronomers to calculate star positions for particular dates.

27. Among Zhilova's many post-1900 publications are the following: M. V. Zhilova, "Angenäherte Elemente und Ephemeride des Planeten Doris," *St. Petersburg, Bulletin of the Academy of Science* 14, no. 5 (1901) (hereafter *Ac. Sci. Bull.*); "Angenäherte Elemente und Ephemeride des Planeten Protogeneia in die Opposition. 1902–1909," *Ac. Sci. Bull.* 16, no. 3 (1902); "Ephemeride des Planeten Doris (48)," *Astronomische Nachrichten* 160 (1902): 149; "Angenäherte Ephemeride des Planeten Doris," *Ac. Sci. Bull.* 21, no. 3 (1904); "Malyi planety," *Ac. Sci. Bull.* 1909, 1911; "Positions des petites planètes déduites par M. Shilow des photographies prises par G. Tikhoff," *Ac. Sci. Bull.* 4 (1912); "Sur l'exactitude des éléments des planètes (48) Doris, (147) Protogeneia et (196) Philomela," *Ac. Sci. Bull.* 4, no. 47 (1912); "Ephemeriden des Planeten (196) Philomela (1914–1925)," *Izvestiia Nikolaevskoi glavnoi astronomicheskoi observatorii* 6, no. 4 (1914); "Malye planety," *Tserkuliary Nikolaevskoi glavnoi astronomicheskoi observatorii* 24 (1917); "Efemerida maloi planety (48)," *Izvestiia glavnoi astronomische observatorii* no. 86 (1918); "Ephemeriden der kleine Planeten (48) Doris, (147) Protogenia," *Berlin-Dahlen, Kleine Planeten . . . Oppositions-Ephemeriden* (1925); "Efemerida maloi planety (196) Philomela dlia protivostoianii s 1926 do 1937 goda," *Izvestiia glavnoi rossiiskoi astronomicheskoi observatorii* 10, no. 4 (1926); "Photographische Beobachtung von kleinen Planeten (48) Doris, (196) Philomela, (566) Stereoskopia," *Astronomische Nachrichten* 232 (1928); "Efemeridy malykh planet (48) Doris, (147) Protogeneia i (196) Philomela na 1929 god," *Berlin-Dahlen, Kleine Planeten . . . Oppositions-Ephemeriden* (1928) (extracts from *Berliner astronomische Jahrbuch*).

28. Nadejda Alexandrovna Polovtzeva (1865–1920)—Genealogy, http://www.geni.com/people/Nadejda-Alexandrovna-Polovtseva/6000000007872512514 (Aug. 2014); Alexei Alexandrovich Comte Bobrinskiy (1852–1927)—Genealogy, http://www.geni.com/people/Alexei-Androvich-Comte-Bobrinskiy/6000000004578414425 (Aug. 2014); Miroslav Marek, "Genealogy of the Bobrinsky family," http://genealogy.euweb.cz/ascania/ascan13.html (Aug. 2014).

29. N. Bobrinsky, "Ephemeride des Planeten (300) Geraldina," *Astronomische Nachrichten* 156 (1901): 171–72.

30. Neuimina, "Zhenshchiny-astronomy," 231; N. I. Nevskaya, "Leningrad Women-Astronomers in the Early Years of Soviet Power," *Proceedings of the International Conference on the Role of Women in the History of Science, Technology and Medicine in the 19th and 20th C.*, Veszprém, Hungary, 15–19 August 1983, 2 vols., 1:139–44.

31. See bibliography, and E. Maximow, "Angenäherte absolute Bahn des Planeten (209) Dido," *Ac. Sci. Bull.* 15, no. 3 (1901): 331–33.

Chapter Five

Geologists, Paleontologists, and Other Explorers

GEOLOGISTS AND PALEONTOLOGISTS

Of the four Russian women who brought out reports in the geological sciences that are listed in the Royal Society's *Catalogue*, three—Mariia Pavlova, Evgeniia Solomko, and Mariia Tsvetaeva—made significant contributions to the field; little information has been found about the fourth, Vera Goloubtzova, who published only one pre-1901 report. The very productive Mariia Pavlova is undoubtedly the best remembered and the most widely known of the group; over the course of her career of more than four decades at Moscow University, her contributions to early Russian paleontology were many and notable, ranging from major surveys of particular areas to outreach efforts, popularizing, and university-level teaching. Evgeniia Solomko spent less than a decade in geological work, and Mariia Tsvetaeva's participation was mainly restricted to summer field research. Nevertheless these two also made valuable contributions to Russian geology.

Mariia Vasil'evna Pavlova,[1] née Gortynskaia (1854–1938), the daughter of Vasilii Stefanovich Gortynskii and his wife, Aleksandra Mikhailovna, was born on 14 (26) June 1854 in Kozelsk, Chernigovskii district, now the Chernigovskii region of Ukraine. Gortynskii, a graduate of Moscow University, a man of eighth grade in the Russian Table of Civil Ranks, was a state provincial doctor; he later became government medical inspector of Chernigovskii district and rose to seventh grade, or Court Councillor rank, which made him one of Russia's class of educated noble bureaucrats.

Mariia was taught at home until the age of ten, when she entered the Kiev Institute for Noble Girls. Like the Smolny Institute in St. Petersburg, the Kiev Institute functioned to a large extent as a finishing school for young ladies but also offered some instruction in science and arithmetic. Mariia completed the six-year course there in 1870, with qualifications for the work of house governess in a variety of subjects. Two or three years later she married Illich-Shishatskii, her senior by many years, who is reported to have been a district doctor in Astrakan.[2] She also taught for a period about this time. When her husband died in 1880, she decided to return to studies and traveled to Paris to attend classes at the Sorbonne.

Long interested in the natural sciences and encouraged in this area by her father, to whom she was very close, she studied zoology, botany, geology, and paleontology and was able to work in the Paris Muséum d'histoire naturelle. While studying under the guidance of Albert Gaudry, a widely recognized expert on fossils of the tertiary period and a convinced Darwinist, her main interest settled on paleontology.

Gaudry's teaching provided the foundation on which Gortynskaia built her brilliant future career, and the friendships she formed while in his laboratory shaped her broad, international outlook, essential for success in her later work. She kept in regular contact with Gaudry after she returned to Russia, and over the course of twenty years, until his death in 1908, they exchanged critiques and reports, discussed publications, and talked about such matters as possible geological excursions.[3] Another long-term contact she formed at this period in her life was with American paleontologist Henry Fairfield Osborn (1857–1935), head of the vertebrate paleontology department of the American Museum of Natural History, whose innovative work in the preparation of museum sites for public displays she much admired. Her most important friendship begun during her time in Paris was undoubtedly that with Aleksei Petrovich Pavlov (1854–1929), her future husband, whom she first met at the home of zoologist M. A. Menzbir, another Russian who was then broadening his scientific horizons in Paris. Making the most of her stay in western Europe, Mariia Gortynskaia visited geological collections in museums in London, Munich, Vienna, and Paris. She also made several excursions to collect vertebrate fossil bones, material that went to enrich Russian museum holdings.

After completing her studies at the Sorbonne in 1884 and receiving the grade of specialist in paleontology, she returned to Ukraine. She spent some time with her family but by 1885 had moved her place of residence to Moscow. In May 1886 she married Aleksei Pavlov, who in January of that year had been made professor extraordinary at Moscow University.

Mariia Pavlova's work at Moscow University began as an unpaid volunteer in the geological museum. She promptly brought out her first scientific publication, a report of her study of ammonite fossils collected by Aleksei Pavlov in Povolzh' (Volga region), her one publication on material from that early period (see bibliography). All her many subsequent studies concerned tertiary mammalian fossils.

The first issue of her celebrated *Studies in the Paleontological History of Hoofed Animals* appeared in1887. In this series, which ran to nine issues published over the course of nineteen years (1887–1906), she traced evolutionary changes in large groups of animals arising from a common ancestor using materials that had been assembled in Russia as well as in western Europe and America. She based her picture of evolutionary trends on the study of as large a population of fossil remains as possible, integrated with Darwinian theory; here she was following the methods used by the outstanding young Russian paleontologist Vladimir Kovalevskii, now acknowledged as one of the founders of Russian paleontology. Focusing in particular on the hippopotamus and horse families, she helped to confirm Kovalevskii's ideas on mammalian evolution in general and especially on the various evolutionary branches of the horse family, a topic then of great interest among biologists and paleontologists. Pavlova's contributions in this area included her work in the early 1890s on the role of *Hipparion*, the small, three-toed horse that represents one of the secondary evolutionary lines of the family. *Hipparion* has special significance since it is considered to be the form that migrated from North America to Asia across the Bering Land Bridge during the Miocene period, bringing the horse family to Asia and Europe. Pavlova placed strong emphasis on the need to base the construction of phylogenetic lines on more than one feature; for *Hipparion* she pointed out the importance of tooth-structure development as well as foot development in demonstrating that *Hipparion* was not the direct ancestor of the modern horse but only its relative.

Although Mariia Pavlova's access to the geological museum at Moscow University in 1886 had depended on help from Aleksei Pavlov and his colleague mineralogist Vladimir Vernadskii, within a decade she was well recognized within the geology community, a valued member of the group, known for her success in reaching out to smaller local organizations and enriching the main Moscow collections. In 1897, when Russia for the first time hosted an International Geological Congress, she and Mariia Tsvetaeva (see below) both took part in the work of the congress and also served on the Organizational Committee, the first Russian women to occupy such roles. Excursions to the Moscow University geological museum were especially popular with the congress delegates.[4]

After Pavlova had completed her long series on the paleontological history of the ungulates, she returned to her study of the *Proboscidea*, the elephants and their extinct allies, work she had begun in the 1890s. Her monograph *Les éléphants fossils de la Russie* appeared in 1910. This was followed in 1913 and 1914 by her two-volume *Mammifères tertiaires de la nouvelle Russie*, coauthored with Aleksei Pavlov. Much of the material for the latter was based on her own collecting efforts and her many visits to distant museums specializing in local materials, where she often found important holdings. These visits, to regions from eastern Siberia to the northern Caucasus and Ukraine, also helped local workers by making clear the wider significance of their holdings and pointing toward particularly profitable future paths to follow.

As part of her overall effort to introduce paleontology to a wider audience, she published several popular works, including translations into Russian of foreign publications such as Henry Neville Hutchinson's *Extinct Monsters* (1893) and Melchior Neumayer's *Die Stämme des Thierreichs* (1889). Her own short work on the causes of extinction of animals in past geological epochs, coauthored with Aleksei Pavlov, came out in 1910.

Pavlova's teaching career before the Soviet era was confined to the Moscow Higher Courses for Women, where a Department of Paleontology was started in 1910, and the Shaniavskii Moscow People's University.[5] Most of the Moscow-educated women geologists and paleontologists from before the revolution were her students; although few in number, several of them went on to successful careers. In 1916 Moscow University awarded Pavlova a doctorate in zoology, *honoris causa*. About a year later, with the postrevolution integration of the Higher Courses for Women into Moscow University, she became one of the faculty as professor of paleontology, a change that inevitably brought an increase in her time spent on teaching and administration at the cost of research and writing.

Many honors, from abroad as well as in Russia, came in the 1920s. In 1921 she became a member of the Ukrainian Academy of Sciences, in 1925 a corresponding member of the division of physical science of the Academy of Sciences of the USSR, and in 1930 an honorary member; she was awarded the title Distinguished Scientist of the Russian Federation in 1928. In 1924 she became a member of the Linnean Society of Lyons, and in 1926 the French Geological Society made a joint award of its prestigious Albert Gaudry Medal to her and Aleksei Pavlov.[6] Moscow University's geological museum, which was founded by Mariia and Aleksei Pavlov, is now the A. P. and M. V. Pavlov Museum of the Moscow Geological Research Institute; its large paleontological section contains the materials collected by Pavlova that formed the basis for its fossil vertebrate exhibits.

After the death in 1929 of Aleksei Pavlov, her close colleague for more than forty years, Mariia Pavlova's spirits and energies gradually declined. Nevertheless, she continued to amass data and to publish, immersing herself in work almost up to her death from pneumonia, at the age of eighty-four, on 23 December 1938 in Moscow.[7] In 2004 the Russian Academy of Sciences, Vernadskii State Geological Museum, Moscow, celebrated the 150th anniversary of the birthdays of academician Aleksei P. Pavlov (1854–1929) and honorary academician Mariia V. Pavlov (1854–1938).[8] Mariia Pavlova was undoubtedly one of that small, select band of leading eastern European women scientists of her generation.

MARIIA KUZ'MINICHNA TSVETAEVA[9] (b. 1854), a contemporary of Mariia Pavlova, was a student and then a coworker of Moscow geologist Sergei Nikolaevich Nikitin for more than twenty years. She was never free to put all her time and effort into geological work because of her pressing need over many years to earn money to support an old, sick mother and a blind uncle; but nevertheless she built a very creditable record of field research, paleontological studies, editorial work, and service on geological committees.

Little information has been found about her family background or her childhood years. In 1872, at age eighteen, she finished gymnasium studies, and immediately after that started teaching natural science in the Fourth Moscow Gymnasium for Girls; within a few years she undertook teaching responsibilities at a second gymnasium as well. Between 1873 and 1877 she also completed the full natural history course in the physical-mathematical sciences division of the recently established Moscow Higher Courses for Women, and appears to have supplemented these studies with further instruction at the Third Moscow Gymnasium for Boys.

At the Higher Courses for Women she studied a broad spread of subjects that included chemistry, physics, and the biological sciences, as well as paleontology, geology, and mineralogy, the latter two fields taught by Sergei Nikitin. She became Nikitin's student and field assistant, working closely with him during summers when she was free from teaching. She accompanied him on field trips, going on foot through almost the whole Volga basin, studying the geological structure of central Russia and collecting paleontological material. Her structural work was incorporated into Nikitin's many papers, and her own publications were limited to paleontological reports. These appeared as two monographs discussing her large collections of nautiloids and other cephalopod fauna of the Carboniferous limestones of central Russia. Tsvetaeva's monographs, which contained a vast amount of new material, were a notable contribution to Russian paleontology and remained important references for many decades (see bibliography).

Between 1884 and 1896 Nikitin, with Tsvetaeva's collaboration, prepared and edited the lengthy Russian geological reference list *Russkaia Geologicheskaia Biblioteka: Bibliothèque Géologique de la Russie*, a serial publication brought out in the *Trudy Geologicheskago Komiteta. Mémoires du Comité Géologique*.[10] Nikitin was a notably active member of this recently established committee, and a frequent contributor to its *Mémoires*. Although she was never formally named as his editorial assistant, Tsvetaeva was thanked by Nikitin for her collaboration in the *Bibliothèque Géologique* work.[11]

Although a somewhat shy, unassuming person, Mariia Tsvetaeva was painstakingly conscientious and confident in her work. In her own time she was well known and respected in the Russian geological community, despite her lack of academic degrees. By no means antisocial, she belonged to the Moscow Naturalists' Society and the Moscow Mineralogical Society; she and Mariia Pavlova were the only two women who served on the Organizational Committee for the Seventh International Geological Congress, which met in St. Petersburg in 1897. A notable highlight of this congress was the excursion from Moscow to the Urals region, an especially popular event arranged by the Organizational Committee and one in which Tsvetaeva took part.

No indication has been found of geological work by Tsvetaeva after 1897. The death of Nikitin at the relatively early age of fifty-nine in 1909 would, at least by then, have limited her research possibilities, even if she had remained active up to that time. She has been called Russia's first woman geologist, the more specific designation of paleontologist being reserved for Mariia Pavlova.[12]

EVGENIIA VIKTOROVNA SOLOMKO,[13] later SOLOMKO-SOTIRIADIS (1862–1898), the first Russian woman to be awarded a doctoral degree in geology, was born on 7 October 1862 in the city of Yaroslavl', about 160 miles northeast of Moscow. She was taught at home during her earliest years, and then attended the Catherine Institute for Girls in Moscow. By the age of sixteen she had completed her studies there with high standing and moved on to the recently opened St. Petersburg Higher Courses for Women, where for five years (1878–1883) she concentrated on geological sciences in the physical-mathematical division. Able and enthusiastic, she became a student of the celebrated petrographer and mineralogist Aleksandr Aleksandrovich Inostrantsev. She carried out laboratory work in petrography and took part in Inostrantsev's field trips in Olonetz Province, about one hundred miles northeast of St. Petersburg, in Karelia on the Russia-Finland border, in the Caucasus, and in east-central Ukraine, where she carried out her first research, a detailed study of a volcanic rock found in Poltava Province.[14]

Directed by Inostrantsev and Pavel Nikolaevich Veniukov, a specialist in Devonian fauna, she then moved on to paleontological studies, beginning with an examination of a difficult group of extinct Devonian hydrozoa, reef-building animals (see bibliography). This was followed by an even more challenging study of a group of Jurassic and Cretaceous corals of the Crimea. Since this group had received little attention from Russian workers, she went abroad for advice and direction, first paying a brief visit to Russian-German geologist and mineralogist Constantin von Grewingk at Dorpat University. Zurich professors Albert Heim and paleontologist and stratigrapher Karl Meyer-Eyman provided assistance, as did the famous Munich paleontologist Karl von Zittel, who became her principal advisor. She studied and classified the many coral collections in the Munich Paleontological Museum and listened to von Zittel's lectures.[15] Her Zurich doctorate was awarded in July 1887, following presentation of a dissertation on her coral studies.[16]

Later the same year Evgeniia Solomko married the Greek director of a commercial college in Odessa and appears to have given up geological work. She died in 1898, at the age of thirty-seven.[17]

VERA GOLOUBTSOVA (fl. 1880s), a student at the St. Petersburg Higher Courses for Women in the mid-1880s, carried out a mineralogical study under the guidance of petrographer Frantz Levinson-Lessig. Working in the university's mineralogical laboratory, she investigated a specimen of porous, black volcanic rock in the university's mineralogical and petrographic collection. This lava rock sample, from Tounka (Tunka) at the southwest end of Lake Baikal, Irkutskaia Province, southern Siberia, was of much interest since the distant region of its origin was then largely unexplored.

Goloubtsova's microscopic study of the crystalline structure indicated that the rock belonged to a group of basaltic lavas; chemical analysis showed that as well as silicon and aluminum oxides, oxides of calcium and iron predominated. Her results were communicated to the Belgian Geological Society, Brussels, by Levinson-Lessig (see bibliography).

ARCHAEOLOGISTS AND ANTHROPOLOGISTS

A great deal of archaeological and anthropological work was being carried out in imperial Russia throughout the nineteenth century, and women were not entirely absent from the scene; a few even played important roles. Only one, V. M. Chebysheva, published reports of her work in a scientific journal indexed by the Royal Society, but at least two others—Anna Mikhailovna Raevskaia and Countess Praskovia Sergeevna Uvarova—are fairly well documented in the historical literature; their stories, sketched here, may help to provide a glimpse into a richer picture of women's work in the field.

In the summer of 1879 V. M. Chebysheva took part in excavations of Viking-age tumuli, or barrows, found in the Smolensk district. Some two hundred miles west of Moscow, Smolensk had been a key stronghold controlling portages between the Western Dvina and the Dnieper rivers, the ancient trade route connecting the Baltic in the north with Kiev and the Byzantine Empire in the south. Its tumuli form one of the largest early medieval gravefields in Russia. Chebysheva's excavations were seventy-five miles east of Smolensk, near Dorogobuzh, a town straddling the Dnieper River, at a pre-Mongol fortress defending the eastern approaches to Smolensk. Her reports appeared in two papers in the *Bulletin* of the Moscow Natural Science Society in 1886 (see bibliography).

Anna Mikhailovna Raevskaia[18] (1820–1883) had interests that ranged over a wide period of prehistory. A lieutenant-general's daughter, she had a good, private education that included tuition in mathematics from Mikhail Ostrogradskii (1801–1862), one of imperial Russia's leading mathematicians; Ostrogradskii thought well of Anna Mikhailovna's mathematical abilities. At the age of eighteen she married General Nikolai Nikolaevich Raevskii, commander of the Black Sea shoreline, but after four years of marriage and the birth of two sons she was widowed.

Already very interested in archaeology, anthropology, and ethnography, she spent five years travelling throughout Italy, visiting its rich collections of antiquities. She met many specialists in the field, Swiss and German as well as Italian, with whom she corresponded after her return to Russia about 1850. These men helped her buy archaeological, anthropological, and fossil collections; she also bought copies of famous items from European museums and private collections. Over the years she often presented these materials to Russian museums and the St. Petersburg Academy of Sciences, with which she kept in close touch.

Anna Raevskaia organized archaeological and anthropological digs in the St. Petersburg region and in the Baltic lands, the work in the latter area being supervised by her friend academician Karl Ernst von Baer, an outstanding naturalist. Despite the reluctance of Count Aleksei Uvarov, president of the Moscow Archaeological Society, she was elected a corresponding member of the society in 1872. Then about fifty-two, she already had a quarter-century of archaeological and anthropological studies behind her.

Raevskaia was well regarded by her contemporaries. Fellow anthropologist Dmitrii Anuchin described her as "a master-spirit of a woman" with "an analytical mind which is so rare among females."[19] She died in 1883, when she was about sixty-three. Her collections were donated to the Moscow Natural Science Society.

Perhaps the most celebrated woman archaeologist of late nineteenth- and early twentieth-century Russia was Countess Praskovia Sergeevna Uvarova,[20] née Princess Shcherbatova (1840–1924). The second child and oldest daughter among nine children of Prince Sergei Aleksandrovich Shcherbatov (1800–1869) and his wife, Praskovia Borisovna Sviatopolk-Chetvertinskaia (1818–1899), she was born on 9 April (28 March) 1840 in the city of Bobriki, in the Kharkov region of south-central Russia. Her earliest years were spent on the Shcherbatov country estate of Lebedianskii, but when the time came for her oldest brother, Boris, to begin his education the family moved to Moscow. Praskovia was well taught at home by a number of very competent tutors. She became fluent in several Western languages, studied art under Aleksei Savrasov, and received music lessons from Nikolai Rubinstein, a brother of the pianist Anton Rubinstein; she also read widely in Russian literature. All these subjects were considered appropriate for high-born girls of her time; science would seem to have been absent from her curriculum.

In 1859 she married Aleksei Sergeevich Uvarov (1825–1884), a son of the well-known minister of education Sergei Uvarov, past president of the Russian Academy of Science and an enthusiastic archaeologist. By the time of his marriage Aleksei Sergeevich was already a widely recognized historian and archaeologist. As was the case with Anna Raevskaia, Praskovia Uvarova's interest in antiquities was awakened by a long visit to Italy. The Uvarovs traveled extensively throughout that country just after their marriage, visiting the ancient ruins and museums, viewing private collections, and admiring the architectural treasures of the old cities. Over the course of the trip they started to develop a plan to collect the many rare valuable objects scattered here and there across their own country.

Over the next two and a half decades, until her husband's death, Praskovia Uvarova joined him in his archaeological work, serving as a very competent assistant, accompanying him on all his many scientific excursions, and helping with his research. As the principal organizer of the All-Russian Archaeological Congresses, and of the regular meetings of the Moscow Archaeological Society (the founding of which was perhaps Count Uvarov's most important achievement), Countess Prascovia became well known in the Moscow scientific community.[21] Nevertheless she was not a member of the society; in complete contradiction to his own close working collaboration with his wife, the Count had little tolerance for women entering into the realm of higher studies. Although an exception was made in 1872, when St. Petersburg-based Anna Raevskaia (see above) became a corresponding member, he was not willing to grant his wife the honor.

In 1885, shortly after her husband's death, Praskovia Uvarova was elected the society's first honorary member; three months later she became president. With official status, and the freedom of a very rich woman to follow her own interests, she devoted her time and impressive energies to the work of the society. As president of the Commission for the Conservation of Ancient Monuments, she led this body that undertook the task of preserving hundreds of important sites, priceless treasures that attracted the attention of Russia's leading archaeologists, art historians, and cultural historians. She herself traveled throughout the country, going as far south as Turkestan and the wild country along the Persian border. She led several major digs sponsored by the Archaeological Society, particularly in the Caucasus region, and also spent considerable time working on the society's collections in Moscow. In addition she organized exhibitions at appropriate times, and took a leading role in preparations for the archaeological congresses of the Aleksandr III Russian History Museum, now the State Historical Museum.[22]

She published an impressive collection of more than eighty scientific works and edited many more.[23] Elected an honorary member of the St. Petersburg Academy of Sciences in 1895 and an honorary professor at several Russian universities, including Moscow University in 1910, she received in 1916 the further honor of a grand celebration marking her half century of work in the Moscow Archaeological Society; thirty of Russia's most distinguished archaeologists attended the gathering.

After the October Revolution of 1917, Countess Uvarova moved to Yugoslavia. She died at Dobrna (near Celia, now in Slovenia) on 30 June 1924, at the age of eighty-four, in straitened circumstances. Of her seven children, born between 1859 and 1869, four outlived her.

NOTES

1. D. V. Nalivkin, *Nashi pervye zhenshchinye-geologi* [Our first women geologists] (Leningrad: Nauka, 1979), 13–36; A. A. Borisiak and V. V. Menner, "Maria Vasilevna Pavlova," *Vestnik Rossiskii Academii Nauk* [Bulletin of the Russian Academy of Sciences] 6 (1939): 78–80; Anne C. Hughes, "Pavlova, Mariia Vasil'evna (1854–1938)," in *The Biographical Dictionary of Women in Science: Pioneering Lives from Ancient Times to the Mid-20th Century*, ed. Marilyn Ogilvie and Joy Harvey (New York and London: Routledge, 2000), 2:992–93; Z. A. Bessudnova, "Pavlova, Mariia Vasil'evna (1854–1938)," in I. A. Starodubtseva, Z. A. Bessudnova, S. K. Pukhonto, et al., *Pavlovskaia Geologicheskaia Shkola* [Pavlov geological school] (Moscow: Nauka, 2004), 21–26.

2. Nalivkin, *Nashi pervye zhenshchinye-geologi*, 17.

3. Svetlana Serova, "La paléontologie en lettres: la correspondance d'Albert Gaudry avec Maria Pavlova (1888–1908)," talk presented at "Les français dans la vie intellectuele et scientifique en Russie au XIX siècle," Singer-Polignac Foundation Conference, 16–17 Sept. 2011, online at http://www.singer-polignac.org/fr/missions/551 (Aug. 2014).

4. This Seventh International Geological Congress (St. Petersburg, August–September 1897) was a notably successful event, particularly for the excursions arranged by the Organizational Committee. Moscow was the starting point for the first excursion, to the Urals, preliminary to the congress itself, and so was the first objective of excursionists after they entered the country. Of the approximately six hundred delegates, about 135 took part in the Urals excursion, which included visits to the rich mining regions on both sides of the mountains. Other excursions included visits to Finland and the Caucasus. See Charles Palache, "The Geological Congress in Russia," *American Naturalist* 31, no. 371 (1897): 951–60.

5. In an effort to improve public education, the Russian Duma passed a bill in 1908 establishing the Shaniavskii Moscow City People's University, the necessary funding being provided from the estate of the relatively liberal-minded General A. L. Shaniavskii (1837–1905); the institution functioned until 1918. Anyone over the age of

sixteen was admitted to either of its two divisions, general and higher-level. Within a few years several thousand students were enrolled, and the higher-level courses were being taught by a number of eminent professors—see *The Great Soviet Encyclopedia*, 3rd ed. (1970–1979).

6. "Tableau des Membres de la Société Linnéenne," *Annales de la Société Linnéenne de Lyons*, n.s., 71 (1924): xli; "Lauréats du Prix Albert Gaudry," *Compte rendu sommaire et Bulletin de la Société Géologique de France* 30, s. 4 (1930): 4.

7. Mariia Pavlova's many post-1900 publications include the following: "Ossements fossils trouvés dans les environs de Krinvoi Rog, gouvernment de Kherson," *Bulletin de la Société Impériale des Naturalistes de Moscou* (hereafter *Moscou Soc. Nat. Bull.*) 1 (1902): 73–98; "Protohippus en Russie," *Moscou Soc. Nat. Bull.* 2 (1903): 173–82; "Mastodon angustidens Cuv. et Mastodon cf. longirostris Kaup, de Kertch," *L'Annuaire Géologique et Minéralogique de la Russie,* Warsaw, 4, livre 6 (1903): 121–39; "Rhinoceros schleiermacheri Kraup des environs d'Ananiew," *Annuaire Géologique et Minéralogique de la Russie*, Warsaw, 7, part 5 (1905); "Revue de la littérature russe pour la paléontologie des mammifères (1901–1906)," *Annuaire Géologique et Minéralogique de la Russie*, Warsaw, 9 (1907): 139–46; "Quelques carnivores fossils du gouvernement de Kerson et de Bessarabie," *Mémoires de la Société des Naturalistes de la Nouvelle-Russie*, Odessa, 32 (1908): 1–22; "Les éléphants post-tertiares de diverses localités en Russie," *Annuaire Géologique et Minéralogique de la Russie*, Novo-Alexandria, 11 (1909): 167–74; "Les éléphants fossils de la Russie," *Nouveau Mémoires de la Société Impériale des Naturalistes de Moscow* 17, no. 2 (1910); with A. P. Pavlov, "Mammifères fossils de Russie," *Nouveau Mémoires de la Société Impériale des Naturalistes de Moscow* 17, livres 2–4 (1910–1915); with A. P. Pavlov, [Causes of extinction of animals in past geological epochs] (Moscow: Gosudartsvennoe izdatel'stvo, 1910–1919?); "Mammifères fossils du Musée de Troitzkoseavsk-Kiakhta," *Société Impériale Russe de Géographie, Travaux de la sous-section Troick-Kiacht, section du Pays d'Amour* 13 (1911): 21–59; "Les données historiques sur le développement des connaissances paléontologiques des mammifères tertiaires et post-tertiaires trouvés en Russie," *Moscou Soc. Nat. Bull., Section Géologique* 31 (1922): 117–48; "Nouvelles données scientifique sur la position de Hipparion," *Moscou Soc. Nat. Bull., Section Géologique* 32 (1924): 313–28; "Cimetière des os des mammifères post-tertiaires trouvés sur la rive gauche du Volga, entre Senguileï et Novodevitchie. 1 Camelidae," *Académie des Sciences de l'Ukraine, Mémoires de la Classe des Sciences Physiques et Mathématiques* 3, fasc. 2 (1926); "Mammifères fossils du gravier de Tirespol, gouvernement de Kerson," *Mémoires de la Section Géologique de la Société des Amis des Sciences Naturelles,* Moscow, 3 (1925): 1–74; *Paléozoologiia 1* (Moscow: Gosudartsvennoe izdatel'stva, 1927); "Mammifères post-tertiaires trouvés sur les bords du Volga près de Senguileï et quelques formes provenant d'autre localities," *Annuaire de la Société Paléontologique de Russie* 9 (1931): 1–42; "Les restes des ossements des mastodontes provenant de l'ancien gouvernement de Kerson," *Travaux de l'Institut Paléozoologique de l'Académie des Sciences de l'URSS* 1 (1932): 69–80; "*Camelopardalis parva* Weith. des couches sarmatiques du Caucase," *Travaux de l'Institut Paléozoologique de l'Académie des Sciences de l'URSS* 2 (1933): 269–71. See bibliography for early papers.

8. *Materialy nauchno-prakticheskoi konferentsii "Problemy regional'noi geologii–muzeinyi rakurs"* [Proceedings of the conference on theoretical and applied geology—"Problems in regional geology—the museum perspective"], ed. G. V. Kalabin (Moscow: Akropol', 2004).

9. Nalivkin, *Nashi pervye zhenshchinye-geologi*, 37–41; M. R. S. Creese, "Fossil Hunters, a Cave Explorer and a Rock Analyst . . . ," in *The Role of Women in the History of Geology*, ed. C. V. Burek and B. Higgs (London: Geological Society Special Publication No. 281, 2007), 38–49.

10. S. N. Nikitin, *Russkaia Geologicheskaia Biblioteka: Bibliothèque Géologique de la Russie* (St. Petersburg: Eggers; corp. author, Geologicheskii Komitet, 1885–1891).

11. *Bibliothèque Géologique de la Russie . . . , 1885–1892*, ed. S. N. Kikitin, vol. 1 (St. Petersburg, 1885), preface, v–vii, on vii.

12. Nalivkin, *Nashi pervye zhenshchinye-geologi*, 39–40.

13. Nalivkin, *Nashi pervye zhenshchinye-geologi*, 42–44; "Evgeniia Solomko—Sotiriadis," *Ezhegodnik po Geologii i Mineralogii Rossii* [Yearbook on Russian geology and mineralogy] 3, sect. 1 (1898–1899): 137; Creese, "Fossil Hunters," 46.

14. Eugénie Solomko, "Étude sur une roche crystalline, trouvée dans le village Issatehki," unbound manuscript, St. Petersburg, 1884. Listed in *Compte rendu des séances de la Société de Géographie et de la Commission Central*, no. 1 (1884): 596, following its presentation to the Paris Société de Géographie by St. Petersburg geologist Pavel Veniukov.

15. Evgenia Solomko could not attend von Zittel's lectures in the same room as his male students but instead listened from an adjoining room through an open door. About fifteen years later von Zittel accepted the Scottish geology student Maria Ogilvie Gordon as his private student. She studied with him for three years and became the first woman to be awarded a doctorate in geology by Munich University. See Mary R. S. Creese, *Ladies in the Laboratory? American and British Women in Science, 1800–1900* (Lanham, Md.: Scarecrow Press, 1998), 294–96.

16. Eugenie Solomko, "Die Jura- und Kreidekorallen der Krim," inaugural dissertation, philosophical faculty, University of Zurich, 1887. See also bibliography.

17. One might speculate that Evgeniia Solomko's husband was Georgios Sotiriadis (1852–1941), a Greek philologist, archaeologist, cartographer, and later professor at the University of Athens. Georgios Sotiriadis studied at the University of Zurich in 1883–1884, and held a doctoral degree. Early in his distinguished career he worked in Odessa and in colleges in several other cities of the region.

18. O. A. Valkova, "The Conquest of Science: Women and Science in Russia, 1860–1940," *Osiris* 23 (2008): 139–40.

19. Dmitrii N. Anuchin, "Anna Mikhailovna Raevskaia," *Izvestiia obshchestva liubitelei estestvoznanii antropologii i ethnographii* 90, no. 3 (*Trudy antropologicheskogo otdela* 18, no. 3 [1896]: 513), quoted in Valkova, "Conquest of Science," 140.

20. Valkova, "Conquest of Science," 148–49; A. I. Sobolevskii, "P. S. Uvarova. 1840–1924," *Izvestiia Rossiiskoi Akademii Nauk* 19, series 6, no. 6–8 (1925): 141–44. Uvarova's memoirs, *Byloe, davno proshedshie schastlivye dni* [The past: Happy days long gone], written after she moved to Yugoslavia following the 1917 October Revolution, was published in 2005 (Moscow: Izdatel'stvo im. Sabashnikovykh).

21. The first All-Russian Archaeological Congress was held in 1869, and these meetings continued at three-year intervals until the First World War and the revolution. They were held in cities throughout the length and breadth of the Russian Empire, with the aim of discussing and making public reports of the scientific work being carried out in archaeology and ethnography.

22. One of her daughters may have helped her in some of her exhibition-organization work, but this has not been confirmed.

23. Uvarova's publications include the following: *Katalog riznitsy Spaso-Preobrazhenskago monastyria v IAroslavie* [Catalogue of the sacristy materials in the Spaso-Preobrazhenskii monastery in Jaroslavl] (Moscow: A. I. Mamontov, 1887); *Trudy V-go arkheologicheskago siezda v Tiflisie 1881* [Proceedings of the Fifth Archaeological Congress, Tiflis, 1881] (Moscow: A. I. Mamontov, 1887); *Materialy po arkheologii Kavkaza, sobrannye ekspeditsiiami Imperatorskago Moskovskago Obshchtestva* [Material on Caucasus archaeology collected by the Moscow Archaeological Society], ed. Praskovia Uvarova, serial publication, 14 vols. (Moscow: A. I. Mamontov, 1888–1916); [Historical note on the work of the Moscow Archaeological Society during its first twenty-five years] (Moscow, 1890); [Regional museums] (Moscow, 1891); *Trudy desiatago arkheologicheskago siezda v Rigie: 1896* [Proceedings of the Tenth Archaeological Congress in Riga, 1896] (Moscow: G. Lissner and A. Geshel, 1900); [The Trieste museum] (Moscow, 1900); *Archaeologie*, vol. 5 of *Die Sammlungen des kaukasischen Museums*, ed. Gustav Radde (Tiflis: Kanzelie des Landeschefs, 1899–); *Kavkaz: Putevyia zamietki* [Caucasus: Travel notes] (Moscow: A. A. Levenson, 1887; later ed., 1904); [The enamels of the Poretskii museum] (Moscow, 1909); [The protection of ancient monuments] (Moscow, 1914); [The architecture of the southwest provinces] (1920).

Chapter Six

Contributors from Imperial Russia's Western Lands

Polish Women

The borders of Poland have changed many times over the centuries and by the early 1790s, after a particularly chaotic period in the region's history, Poland was partitioned among its three powerful neighbors, Austria-Hungary, Prussia, and Russia. Throughout the upheavals of the Napoleonic Wars these three basic divisions remained, although their borders underwent several changes and adjustments. One major change was the westward expansion of the zone held by imperial Russia to include the region known as Kingdom of Poland or Congress Poland, with its capital city, Warsaw (see fig. 5).

Figure 5. Imperial Russia's western lands, showing the regions of partitioned Poland.

Locations of Poland's leading educational centers changed over the years, but by the last decades of the nineteenth century Warsaw was especially prominent and was the city with which most of the women scientists discussed here were associated. Over the course of the century measures were taken by each of the occupying powers to replace Polish cultural influences and the Polish educational system, including language of instruction, with their own. Warsaw University, for many years the Russian University of Warsaw, began to function as a Polish institution about the end of the First World War.[1]

Nineteenth-century Warsaw was by no means lacking in its strongly patriotic families, usually members of the upper classes and nobility from whom most of the early women scientists were drawn. The city had many illegal underground organizations, including private, fee-paying schools where the language of instruction and the predominating cultural influences were Polish. Among these private schools were several boarding institutions for girls that provided secondary education, although frequently not to university-entrance level. Not until about 1900 were women able to enroll at universities in Polish lands, but as early as 1871 a Polish woman became a student at the University of Zurich, and by the 1880s, a time of increasingly severe tension between the Polish upper and middle classes and the St. Petersburg government, many others were going to Switzerland for higher education.

As was the case in the rest of the empire, science was the most admired area of higher education, particularly pure science and research, and when they finally had access to universities, women as much as men gravitated to scientific training rather than to the liberal arts. Maria Sklodowska-Curie dominates any picture of late nineteenth- and early twentieth-century Polish women scientists, to such an extent that others have languished in the shadows,[2] but the contributions of a further twenty-one women from this period identifiable as being of Polish background are listed in the Royal Society's *Catalogue*. Of the pre-1901 papers they authored, more than half reported work in the biological sciences, carried out in western European laboratories; like the medical women in other parts of the empire, they concentrated on neurology and physiology (see fig. 3).

Among those who published a notable body of research in these areas, two of the most prominent were Jósefa Joteyko and Michalina Stefanowska. Both acquired their university education in Geneva in the late 1880s and then spent several years at the Solvay Institute in Brussels, where they collaborated fairly extensively in the newer areas of neurological research. Both spent several decades in the West and then went back to Poland for the final stages of their working lives, Stefanowska more than Joteyko succeeding in reintegrating successfully into Polish scientific society.

Jósefa (Joséphine) Franciszka Joteyko[3] (1866–1928), the oldest of four children of landowner Lucjan Joteyko and his wife, Karolina, née Odrowaz-Kurzanska, was born on 29 January 1866 at her family's large estate of Poczujki, near Kiev, Ukraine. With her sister, Zofia, and her brothers, Mieczysław and Tadeusz, she spent her earliest years in the pleasant wooded surroundings of Poczujki, but about 1873 the Joteykos, members of Poland's intellectual elite, leased the estate and moved to the capital city.[4]

Educated privately in her earliest years, Jósefa, along with her sister, Zofia, attended the Sierakowskiej boarding school, one of Warsaw's secondary schools for girls. However, her parents having decided that the science instruction provided was unsatisfactory, she was withdrawn after a few months and then tutored privately. By 1884 she had reached university-entrance level in science, spoke four languages, and had acquired a strongly patriotic outlook. Not having met the Russian-language requirements for certification as a private teacher, and since women were not admitted to Warsaw University, she, like others of her generation, decided to continue her education abroad. With the encouragement of her mother, a forward-looking woman, she went to Geneva, the home since the 1848 revolution of her mother's uncle, the Polish writer Zygmunt Milkowski, who published under the name Teodor Tomasz Jez. She enrolled at Geneva University in 1886 for studies in a range of physical and biological sciences and was awarded a bachelor's degree in 1888. The following year she moved to the Free University of Brussels to begin medical studies, which she continued at the University of Paris faculty of medicine. Her University of Paris MD was awarded in 1896.

The three western universities where Joteyko spent ten years as a student had given her not only a scientific and professional training, but thanks to their generally liberal atmosphere, very different from that in Poland, had introduced her to new ideas in democracy and social progress. She enthusiastically embraced the view that advances in the social sciences (including psychology and pedology), and then in society itself, would be achieved only if grounded in rationalism based on quantitative experimental work; that is to say, by the same methods as those that were being used in the physical and biological sciences. Her doctoral studies, carried out under the guidance of Paris physiologist Charles Richet, concerned muscle exhaustion and oxygen requirements during recovery, work that laid the foundation for her future research on the effects of fatigue, both physical and mental, on child development. A brief note, published jointly with Richet, summarized the findings discussed in her 1896 dissertation, "La fatigue et la respiration élémentaire du muscle." This was promptly followed by several other short papers on muscle and nerve-center fatigue, work that was reported more fully in her seventy-two page book, *Recherches sur la fatigue névro-musculaire* . . . .[5]

Joteyko had a private medical practice in the Paris suburb of Neuilly for two years after graduating. By 1898, however, having decided to concentrate on research, she moved to the Kasimir Laboratory of psycho-physiology at the Free University of Brussels, where she led research until 1914 while also working at the Solvay Institute of Physiology, Brussels, during much of the same period.[6] From 1906 until 1914 she taught pedology (the physiological and psychological study of the child) at training colleges for women teachers in Mons and Charleroi, cities to the south of Brussels. Beginning about 1901 she held a lectureship in experimental psychology at the Free University of Brussels.

Her vigorous efforts to promote the scientific investigation of early childhood development continued with the establishment of the Brussels International Faculty of Pedology, which she directed until 1914, and the founding of the Brussels *Revue Psychologique*, which she edited from 1912 to 1914. With the outbreak of the First World War she moved to Paris and served for a time as lecturer in experimental psychology at the Collège de France; at the same time she continued her joint research with her Polish student Maria Grzegorzewska, then enrolled at the Sorbonne. In 1916 she and Grzegorzewska, later a pioneer in special education, founded the Polish League of Education.[7]

Over the course of about three decades, from her graduation in 1896 on, Joteyko brought out some 260 notes, papers, and books, an impressive publication record. These discussed the two subjects that were her special interests, the investigation of fatigue, both physical and mental, and child development. Most of her reports were of single authorship, although she also had a productive, decadelong collaboration with her compatriot Michalina Stefanowska (see below), a neurologist and physiologist who worked as a colleague at the Solvay Institute for a number of years. Their joint work resulted in close to twenty coauthored publications, including their 1909 book, *Psycho-physiologie de la doleur* (Psycho-physiology of pain).[8]

Joteyko's results from her studies on the behavior of muscles and nerve centers, including oxygen requirements, appeared in leading French and Belgian journals over more than two decades. The work was largely quantitative, several techniques and types of measuring equipment being by then available, although the understanding of muscle physiology was still at an early stage.[9] Other studies included her work, much of it carried out jointly with Stefanowska, on anesthesia of nerve centers and the effects of various chemicals and intermittent electrical current on muscles.[10] About this time she also contributed an article on fatigue to the *Dictionnaire de Psychologie*, brought out in 1903 under the editorship of her former research director, Charles Richet; a further study on fatigue in military training appeared as a hundred-page report in a Solvay Institute of Sociology series in 1905.[11]

Much interested in the question of vegetarian diet, she and one of her students, Varia Kipiani, carried out a detailed study using a number of quantitative methods on the effects of a vegetarian diet on muscle strength and endurance. Their book, *Enquête scientifique sur les végétariens de Bruxelles* (1907), aroused considerable interest in Britain and the United States as well as on the Continent, vegetarianism being a matter of much discussion and research among leading physiologists and physiological chemists at that time. Joteyko and Kipiani found that while there was little difference in strength between groups of vegetarians and carnivores, vegetarians surpassed the carnivores by a wide margin in endurance. Joteyko and Kipiani's conclusion that a vegetarian diet would reduce fatigue and therefore frequency of accidents was also in agreement with those of several other research workers. The book was very favorably reviewed by the well-known economist and health campaigner Irving Fisher in the American journal *Science* in 1907. Joteyko's paper at the 1910 World Vegetarian Congress in Brussels presented her findings from a study of 170 child vegetarians.[12]

By 1909 Joteyko was studying the effects on child development of excessive schoolwork, a subject she discussed further in her 1922 report on the results of experimental research into the intellectual level of students in the first three years of lycée education.[13] Her 1917 work on the science of labor and its organization brought together in one volume a series of articles published in 1916 and 1917 in the *Revue Psychologique* and other French-language journals; it was republished in English in Britain and the United States in 1919 and reissued in 1974.[14]

A prominent figure in her field, she regularly participated in conferences and was a member of several leading biological and chemical societies, including the Société Belge de Neurologie, in which she served as vice president in 1904–1905 and president in 1905–1906. Honors included *lauréat* of the Académie des Sciences, Paris, and of the Académie Royale de Médecine, Brussels, and the Lallemand Prize (Académie des Sciences, Paris), shared with Garnier and Gololiau in 1903. A member of the Académie Royale de Médecine de Belgique and a corresponding member of the Société Royale des Sciences Médicales et Naturelles de Bruxelles, she served as presiding officer at the First Belgian Congress of Psychiatry in Liège in 1905.

Throughout her more than thirty-year career in France and Belgium she had kept in fairly close contact with scientific developments in Poland, and with the end of the First World War and the establishing of Polish independence, she returned to Warsaw. The country's long period of division into three political and administrative sectors had led to three distinct systems of education. Eager to contribute to the work of reconstructing Polish educational theory and practice, Joteko sought to assist in the modernization of ideas on child development and educational theory by incorporating new Western findings, particularly in such areas as experimental psychology and pedology. For a few years she held lectureships in general psychology and pedology at the Free University of Warsaw and also taught pedagogical psychology in the faculty of medicine of Warsaw State University and at the State Institute of Pedology. In 1927 she founded the quarterly serial *Archives polonaises de la Psychologie (Polskie Archiwum Psychologii)*, several volumes of which included a section on pedological psychology. Publication continued until 1939, Joteyko serving as editor of the first two volumes.

She was unable to establish a Polish scientific school under her direction, but with her students and collaborators she did begin a relatively short-lived movement that enriched the discussions then going on in the country's pedagogical community. The time was one of considerable struggle between research scientists and party political men; Joteyko's Western outlook was in conflict with the emerging Soviet psychology and pedology of the 1920s.

One of the most distinguished of Poland's women physician-scientists who completed her training before the end of the nineteenth century, Jósefa Joteyko was a pioneer in the field of developmental psychology and a vigorous supporter of research in pedology, as well as a physiologist and dedicated experimental psychologist. She died in Warsaw, in somewhat straitened circumstances, on 24 April 1928, at the age of sixty-two.[15]

Michalina (Micheline) Stefanowska[16] (1855–1942), physiologist, neurologist, and plant physiologist, the daughter of Ferdinand Stefanowski and his wife, Joanna, née Sienkiewiczów, was born on 20 November 1855 in Grodno, about 160 miles northeast of Warsaw. After completing gymnasium studies in 1872 she stayed on in Grodno as a teacher of natural science and geography in a junior gymnasium for girls; this was followed by a similar post in a girls' junior gymnasium in Lodz, then one of western Poland's fast-growing cities.

In the mid 1880s, when she was about thirty, Michalina Stefanowska enrolled at the University of Geneva, where she studied biology, concentrating on zoology. Well known throughout Europe for the solid reputation it had already established in developmental biology, Geneva was the location of several eminent naturalists, including Carl Vogt (1817–1895) and Hermann Fol (1845–1892). Stefanowska's research project was carried out under Fol's direction with further assistance from Emil Frey-Gessner, entomologist and conservator of the Geneva Natural History Museum. A detailed investigation of the physiological effects of variation of light intensity on the eyes of arthropods and insects, it was published in 1892 (see bibliography). The study won Stefanowska the university's Davy Prize.[17] Her doctorate in biological science was awarded in 1889.

She continued her studies in Paris during the period 1891–1897, expanding her area of interest to include the developing and very active field of experimental psychology. Then followed nine years of research (1897–1906) at the Solvay Institute of Physiology, Brussels, and at the Free University of Brussels, where she carried out an impressive amount of research that resulted in many publications. By

1903 she had more than a decade of advanced study and research experience behind her, recognized by two noteworthy awards.[18] That year she accepted a second position, that of *Privatdozent* in physiology at the University of Geneva.

Her publications between 1897 and 1900, many brought out in the Solvay Institute's *Travaux*, reported her investigations of nerve-cell organization, the mechanism of neurotransmission, and the effect of anesthesia and electrical stimulation on cortical circuits.[19] A number of her early papers resulted from joint studies with her compatriot physiologist Jósefa Joteyko, also then at the University of Brussels;[20] particularly noteworthy were their monographs, *Recherches algésimetriques* (Studies on the measurement of pain, 1903) and *Psycho-physiologie de la douleur* (Psychology and physiology of pain, 1909).

On taking up her appointment in Geneva in 1903 Stefanowska joined in research at the university's Botanical Institute and at the École Cantonale d'Horticulture, both places of much activity in plant-physiology studies. Her work, focused on weight gain in plants, animals, and fungi, was reported in about ten communications, mainly in the Paris Académie des Sciences *Comptes rendus*, between 1903 and 1905.[21]

In 1908, when she was about fifty-three, she went back to Poland and for four years held a lectureship in physiology and neurology in the humanities faculty of one of the Wydz higher education institutions in Warsaw. From 1912 until 1917, throughout most of the First World War and the German occupation, she headed the Orzeszkowa Gymnasium for girls in Lodz, about eighty miles southwest of Warsaw, while also lecturing in physiology and neurology at the University of Poznan. In 1922 her Poznan appointment was changed to assistant professor of physiology and neurology, and the following year she became professor of anthropology, a position she held until 1939.

Michalina Stefanowska was the second woman, after Maria Sklodowska-Curie, to become a member of the Polish Academy of Learning (Polish Academy of Arts and Sciences); she joined as a corresponding member in 1931. Other honors included membership in the Poznan Society of Friends of Learning.

Remembered mainly as a pioneering neurologist, physiologist, and experimental psychologist, she came, through her work in psychoneurology, to a consideration of the phenomenon of hypnotism. She saw the new and rapidly growing branch of animal-behavior studies, a field in which hypnotism could have a role, as offering an ethological approach to human behavior.[22]

With interests that ranged widely across the life sciences and science in general, her years of experience in western Europe, and her background in both teaching and research, she recognized the ever-growing educational value of the natural sciences. Joining in the effort to introduce up-to-date Western scientific ideas to the general public as early as the mid-1890s, she prepared translations into Polish of important works by French scientists. The first of these was the book *Qu'est-ce que le ciel?*, by the astronomer, popularizer of astronomy, and prolific author Camille Flammarion (1842–1925); over the years this much-admired work was translated into a great many languages, including Hindi in 1931 and Chinese in 1974. Her translation of the comprehensive work on physical geography by Élisée Reclus (1830–1905) appeared in 1894. Additional publications directed at the general reader included her book on marine life, which long remained popular, and a short description of insect life in bee colonies.[23]

One of Poland's notable contributors to the biological sciences in the late nineteenth and early twentieth centuries, Michalina Stefanowska lived through turbulent times when her country suffered successive foreign invasions. She died in Cracow on 15 December 1942, at the age of eighty-seven, during a very difficult period when Poland was once more caught up in heavy fighting as the Second World War continued. In the summer of 2009, as part of its celebration of the 450th anniversary of its founding, the University of Geneva arranged a display on the outside walls of one of its buildings of portraits of about one hundred of its former lecturers. Among those portraits, some of the very famous and some of the almost forgotten, was Michalina Stefanowska's.[24]

Wanda Szczawinska[25] (1866–1944) was the youngest of three sisters, all of whom were prominent in the work of improving living conditions of women and children in the poor sections of Polish society in the early years of the twentieth century, the period after the First World War when Poland was once again an independent state.

Following the example of her sister Jadwiga, Wanda Szczawinska enrolled at the University of Geneva in the mid-1880s. Studying under the direction of zoologist Carl Vogt (1817–1898), she extended the investigations of her compatriot Michalina Stefanowska on the movement of pigment granules in arthropod eyes caused by changes in light conditions. Stefanowska had focused on arthropods and insects (see section on Stefanowska, above); Szczawinska, using the same procedures, examined Crustacea and three spiders not examined by Stefanowska. These were the small crustaceans *Grammarus* (sometimes designated "shrimp") and *Astacus fluviatilis* Fabricus (*Astacus astacus* Linnaeus, the European crayfish), and the spiders *Lycosa hortensis*, *Epeira diadema*, or *Aranea diadema,* known as the "cross spider," and *Linyphia triangularis*, one of many "money spiders." Her findings, which formed her dissertation for a doctorate in natural sciences, awarded in 1891, were published in *Archives de Biologie* in 1890 (see bibliography) and in booklet form in 1891.[26]

She then moved to France and began studies on the nerve systems of cartilaginous fishes, working under the guidance of the distinguished biologist Yves Delage of the Sorbonne and the Roscoff Biological Station in Brittany; Henri de Lacaze-Duthier, also of the Sorbonne and the Roscoff laboratory, and Jean Joyeux-Laffuie, professor at the University of Caen, provided her with materials.[27]

Szczawinska examined the nerve systems of fishes of the genus *Raja*: *Raja clavata*, the Thornback ray or skate, and *Raja punctata*, the Atlantic starry skate (or ray), also found in the Mediterranean. For contrast she studied the nerve system of one modern bony fish, *Cyprinus carpio*, the common carp; results were published in *Archives de Biologie* in 1898. She also brought out a review of recently published research on nerve-cell structure in 1896 (see bibliography).

While pursuing these neurological research interests Szczawinska was also fulfilling requirements for qualification as a physician. Under the guidance of Professor Grancher she worked in Paris clinics specializing in infancy and childhood diseases of the digestive tract; she quickly realized that correct early assistance in this area provided long-term benefit. In 1902, when she was about thirty-six, she was awarded an MD by the University of Paris, after presentation of a thesis on cytotoxic serums, that is, serums made from the blood of laboratory animals that can cause death.[28]

She stayed in Paris for another nine years, taking a prominent part in the provision of infant care, the cause she championed with vigor and persistence after she returned to Warsaw in 1911. In 1912, after her series of lectures under the general title "Child Health," she organized the Baron Lenval Paradin child health institute, with an outpatient clinic that provided medical examinations for infants, and she also took on the leadership of the Kropla Mleka (Drop of Milk), an organization until then run by the Sisters of Mercy. In 1914, largely as the result of her efforts, the widely known Drop of Milk Company was founded. Staffed by women doctors, this company was active until the outbreak of the Second World War, providing practical and material help for poor children and their families; Wanda Szczawinska led and managed the company throughout its existence.[29] During the First World War years she also worked in Warsaw hospitals, substituting for male doctors called for military service. The Drop of Milk Company was a much-respected organization that provided vital assistance to a great many needy children in their most vulnerable years; its value was recognized by the award of a gold medal at the Twelfth Congress of Polish Naturalists and Physicians in 1925. The present Polish Mothers and Children Institute (Instytut Matki i Dziecka), founded in 1951, has its roots in Kropla Mleka.

Active in women doctors' organizations from her return to Warsaw in 1911, Wanda Szczawinska became the first secretary of the Polish section of the Medical Women's International Association when that was founded in 1924. In 1926 she held the office of president. She died in 1944, when she was about seventy-eight, lonely and forgotten, in a home for the elderly in Warsaw, a city then caught in the chaos of the final months of German occupation near the end of the Second World War. She has been described as the last of the women doctors who undertook the task of serving the poorer sections of the Warsaw community and the first to fight for the professional rights of women doctors.[30]

Rozalia Nusbaum-Hilarowicz,[31] née Głębocka (1859–1933), was born in Warsaw on 24 October 1859. She studied biological sciences at the University of Geneva for a time and then returned to Warsaw, where she taught in Polish underground schools for several years. In 1886 she married zoologist Jósef Nusbaum, one of the leading figures in the Warsaw Floating University, an underground enterprise that operated during the period of Russian control of the region. In 1892 Jósef Nusbaum joined the teaching staff at the University of Lvov and Rozalia moved with him to Lvov. There she continued her education, attending lectures in philosophy while also teaching science and mathematics in girls' gymnasia. About 1907 Nusbaum changed his name to Nusbaum-Hilarowicz, and Rozalia's son, born in 1890, was known as Henryk Hilarowicz.

She published a number of critical essays discussing ideas and developments in physiology, psychology, philosophy, and pedagogy. Her lengthy paper in 1898 in the Polish journal *Kosmos* (see bibliography) examined the hypothesis put forward by Italian physician and criminologist Cesare Lombroso (1835–1909) that those who commit serious offenses are sick persons and should not be designated criminals. She discussed Lombroso's ideas in the light of recent neuro-anatomical research, an area in which there was much activity about that time, particularly in centers such as the Paris teaching hospitals.

Rozalia Nusbaum-Hilarowicz died in Lvov on 1 February 1933, in her seventy-fourth year, having outlived her husband by sixteen years. Her son, Henryk, a surgeon, was put to death in 1941, one of many Polish professionals and academics who suffered this fate during the period of German occupation.

Little has been uncovered about the background and life story of Z. GRUZEWSKA,[32] later Mme GATIN-GRUZEWSKA (fl. 1899–1940). Evidence available so far suggests that she did not continue formal academic studies beyond the *licence* ès *sciences* level, which is approximately a master's degree. Nevertheless, in her long career in biochemical research she often worked with senior scientists and also brought out a substantial number of research reports under her name only. Over the years she clearly acquired considerable expertise in research in physiological chemistry and biochemistry, but despite her impressive publication record, she remains a relatively unrecognized figure in the history of Polish women scientists of the early twentieth century. Not belonging to the group of women physicians and graduate medical scientists now becoming increasingly well known to historians, and without a published doctoral dissertation, she would seem to have fallen through the main catch nets used by historians; her name has been found only in the technical literature.

Sometime between her first publication in 1899 (see bibliography) and her second in 1903, she married, becoming Mme Gatin-Gruzewska.[33] After the four-year pause between her first and second papers she published steadily, often bringing out several reports, either joint or single-author, until the start of the First World War. After the disruption this brought, her contributions reappeared in the pages of the biological journals, continuing until the late 1930s.

In 1903 she worked for a time in the physiological laboratory of Bonn University, beginning a series of studies on glycogen (the "storage depot" of polysaccharide nutrient reserves in animal cells). Continued the following year at the physical-chemical laboratory and the Chemical Institute of the University of Göttingen, these studies gave rise to several publications, including a lengthy discussion on pure glycogen, a report on her molecular-weight determination for the compound, and a joint paper with colloid chemist Wilhelm Biltz on an ultramicroscopic study of pure glycogen in solution.[34]

From 1904 her place of work was the physiological laboratory at the Sorbonne. During the nine prewar years, as well as extending her glycogen work, she carried out a number of investigations on the hydrolysis of plant starches; much of this research was published under her name only.[35] In 1907 she was proposed for membership in the Société Chimique de France.[36]

Later work included studies on the mucilaginous material extracted from the seaweed *Laminaria flexicaulis*, acidity determinations of mineral waters, and research on gastric secretions carried out jointly with Paul Carnot, professor of clinical medicine at the University of Paris.[37] Overall she published about seventy papers and notes.

STANISLAWA POPLAWSKA, Mme,[38] was born in Warsaw in 1862. After completing studies at Gymnasium IV in Warsaw, she went to Switzerland for medical training, enrolling at Zurich University in the winter semester of 1885. While there she carried out investigations on eye injuries at the university eye clinic under the direction of Zurich ophthalmologist Otto Haab. Her work, microscopic examinations of injured eye tissue and studies of resulting inflammatory conditions, was reported in two papers published in 1890 and 1892 (see bibliography); one of these was accepted as an inaugural dissertation by Bern University, which awarded her an MD in 1890. In the early 1890s she spent some time in the small, recently established Zurich laboratory of Russian-Swiss neuropathologist Constantin von Monakow, a facility in which there was very considerable activity at that time and which later developed into the Zurich University Brain Research Institute. She and neurologist Ol'ga Leonova (see chapter 1) were the first two women to carry out research in that laboratory.

After returning to Warsaw she worked under the direction of ophthalmologist Zygmunt Kramsztyk, man of letters and a leading activist in the Warsaw medical community. In 1902, on the opening of the Starozakonnych Hospital in Warsaw, Kramsztyk became its head doctor and director; patterned after modern hospitals and clinics in western European countries, it was the most up-to-date medical facility in the city.[39] Stanislawa Poplawska's appointment as head of the Department of Ophthalmology in 1907 was the first appointment of a female physician to a hospital staff position in Warsaw.

Active in social work and women's rights organizations, she was president of the Polish Association for Equal Rights for Women, and spoke out strongly on such matters as the necessity of equal access to education and equal pay for women.

Anna Tomaszewicz-Dobrska[40] (1854–1918), physician and social activist, was one of at least two daughters of Wladislav Tomaszewicz, who served with the rank of colonel in the Russian police force, and his wife, Jadwiga, a daughter of the Polish noble family Kołaczkowskich. Despite Jadwiga Tomaszewicz's heritage on her mother's side, the background and choice of profession of Wladislav Tomaszewicz meant that his family was subjected to extreme social ostracism by the Polish community, a circumstance thought to have contributed to Anna's determination to become a doctor and establish her own separate identity.[41]

She was born in Mława, about 112 miles north-northwest of Warsaw, on 13 April 1854, but moved with her family to Warsaw in 1866. In 1869, at age fifteen, she graduated with honors from the Paszkiewiczowej school, one of many private secondary schools for girls in Warsaw at the time. Having convinced her parents of her determination to study medicine at the University of Zurich, which had begun to accept women students in 1867, she continued studies in physics, chemistry, Latin, French, and German for two more years, guided by private tutors who brought her to university-entrance level. She enrolled in the medical faculty at the University of Zurich in the winter semester of 1871, becoming the first Polish woman and the fifteenth woman overall to take and complete the medical course.[42] While still an undergraduate she carried out research in the laboratory of neurologist and neuropsychiatrist Eduard Hitzig on the effects of trichloracetaldehyde and trichloracetic acid on the nerve system. The work, during the course of which she contracted typhus, was reported in a fairly lengthy paper in Pflüger's *Archiv für die gesammte Physiologie* in 1874 (see bibliography). Her doctoral research, done under the guidance of physiologist and speech scientist Ludimar Hermann, concerned aspects of the physiology of the inner ear.[43] After receiving her degree in August 1877 she had short periods of further study in Vienna and Berlin before returning to Warsaw, where she expected to begin practicing as a physician.

The reception of Poland's first female physician by the Warsaw Medical Society was less than enthusiastic. Despite the favorable opinion of Professor Henryk Hoyer, the society's president and the country's leading biologist, who examined her credentials, the members voted against her admission to the society by a slim margin, thus denying her the right to practice. There remained the possibility of having her diploma certified by the St. Petersburg Medical-Surgical Society, which would give her the right to practice throughout Russia, including the Russian region of Poland. However, when she arrived in St. Petersburg, she discovered that the proclamation issued by Tsar Aleksandr II in June 1873 prohibiting further enrollment of Russian women at Zurich University and denying future graduates the right to practice in Russia applied equally to her. A way around these difficulties appears to have come by chance. The sultan of Turkey, a friend of Tsar Aleksandr, was soon to spend some time in St. Petersburg and would be accompanied by his harem, whose members would require the services of a female physician fluent in French, German, and English. Anna Tomaszewicz being the only medically qualified woman available who met these requirements, the authorities decided to remain blind to the awkward place of origin of her diploma, allowed her to take the St. Petersburg Medical-Surgical Academy's final examinations, and issued the necessary certification. This also meant that she could practice in Warsaw.

In 1881 she married Konrad Dobrski (1849–1915), one of Warsaw's notable late nineteenth- and early twentieth-century physicians and social activists. Dobrski was also a well-known writer, translator, and, from 1877 until 1880, coeditor of the Polish public health journal *Zdrowie* (Health).

Anna Tomaszewicz-Dobrska's private practice in gynecology and pediatrics was small at first but with time expanded considerably. Latterly she was one of Warsaw's most popular physicians, drawing patients from all levels of society; with the indigent poor she often waived her fees. In 1882 she became physician in charge of one of the recently established, largely philanthropic maternity centers for poor women. A primitive facility in a woefully inadequate building, it improved somewhat when moved to a better site in 1889. Tomaszewicz-Dobrska remained the head of this clinic until it closed in 1911, and over the course of her three decades there was able to bring about remarkable improvements in professional standards. There is little doubt that these changes were furthered by her program of lectures for midwives. She is reported to have published a full and detailed account of the work of the clinic and shelter-house after its 1911 closing, but the existence of this publication has not, so far, been verified.[44] Additional medical work included her service as physician to a Warsaw hostel for women teachers. Several charitable organizations also benefited from her considerable energy and commitment, among them the Association of Summer Camps for Children; she is said to have carried out a research project on the value of these camps, but the work remains unpublished.[45] The last decades of the nineteenth century and the opening years of the twentieth saw great cultural and political ferment in eastern Europe, and Anna Tomaszewicz-Dobrska along with her husband Konrad Dobrski was much involved in developments in Poland. Very active in the Polish Culture Society from its founding in 1906 by writer and philosopher Aleksander Świętochowski, she served on its central board until 1913, when it ceased to exist. Among her friends and correspondents was the celebrated Polish novelist and essayist Eliza Orzeszkowa (1841–1910), whose many powerful works dealt with conditions across the whole range of Polish society. A champion of equal rights for women, Tomaszewicz-Dobrska helped organize the First Assembly of Polish Women and served as its vice president. She was also a strong supporter of the Polish Drop of Milk Company, Kropla Mleka, which established clinics to help needy women and their children (see also the discussion of Wanda Szczawinska, above).

Anna Tomaszewicz-Dobrska had one child, her son Ignacius.[46] She died in Warsaw, at the age of sixty-four, on 13 June 1918, shortly before Poland gained its independence.[47]

Only minimal information has been collected about the remaining two Polish medical women in this group, Nathalia Kleindienst and Eugénie de Przedniewicz.

Nathalia Kleindienst,[48] the daughter of Frau M. Kleindienst, was born in Warsaw in 1867. After secondary schooling at a Warsaw gymnasium she went to Switzerland for medical studies, and enrolled at the University of Zurich in the spring semester of 1887. Later she moved to the Institute for Experimental Pharmacology in Bern, where she spent four semesters. Working mainly under the direction of Rudolph Demme, professor of pharmacology, and surgery professor Charles Girard, she studied the use of amyl alcohol as an anesthetic. Although the possible use of this compound in dentistry had been investigated some thirty-five years earlier, interest had declined. Using both warm-blooded and cold-blooded animals, Nathalia Kleindienst examined its effect on various vital signs. She received her doctoral degree (pharmacology/medicinal chemistry) in 1892 after presenting her results, which were published in the *Deutsche Zeitschrift für Chirurgie* (see bibliography).[49]

Eugénie Antoine de Przedniewicz, Mlle (fl. 1890s–1900) received an MD from the University of Paris faculty of medicine in 1896, after completing her research on the subject "infection and symmetry."[50] A second study by her, on the role of phosphorus in nutrition, was presented at the Thirteenth International Congress of Medicine in Paris in August 1900 (see bibliography).

Papers in the natural sciences by four late nineteenth-century Polish women are listed in the bibliography; three were botanists—Gabriela Balicka-Iwanowska, Mathilde Goldflus, and Marie Maliniak—and one was an entomologist, Mariia Ivanovna Pavlova.

Gabriele Iwanowska,[51] later Balicka-Iwanowska (1871–1962), was born in Warsaw. She studied under the direction of Robert Chodat at the University of Geneva's Botanical Institute in the late 1880s and early 1890s, carrying out a number of investigations concerning the systematics of the genus *Iris* and its close relatives. This research, based on anatomical work, appeared in a series of joint reports with Chodat between 1890 and 1892; her doctoral dissertation (1893) presented a full account.[52]

While in Chodat's laboratory she undertook at least two additional studies, one on marine algae, a special interest of Chodat, and the other on plants of the genus *Tremandra* of the Tremandraceae family, rare flowering shrubs from western Australia. The results of her structure study of blue algae, members of the division Rhodophyta collected from the Mediterranean coast, were published in 1895 (see bibliography).

In 1891 Gabriele Iwanowska married Zygmunt Balicki (1858–1916), a fellow student who later became a well-known figure in Polish political movements. Balicki studied law and social sciences at the Universities of Zurich and Geneva, and received a doctoral degree from the University of Geneva in 1896.

About that time Gabriele Balicka-Iwanowska moved to the Institute for Plant Physiology in Munich, where, in collaboration with distinguished morphologist Karl Gobel, she continued her plant classification studies using materials from the Munich botanical garden. In 1897 she reported on the morphology of the unusual annual *Thelygonum cynocrambe* L.; a prostrate plant native to Mediterranean regions, and one of only three species in the *Thelygonum* (now *Theligonum*) genus, its classification was then undetermined. An additional paper, which concerned a group of gamopetalous or fused flower-petal plants, appeared two years later (see bibliography).

In 1898 Zygmunt Balicki returned to Poland, where he became very active in political life, taking part in the founding of the National League and the National Democratic Party. Most likely Gabriele also went back to Poland about then.

For some years she continued to pursue her research interests in the laboratory of the distinguished plant physiologist and agrochemist Emil Godlewski Sr. at the Jagellonian University in Cracow; she also taught plant physiology at the University of Warsaw. Undoubtedly influenced by her husband's political and sociological activities, she took a special interest in educational issues; after the First World War she served as a delegate to the lower house of the Polish parliament.

Gabriele Balicka-Iwanowska's work covered a wide range of plant science, including anatomy, cytology, and physiology as well as the role of minerals in plant growth and plant protein metabolism—a creditable record for a woman of her time. She died in Cracow on 20 February 1962, when she was about ninety-one.

MATHILDE GOLDFLUS[53] (fl. 1890s), another student of Robert Chodat, was awarded a natural science doctorate from the University of Geneva in 1898. As well as her dissertation research on the structure and function of epithelial cells in plants of the Compositae family (see bibliography), she carried out joint studies with Chodat on the ovule of composite plants, and with Chodat and Swiss-Italian algologist Agilulf Preda on culture and development of blue algae (see bibliography).

MARIE MALINIAK[54] (fl. 1890s) carried out investigations in the botanical laboratory of the University of Warsaw in the late 1890s. She may also have been associated with the Jagellonian University in Cracow. Her research report on the formation of plant proteins was published in 1900 (see bibliography).

Entomologist MARIIA IVANOVNA PAVLOVA[55] (1867–1897) graduated from Warsaw Gymnasium III and then attended the Higher Courses for Women in St. Petersburg; under the guidance of zoologist Solomon Herzenstein, custodian of the zoological museum of the Imperial Academy of Sciences, she began the studies of insect anatomy that were to continue for the rest of her short life.

After completing the required training, she returned to Warsaw Gymnasium III as teacher of natural science and mathematics and, despite worsening pulmonary difficulties, at the same time continued entomological research at the Zoological Institute of Warsaw University, setting herself an ambitious program of demanding work. The detailed and wide-ranging discussion of insect blood-vessel and nerve systems she published in 1895 was described as a model presentation, especially notable for her own outstanding illustrations; illness prevented completion of her planned second part.[56] She also brought out two short papers on insect blood and nerve circulation in *Zoologischer Anzeiger* in 1895, and a longer work on the embryology and development of the *Mantidae* in 1896 (see bibliography). By then worsening health had caused her to give up her teaching position and leave Warsaw. She spent the winter of 1896–1897 at Mentone, hoping that the Mediterranean climate would reduce the distress caused by her lung disease. She died in Paris on 26 June 1897, when she was about thirty.

Of the six nineteenth-century Polish women contributors to research in chemistry noted here, two were students of Friedrich Kehrmann at the University of Geneva, one worked with Arthur Hantzsch and Heinrich Goldschmidt at Zurich Polytechnicum, two trained in Poland, and one in Leipzig. Only minimal information about them is offered. All six carried out research in organic chemistry, mostly in the late 1880s and in the 1890s.

WANDA HABERKANT[57] (1871–1930), a daughter of Jan Adam Kaberkant and his wife, Helena, née Tydelska, was born on 10 May 1871 in the town of Dabie, about seventy-five miles west of Warsaw. Pastor of the long-established evangelical Protestant church in the parish of Kalisz, Adam Haberkant was a leading figure in community life and educational development. Wanda attended a state gymnasium in the town of Kalisz, about thirty-five miles south of Dabie, and graduated in 1887. She may then have taught for a few years, but by the early 1890s had enrolled at the University of Geneva, where she studied under the direction of Friedrich Kehrmann. Her research on naphthalene-picric acid complexes was published in 1897 and 1898 in two short papers coauthored by Kehrmann (see bibliography). She received a doctorate in 1897, but no evidence is available of further work in chemical research. After returning to Poland she began a career in education, first as a teacher and from 1923 as a schools inspector. She died in Warsaw on 16 March 1930, at the age of fifty-eight.

MATHILDE IDZ'KOWSKAIA (fl. 1890s) studied with Egor Wagner at the University of Warsaw in the early 1890s, joining in his lengthy series of investigations on the oxidation of alkenes (unsaturated hydrocarbons) with potassium permanganate. She coauthored two substantial papers on this work, published in Polish and Russian journals in 1897 and 1898. After moving to the University of Geneva she collaborated with Friedrich Kehrmann in research on organic nitro compounds, work published in 1899 (see bibliography).

S. V. BUSHMAKINA, Mlle (fl. 1890s) was also a student of Egor Wagner at the University of Warsaw. Their joint work on the potassium permanganate oxidation of isosafrole to piperonal, a compound used in the perfume industry, was reported in a Warsaw journal in 1891 (see bibliography).

NATALIA POLONOWSKA (fl. 1880s) studied at the Federal Polytechnicum in Zurich, then essentially the science and mathematics division of the University of Zurich. Her research, guided by Heinrich Goldschmidt and Arthur Hantzsch, was reported in four papers and notes, two of them coauthored by Goldschmidt (see bibliography).

R. SRZEDZIŃSKA (fl. 1890s) published a short note on alkaloids with Warsaw medicinal chemist Adam Jaworowski in the journal *Kosmos* in 1895 (see bibliography).

Also in this group is MARIIA MAZUROWSKA (pseud. Marjan Orlowsky). Two papers by her reported work done under the direction of Gustav Wiedemann in the laboratory for physical chemistry at the University of Leipzig (see bibliography).

A publication by one other Polish woman, JÓSEFA CZARNOWSKA, was found listed in the Royal Society's *Catalogue*. Her nineteen-page report on a memoir by Russian geologist and stratigrapher Nikolai Andrusov (1861–1924) appeared in the journal *Kosmos* in 1900. The memoir by Andrusov, professor at the University of Kiev from 1905 to 1914, discussed some of his extensive studies of the geology and sediments of the Aralo-Caspian basin, carried out in the 1890s. Czarnowska's report was a full presentation of the results of Andrusov's detailed and comprehensive investigation of the Kara-Bogas-Gol, the hundred-mile-long, shallow, inundated depression that formed a salty lagoon on the eastern shore of the Caspian (see bibliography). This study, carried out at the request of the Russian Ministry of Agriculture, focused on two areas in particular—the underlying stratigraphy and the current condition of the lagoon. Czarnowska indicated how, after setting out the theories and results of other workers on the region, Andrusov had concentrated on fossil content of early deposits and then gone on to determine current mineral content of Caspian and Kara-Bogas waters, as well as studying the flora and fauna of these waters. She also made a brief mention of an expedition to the region two years later by a team of specialists in various fields, which included Andrusov serving as geologist.

No definitive biographical information about Jósefa Czarnowska has been found; she would appear to have been active in late nineteenth-century literary work, especially in the dissemination of scientific research results.

NOTES

1. Piotr S. Wandycz, *The Lands of Partitioned Poland, 1795–1918* (Seattle: University of Washington Press, 1974).

2. See Mary R. S. Creese, *Ladies in the Laboratory II: West European Women in Science, 1800–1900* (Lanham, Md.: Scarecrow Press, 2004), 81–84.

3. E. Smorscek, "Le centenaire de la naissance de Jósefa Joteyko," *Bulletin de psychologie scolaire et d'orientation* 15 (1966): 45–50; Laura Lynn Windsor, *Women in Medicine: An Encyclopedia* (Santa Barbara, Calif.: ABC-CLIO, 2002), 116; "Josefa Franciszka Joyteko (1866–1928)," in "Femmes scientifique en France," http://newscientist.ulb.ac.be/france/fr_fcel_fr.htm (July 2014); Joy Harvey and Marilyn Ogilvie, "Joteyko, Joséphine (1866–1929)," in *The Biographical Dictionary of Women in Science: Pioneering Lives from Ancient Times to the Mid-20th Century*, ed. Marilyn Ogilvie and Joy Harvey (New York and London: Routledge, 2000),

1:670–71; Esther Pohl Lovejoy, *Women Doctors of the World* (New York: Macmillan, 1957), 172–73, 188; Lipinska, *Les femmes et le progrès des sciences médicales*, 146; "Życie i twórczość naukowa Józefa Joteyko" [The life and scientific work of Jósefa Joteyko], http://www.apk_poczatkowe2.republika.pl/pawlikowska1.doc (July 2014); "Societies and Academies. Paris," *Nature* 69, No. 1783 (1903): 215–16, on 216; see also Ivan Z. Holowinsky, "Pedology in Europe and Developmental Psychology in Ukraine," *School Psychology International (SAGE)* 14 (1993): 327–38. Joteyko's name is sometimes spelled Ioteyko.

4. Tadeusz Joteyko became well known as a composer and writer.

5. J. Ioteyko (Mlle. Dr.), *Recherches sur la fatigue névro-musculaire et sur l'excitabilité* électrique *des muscles et des nerfs* (Brussels: H. Lamartin, 1900). See bibliography for most of Joteyko's pre-1901 journal articles.

6. The Solvay Institute of Physiology was one of three major research facilities established at the Free University of Brussels by Belgian chemist, industrialist, philanthropist, and statesman Ernest Solvay in the 1890s and early 1900s—the Institute of Sociology (1894), the Institute of Physiology (1895), and the École de Commerce Solvay (1903). Solvay was also the founder of the International Institute for Physics and Chemistry and the organizer of the famous Solvay Conferences of the years immediately preceding the First World War. Well ahead of his time in his ideas on social development, he believed that only by applying to the social sciences the methods of observation and experiment used in the physical and biological sciences could advances be made. (Such ideas were then being promoted in other Western countries as well, including the United States, where one of the leaders was experimental psychologist G. Stanley Hall of Clark University in Worcester, Massachusetts.) The Solvay Institute of Sociology brought out several series of publications, including its *Bulletin mensual*, *Notes et mémoires*, *Études socials*, and *Actualités sociales*.

7. For Grzegorzewska, see Alicja Siemak-Tylikowska, "Maria Grzegorzewska (1888–1967)," *Prospects: The Quarterly Review of Comparative Education* 23, no. 3/4 (1993): 625–37, online at http://www.ibe.unesco.org/publications/ThinkersPdf/grzegore.pdf (July 2014).

8. Jósefa Joteyko and Michalina Stefanowska, *Psycho-physiologie de la doleur*, Bibliothèque de Philosophie contemporaine series (Paris: Librairie Félix Alcan, 1909), 251 pp.

9. See for instance, Jósefa Joteyko, "La fatigue intellectuelle et sa mesure," *Revue de l'Université de Bruxelles* 8 (1902–1903): 497–533; Charles Henry and Jósefa Joteyko, "Sur les équations générals des courbes de fatigue" (Paris: Gauthier-Villars, 1903); Charles Henry and Jósefa Joteyko, "Sur une relation entre le travail et le travail dit statique, énergétiquement équivalent à l'ergographe" (Paris: Gauthier, 1903). See also the bibliography for the 1900 paper "Le quotient de la fatigue H/N."

10. Jósefa Joteyko and Michalina Stefanowska, "Influence des anesthésiques sur l'excitabilité des muscles et nerfs" (Brussels: H. Lamartin, 1901), 64 pp.; Jósefa Joteyko, "Effets psychologiques des ondes induites de fermature et du rupture dans l'anesthésie des muscles," *Institut Solvay. Travaux de Laboratoire* 5, fasc. 1 (1902): 1–38; Jósefa Joteyko and Michalina Stefanowska, "De l'excitation des muscles et des nerfs par courants faradiques de fermeture et de l'ouverture," *Journal de Neurologie* 7 (1902): 206–13; Jósefa Joteyko and Michalina Stefanowska, "Dissociation des phénomènes de sensibilité et de motilité dans l'anesthésie par l'éther et rapport sur le mémoire par M. Venneman," *Bulletin de l'Académie Royale de Médecine de Belgique* 16, s. 4 (1902): 373–406; Jósefa Joteyko, "Influence de menthol sur les nerfs cutanés," *Journal de Neurologie* 8 (1903): 223– 32; Michalina Stefanowska and Jósefa Joteyko, "Contributions à l'étude expérimental de la douleur (Recherches algésimetrique)," Laboratoire Psychologique Kasimir, l'Université de Bruxelles, *L'Année Psychologique* (1903): 461–70, also published in *Recherches algésimetriques* (Brussels: Hayez, 1903), 86 pp.; Jósefa Joteyko, "Étude sur la contraction tonique du muscle strié et ses exitants," *Mémoires couronnés et autres mémoires publiés par l'Académie Royale de Médecine de Belgique* 18, no. 1–2 (1903): 1–100.

11. Jósefa Joteyko, "Fatigue," in *Dictionnaire de Physiologie*, ed. Ch. Richet (Paris: Librairie Félix Alcan, 1903); Jósefa Joteyko, "Entraînment et fatigue au point de vue militaire: Avec une préface de Charles Richet," *Institut de Sociologie, Institut Solvay, Bruxelles, Travaux. Actualités socials* no. 5 (1905).

12. Varia Kipiani and Jósefa Joteyko, "Enquête scientifique sur les végétariens de Bruxelles: Leur résistance à la fatigue étudiée à l'ergograph . . . ," (Paris: H. Lamartin, Société végétarienne de France, 1907), 77 pp. For Irving Fisher's review, see "Diet and Endurance in Brussels," *Science*, n.s., 26 (Oct. 1907): 561–63. Jósefa Joteyko, "L'enfance végétarienne: Enquête sur 170 enfants végétariens," *Comptes rendus, Congrès Végétarien International, Bruxelles* (1910), vol. 2.

13. Jósefa Joteyko, "Le surmenage scolaire," *Rapport au Cinquième Congrès Belge de Neurologie et de Psychiatrie* (1909); Jósefa Joteyko, "Une éducation dynamique de la mémoire" (Paris: Libraire Félix Alcan, 1916), excerpted from *L'éducation* 8, no. 4 (December 1916); Jósefa Joteyko, "Les méthodes belges d'éducation technique," *Revue Générale des Sciences* (30 May 1917); Jósefa Joteyko, "Recherches expérimentales sur le niveau intellectuel des élèves des trois premières années au lycée," *Travaux Psychologiques* 1 (1922), Commission Pédagogique du Ministère des Cultes et de l'Instruction Publique (Warsaw); Jósefa Joteyko, "La psychologie

de la form dans ses rapports avec les thèses de l'école active," *Archives Polonaises de Psychologie* (Warsaw) 1, no. 3 (1926–1927).

14. Jósefa Joteyko, *La science du travail et son organisation* (Paris: Librairie Félix Alcan, 1917), translated into English as *The Science of Labour and its Organization* (London: G. Routledge & Sons; New York: E. P. Dutton, 1919; repr. 1974), 260 pp. See also "La productivité et la durée du travail," *Revue de l'Institut de Sociologie* 1, no. 1 (1921); "Le facteur psycho-physiologique dans le travail industriel," *Bulletin de l'Institut Général Psychologique* (Paris) no. 4–6 (1918).

15. Additional French-language publications by Jósefa Joteyko include the following: "Résumé des travaux scientifiques (période décennale 1896–1906)" (Gand: Société coopérative Volksdrukkerij, 1906); *Aide-mémoire de Psychologie Expérimentale et de Pédologie* (Brussels, 1907), 322 pp.; *La fonction musculaire*, vol. 9 of *Encyclopédie Scientifique. Bibliothèque de Physiologie* (Paris: O. Doin et fils, 1909), 420 pp.; *La Fatigue* (Paris: Flammarion, 1920), 331 pp.

16. Alphonse Rebière, *Les Femmes dans la science: Notes recueilles*, 2nd ed. (Paris: Nony, 1897), 271; "Zakład Antropologii. Prof. fizjologii układu nerwowego Dr. Nauk przyrodniczych Michalina Stefanowska (20 XI 1855 – 15 XII 1942)," University of Poznan, Main Library Publications, http://www2.bg.am.poznan.pl/wydawnictwa/_1919_39/115_135.pdf (July 2014); see also the website of Jagellonian University in Cracow for a representative list of deceased biological scientists who were members of the Polish Academy of Arts and Sciences or its predecessor organizations, http://www.khm.cm-uj.krakow.pl/wydzialy.html (July 2014); "Faces à faces: Habillage d'Uni Dufour, 3 juin–30 septembre 2009," www.unige.ch/450/expositions/faces/Liste_site_Unige.pdf (July 2014); see also Albert Moll, *Der Hypnotismus* (Berlin, 1889), republished as *The Study of Hypnosis: Historical, Clinical and Experimental Research in the Techniques of Hypnotic Induction* (New York: Institute for Research in Hypnosis, 1958).

17. Sir Humphry Davy had strong connections with Switzerland, especially Geneva, and with early nineteenth-century Swiss men of science. His discovery of the electric arc was to bring great benefit to science and industry in Switzerland with its rich supply of waterpower-generated electricity. When Davy died in Geneva in 1829, he was buried in the Protestant cemetery there; the Conseil d'État granted a concession for ninety-nine years, later extended at the insistence of the university's faculty of science. In gratitude Lady Davy donated money to found a prize for the encouragement of scientific research. See Ch. Eug. Guye, "Faraday's Connection with Switzerland and Swiss Industrial and Economic Development," supplement to *Nature* 128, no. 3226 (29 Aug. 1931): 349–51.

18. In 1903, she shared the Paris Academy of Sciences Monyton Prize with Joteyko and M. Radzikowski; this followed the Dieudonnée Prize of the Belgian Royal Academy of Medicine, shared with Joteyko in 1901.

19. See bibliography. For a twenty-first-century paper that includes an assessment of the importance and significance of Stefanowska's work on dendritic spines (thread-like appendages of neurons) and the possibility of their temporarily changing shape and number, see the paper by Javier DeFelipe of Madrid, a specialist on neurotransmission and cerebral-cortex organization, "Brain Plasticity and Mental Processes: Cajal Again," *Neuroscience*, Nature reviews, 7 (October 2006): 811–17, especially 814, 815.

20. See above section on Joteyko and especially notes 7 and 9 for joint Joteyko-Stefanowska publications. Stefanowska's early single-author papers include the following: "Étude histologique du cerveau dans le sommeil provoqué par la fatigue," *Journal de Neurologie*, Brussels, 5 (1900): 181–85; "Résistance rationelle variable dans les différents territoires du cerveau," *Journal de Neurologie*, Brussels, 6 (1901): 5–8; "Dans quelles conditions se forment les varicosités sur les dendrites cérébrales?," *Comptes Rendus, 4th International Congress of Psychology*, Paris (1901): 247–48; "Sur les appendices piriformes des cellules nerveuses," *Comptes Rendus, 4th International Congress of Psychology*, Paris (1901): 249–50; "Déshydration du protoplasme vivant par l'éther, le chloroform et l'alcool. Contribution à l'étude du méchanisme de l'anesthésie," *Annales de la Société Belge de Microscopie* 81 (1900–1901): 63–73; [Les terminaisons réelles des cellules nerveuses et leur signification dans les procès psychiques], *Kosmos* (Lwow) 26 (1901): 244–50; "Les appendices terminaux des dendrites cérébreux et leurs différents états physiologiques," *Institut Solvay, Travaux de Laboratoire* (1901): 1–58; "Diversité de résistance des différents territoires cérébraux vis-à-vis du traumatisme et de l'intoxication," *Institut Solvay, Travaux de Laboratoire* (1901):167–75; "La grande hypnose chez les grenouilles en inanition," *Bulletin de l'Académie Royale de Médecine de Belgique*, Séance du 26 juillet 1902: [185]–[227]; "La théorie du neurone dans la dernière période décennale 1896–1906," 2ème Congrès Belge de Neurologie et Psychiatrie, 1906 (Brussels: L. Severeyins, 1906), 99 pp.

21. "Sur la croissance en poids de la souris blanche," *Comptes Rendus hebdomadaires des Séances de l'Académie des Sciences*, Paris, 136 (1903): 1090–93; "Sur la croissance en poids des végétaux," *Comptes Rendus hebdomadaires des Séances de l'Académie des Sciences*, Paris, 138 (1904): 304–6; "Sur la loi de variation de poids du *Penicillium glaucum* en fonction de l'âge," *Comptes Rendus hebdomadaires des Séances de l'Académie des Sciences*, Paris, 139 (1904): 879–81; "Sur la croissance en poids des animaux et des végétaux," *Archives des*

Sciences Physiques et Naturelles 18, période 4 (1904): 474–76; "Le rendement organique de la plante en fonction de temps," *Congrès International de Physiologie*, Brussels (1904); "Sur l'accroissement du poids des substances organiques et minérales dans l'avoine en fonction d'âge," *Comptes Rendus hebdomadaires des Séances de l'Académie des Sciences*, Paris, 140 (1905): 58–60; "Sur la croissance en poids du cobaye," *Comptes Rendus hebdomadaires des Séances de l'Académie des Sciences*, Paris, 140 (1905): 879–80; "Sur la croissance en poids du poulet," *Comptes Rendus hebdomadaires des Séances de l'Académie des Sciences*, Paris, 141 (1905): 269–71; "Recherches statistiques sur l'évolution de la taille des végétaux," *Comptes Rendus hebdomadaires des Séances de l'Académie des Sciences*, Paris, 141 (1905): 600–603; Michalina Stefanowska and Henri Chrétien, "Recherches statistiques sur l'évolution de la taille du lin," *Comptes Rendus hebdomadaires des Séances de l'Académie des Sciences*, Paris, 141 (1905): 900–903.

22. Ethology, the study of character formation and behavior, both inherited (instinctual) and learned, of animals in the wild, is a relatively new branch of zoology. One of the best-known names in the field is that of the Austrian naturalist Konrad Lorenz, whose studies of the behavior of young birds were published in the 1960s. More recent work includes the scientific study of human behavior from an ethological perspective.

23. Camille Flammarion, *Qu'est-ce que le ciel?* (Paris: Ernest Flammarion, 1891–1893), trans. Michalina Stefanowska as *Niebo* (Warsaw: Gebethner i Wolff, 1895); Élisée Reclus, *La terre: Description des phénomènes de la vie du globe* (Paris: L. Hachette, 1868–1869), trans. Michalina Stefanowska as *Zjawiska ziemskie: [1], Lady stale* (Warsaw: Gebethner i Wolff, 1894); Michalina Stefanowska, *Zycie w Oceanie: Opis popularny roslin i zwierzat morskich* [Life in the Ocean: A popular description of marine plants and animals] (Warsaw: Michała Arta, 1905); Michalina Stefanowska, *Co sie dzieje w ulach? Zycie i obyczaje pszczoly zwyczajnej* [What happens in hives? Life and habits of common bees] (Warsaw, 1906), 81 pp.

24. See "Faces à faces" University of Geneva (listed above in note 16).

25. Rebière, *Femmes dans la science*, 250; information in technical papers; Magdalena Bajer, "Melcerowie Szczawińskie Stromengerowie," Strona glówna "Forum Akademickiego," Archiwum z roku 2003, Spis Treści numeru 6/2003, http://www.forumakad.pl/archiwum/2003/06/artykuly/23-rody_uczone.htm (July 2014); Barbara Bruziewicz-Miklaszewska, "Żeby ślad pozostał i pamięć: Medicial [*sic*] Women's International Association-MWIA-1919–1999," *Gazeta Lekarska* no. 4(99), s. 51 (1999): 53 [a review of the medical newspapers for the period], http://www.oil.org.pl/xml/nil/gazeta/numery/n1999/n199904/n19990424 (July 2014); *Biuletyn Federacji Polskich Organizacji Medycznych na Obczyznie* no. 4 (10) (1998), http://www.polmed.org/biuletyn/biul10.htm (Dec. 2013); Zofia Podgórska Klawe, "Kobiety—Lekarze Warszawy" [Women doctors in Warsaw], *Archiwum Historii Medycyny* 28 (1965): 243–50, especially 249–50.

26. Wanda Szczawinska, "Contribution à l'étude des yeux de quelques crustacés et recherches expérimentales sur les mouvements du pigment granuleux et des cellules pigmentaires sous l'influence de la lumière et de l'obscurité dans les yeux des crustacés et des arachnides," dissertation, University of Geneva, 1891 (Liège: Imprimerie Vaillant-Carmanne, 1891), 52 pp.

27. Yves Delage (1854–1920) had wide interests across the biological sciences, from the circulatory systems of microcrustaceans and marine worms to the evolution of sponges and the physiology of the inner ear of humans; he also made important contributions to theories of evolution. Henri de Lacaze-Duthier, Delage's predecessor as director of the Roscoff laboratory, specialized in small marine animals; he founded both the Roscoff laboratory and the Arago Marine Laboratory at Banyuls-sur-Mer, a fishing port about three miles north of the Spanish border, at the eastern end of the Pyrenees. The Lacaze-Duthier Canyon, an underwater canyon situated about twenty-two miles east of Banyuls, a site rich in fauna and flora, is still an important research area. See http://www.larousse.fr/encyclopedie/personnage/Yves_Delage/115937 (July 2014).

28. Wanda Szczawinska, "Sérums cytotoxiques," doctoral thesis, medical faculty, University of Paris, 1902.

29. Szczawinska's book on the Drop of Milk Company, *Życie ogniska* [The spark that starts the flame of life], appeared in 1917 (Warsaw: Druk Piotra Laskauera).

30. Klawe, "Kobiety-Lekarze Warszawy," 250.

31. Grzegorz Brzęk, "Rozalia Nusbaum-Hilarowiczowa," in *Słownik Biologów Polskich* [Dictionary of Polish biologists], ed. Stanislaw Feliksiak (Warsaw: Polish Academy of Science, 1987).

32. Information collected from technical papers.

33. One might wonder if Mlle Gruzewska married botanist Charles-Louis Gatin (1877–1916). She coauthored one paper with him in 1905, but there is no mention of her in his obituary notice. See C.-L. Gatin and Mme. Gruzewska, "Action de quelques diastase animales sur certain mannanes," *Société de Biologie. Comptes Rendus* 58 (20 May 1905): 847–49; Jean Friedel, "Notice sur Charles-Louis Gatin," *Revue Générale de Botanique* 31 (1919): 65–74.

34. Mme Z. Gatin-Gruzewska, "Praezipitationserscheinung des reinen Glykogenes. Vorläufige Mitteilung," *Archiv für die gesamte Physiologie* 100 (1903): 634–35; Mme Z. Gatin-Gruzewska, "Das reine Glykogen," *Archiv*

für die gesamte Physiologie 102 (1904): 569–91; Mme Z. Gatin-Gruzewska, "Das Molekulargewicht des Glykogenes," *Archiv für die gesamte Physiologie* 103 (1904): 282–86; Mme Z. Gatin-Gruzewska, "Die Wanderungen des Glykogenes unter dem Einfluss des elektrischen Stromes," *Archiv für die gesamte Physiologie* 103 (1904): 287–88; Mme Z. Gatin-Gruzewska and Wilhelm Biltz, "Ultramikroskopische Beobactungen an Lösungen reinen Glykogens," *Archiv für die gesamte Physiologie* 105 (1904): 115–21; Mme J. [*sic*] Gatin-Gruzewska, "Composition du foie des chiens nourris en vue de la production de la quantité maximale de glycogène," *Société de Biologie. Comptes Rendus* 58, no. 1 (1905): 423–25; H. Bierry and Mme Z. Gatin-Gruzewska, "Action physiologique de l'adrénaline pure," *Société de Biologie. Comptes Rendus* 58, no. 1 (1905): 902–3; H. Bierry and Mme Z. Gatin-Gruzewska, "L'adrénaline produit-elle la glycosurie par son action sur le pancréas?," *Société de Biologie. Comptes Rendus* 58, no. 1 (1905): 904–5; Mme Z. Gatin-Gruzewska, "Action de l'adrénaline sur la tenure du muscle en glycogène," *Comptes Rendus hebdomadaires des Séances de l'Académie des Sciences* 142 (1906): 1165–67; Mme Z. Gatin-Gruzewska, "Influence de la dilution des solutions d'adrénaline sur la disparation du glycogène dans le foie et dans le muscle," *Société de Biologie. Comptes Rendus* 60, no. 1 (1906): 940–41.

35. Mme Z. Gruzewska and L. Maquenne, "Sur la composition du grain d'amidon," *Comptes Rendus hebdomadaires des Séances de l'Académie des Sciences* 146 (1908): 540–42; Mme Z. Gruzewska, "L'oxydation et l'hydrolyse du glycogène sous l'action du peroxide d'hydrogène," *Société de Biologie. Comptes Rendus* 68 (1910): 274–76; Mme Z. Gruzewska, "Sur les produits d'hydrolyse de l'amidon sous l'action de l'eau oxygénée," *Société de Biologie. Comptes Rendus* 68 (1910): 1084–86; Mme Z. Gruzewska, "Purification rapide de grandes quantités de glycogène et séparation de granules de différents grandeurs," *Société de Biologie. Comptes Rendus* 69 (1910): 526–28; Mme Z. Gruzewska, "L'oxydation et l'hydrolyse de glycogène sous l'action de peroxide d'hydrogène," *Société Chimique de France, Bulletin* 7 (1910): 743–46; Mme Z. Gruzewska, "Quelques propriétés caractèristiques de l'amylose et de l'amylopectine," *Comptes Rendus hebdomadaires des Séances de l'Académie des Sciences* 152 (1911): 785–88; Mme Z. Gruzewska, "Contibution à l'étude de l'amidon. 1. L'amylose et l'amylopectine. La séparation des deux constituants du grain d'amidon et leurs principaux caractères," *Journal de Physiologie et de Pathologie générale*, Paris, 14 (1912): 7–20; Mme Z. Gruzewska, "Contribution à etude de l'amidon. 2. Hydrolyse de l'amidon et de ses constituants par le suc pancréatique de chien et par peroxide d'hydrogène," *Journal de Physiologie et de Pathologie générale*, Paris, 14 (1912): 32–41.

36. *Bulletin de la Société Chimique*, Paris 1, s. 4 (1907): 50.

37. Mme Z. Gruzewska, "Contributions à l'étude de la Laminarine du *Laminaria flexicaulis*," *Comptes Rendus hebdomadaires des Séances de l'Académie des Sciences* 170, no. 9 (1920): 521–23; Mme Z. Gruzewska, "L'action des diastases sur la Laminarine," *Bulletin de la Société de Chimie Biologique* 3 (1921): 490–97; Mme Z. Gruzewska, "A propos du pH de certaines eaux minérales," *Bulletin de l'Académie de Médecine* 92, no. 37 (1924); Paul Carnot and Mme Z. Gruzewska, "Variations de concentration ionique de la bile et du suc pancréatique pendant la sécrétion acide du suc gastrique," *Société de Biologie. Comptes Rendus* 93 (1925): 240–42; Paul Carnot and Mme Z. Gruzewska, "La concentration ionique de la bile et sa teneur en CO_2 pendant la sécrétion gastrique," *Société de Biologie. Comptes Rendus* 94 (1926): 369–73.

38. Sandra L. Chaff, et al., comp. and ed., *Women in Medicine: A Bibliography of the Literature on Women Physicians* (Metuchen, N.J.: Scarecrow Press, 1977), 106; Barbara Bachmann and Elke Bradenahl, "Medizinstudium von Frauen in Bern, 1871–1914," doctoral dissertation, University of Bern, 1990 (Bern: Medizinhistorisches Institut der Universität Bern, 1990); matriculation register, Zurich University, http://www.matrikel.uzh.ch/active/static/26779.htm (July 2014); Klawe, "Kobiety-Lekarze Warszawy," 246; Constantin von Monakow, *Vita Mea. Mein Leben*, ed. Alfred W. Gubser and Erwin H. Ackerkencht (Bern: Verlag Hans Huber, 1970), 205.

39. Also known as the Wolski Hospital, or the New Orthodox Hospital in Czyste, the Starozakonnych Hospital was the successor of the much older Jewish Hospital; efforts to establish it began in the late 1880s. Beginning in 1907 its administration was taken over by the city, as was that of all other Warsaw hospitals. In many ways a model medical center, as its planners had envisaged, its clinics were long used by the Warsaw Medical Academy; its staff also provided further training and retraining courses for physicians and nurses. See http://pl.wikipedia.org/wiki/Synagoga_Szpitala_Starozakonnych_w_Warszawie (July 2014).

40. Matriculation register, Zurich University, http://www.matrikel.uzh.ch/active/static/27484.htm (July 2014); "Outstanding Mława Citizens' Profiles," http://www.archiwum.mlawa.pl/en/index-4.html (July 2014); H. Kulich, [The centenary of the opening of the gates of medical faculties of universities to women], *Archiwum Historii i Filozofii Medycyny* (Warsaw) 63, no. 3–4 (2000): 175–81, English abstract at http://www.ncbi.nlm.nih.gov/pubmed/11768365 (July 2014); Lipinska, *Histoire des Femmes Médecins depuis l'Antiquité jusqu'à nos Jours* (Paris: G. Jacques, 1900), 469; Ursula Dobrzańska, "Trudne przypadki pierwszej polskiej lekarki" [The obstacles facing the first Polish woman physician], http://netbird.pl/a/320/7285,2, machine translation (Jan. 2010); Jerzy Kulikowski, "Medycynierka" (Doktor Anna Tomaszewicz-Dobrska), in *POLITYKA.PL* [Policy] 34, 21 August 2007, http://www.polityka.pl/historia/226843,1,medycynierka.read, machine translation (July 2014); Zbigniew

Filar, *Anna Tomaszewicz Dobrska: Karta z dziejów polskich lekarek* [Anna Tomaszewicz Dobrska: A note from the history of Polish women doctors], Polskiego Towarzystwo Historii Medycyny. Życiorysy no. 4 (1959), 263 pp. See especially 146–47 and English summary, 239–42.

41. See Kulikowski, "Medycynierka."

42. Six Russian women, three British, two Swiss, two Germans, and one American received Zurich medical degrees before Tomaszewicz—see Hannay Rohner, *Die ersten 30 Jahre des medizinischen Frauenstudiums an der Universität* Zürich, *1867–1893*, Zürcher Medizingeschichtliche Abhandlungen, Neue Reihe, no. 89 (Zurich: Juris Druck, 1972), 78–80.

43. Anna Tomaszewicz, "Beiträge zur Physiologie des Ohrlabyrinths," dissertation, medical faculty, Zurich University, 1877 (Zurich: Genossenschafts-Buchdr., 1877).

44. Anna Tomaszewicz-Dobrska, "Sprawozdanie z działalności Przytułku położniczego 2-go za cały 29-letni czas jego istnienia 1882–1911" [Report on the activity of Shelter House (Lying-in Hospital) No. 2 for the entire time of its existence from 17 November 1882 to 1 July 1911] (Warsaw, 1912); Filar, *Anna Tomaszewicz-Dobrska*, 147, 242.

45. Filar, *Anna Tomaszewicz-Dobrska*, 242.

46. Filar, *Anna Tomaszewicz-Dobrska*, 147–48, includes a letter from Anna Tomaszewicz-Dobrska to Ignacius Dobrski, dated 1 March 1915. Filar, 238, also refers to letters from Anna to her sister Zofia Tomaszewicz-Chabrié in Paris (dated March 1895 and January 1915). Zofia's life and circumstances have not been investigated; one might wonder if she was the wife of chemist Camille Chabrié, from 1908 to 1928 director of what is now the École National Supérieure de Chimie de Paris.

47. Sources vary about the exact date of death, ranging from 9 April 1918 to 21 June 1918.

48. Matriculation register, Zurich University, http://www.matrikel.uzh.ch/active/static/26044.htm (July 2014); Bachmann and Bradenahl, "Medizinstudium von Frauen in Bern"; archives, Bern University, http://apps.uniar chiv.unibe.ch/syscomm/images/mata/6412_6423.gif (July 2014); information in technical papers.

49. Nathalie Kleindienst, "Ueber pental als Anaestheticum," inaugural dissertation, medical faculty, Bern University, 1892 (Leipzig: Hirschfeld, 1892).

50. Eugénie Antoine de Przedniewicz, "Infection et symétrie," doctoral dissertation, medical faculty, University of Paris, 1896.

51. Alicja Zemanek, "Pro Memoria. 130 rocznica Gabrieli Balickiej-Iwanowskiej (1871–20 II 1962)" [130th anniversary of the birth of Gabriela Balicka-Iwanowska (1871–20 February 1962)], Krakov, Institut Botaniki, *Wiadomości Botaniczne* [Botanical reports, Cracow Botanical Institute] 45, no. 1/2 (2001): 74; Neumann, *Studentinnen aus dem Russischen Reich in der Schweiz: Das Frauenstudium an den Schweizer Hochschulen*, 156; matriculation register, Zurich University, entry for Siegmund (Zygmunt) Balicki; information in technical papers.

52. Gabrielle Balicka-Iwanowska, "Contribution à l'étude anatomique et systématique du genre Iris et des genres voisins," doctoral thesis, University of Geneva, 1893 (Geneva: Imp. Aubert-Schuchardt, 1893).

53. John Henry Barnhart, *Biographical Notes upon Botanists* (Boston: G. K. Hall, 1965); information in technical papers.

54. Information in technical papers.

55. N. Nassonov, [Memorial for Mariia Iwanovna Pavlova], *Anatomische Hefte, Abteilung 2, Ergebnisse der Anatomie und Entwicklungsgeschichte* 9, *Biographisches Geschichte* 11 (1900): 518–19.

56. Mariia Ivanova Pavlova, [Circulatory and sympathetic nervous system of insects] (Warsaw: Vabotui Zool. Kabineta, Imperial University of Warsaw, 1895), 96 pp.; see also Nassonov, "Memorial," 518.

57. Kalisz online, History, http://www.info.kalisz.pl/historia/ewangelicy.htm (July 2014); "Haberkant, Wanda," [Database of Polish People], http://baza-nazwisk.de/erw-suche.php?data=59376&hits=3560&ds=2171&title=Hab erkant%20Wanda (July 2014); information in technical papers.

Appendix

Additional Biographical Sketches and Notes

Information is given here about four Europeans whose countries of origin had not been determined before publication of volume 2 of this survey, where they would better belong. Notes have been added about five others who were probably European but whose nationalities are still unverified. Also included are sketches of three remarkable women of scientific bent from the Far East, and a note about one from Mexico who reported what then appeared to be an extraordinary sky phenomenon.

A FRENCH PHARMACIST

Louise Rosalie Napias, later Chaboseau-Napias[1] (1878–1952), who was most likely the first French woman pharmacist, was born on 13 February 1878 in the Fourteenth Arrondissement of Paris, a left-bank district; by the early twentieth century, this arrondissement was the home of artists and was sometimes known as "Bohemian Central." Descendent of one of the early social reformers, the *fouriéristes*, Louise was the daughter of Clémence Joséphine Napias, a seamstress and embroiderer. Her father did not live with his family, but her uncle, Dr. Henri Napias (1842–1901), helped her during her early years. A member of the Académie de Médecine, Henri Napias had a notable career in public health and industrial legislation, areas that would later be of special interest to Louise.

She acquired much of her early education as a fee-paying student and by attending classes provided by public instruction societies such as the Société pour l'Instruction Élémentaire, where she won prizes. Her ambition of becoming a pharmacist required overcoming the hostility of male pharmacists who objected to the idea of women entering their profession; limited financial resources were a further handicap. However, she completed her bachelor's-level studies and proceeded on to the *internat*, the required period of supervised practical training in a hospital pharmacy. Throughout these studies she had strong support and encouragement from Blanche Edwards-Pilliet, a pioneering woman physician and a leading social reformer.[2] Her physician uncle also provided important help; largely through his good offices she received a paid appointment as a laboratory assistant at the Pasteur Institute, where she was able to fulfill the research requirements for a pharmacy degree.

An 1898 note she coauthored with Henri Pottevin, head of the Pasteur Institute laboratory, reported work on sugar fermentation, but her major research, directed by biologist Emil Roux, was a biochemical study of the breakdown of starches and sugars by the tuberculosis bacillus. Her results appeared in a sixteen-page report in the institute's *Annals* in 1900 (see bibliography).

During her student years Louise Napias became increasingly involved in social work, joining forces with several prominent activists in the cause, particularly in the areas of public instruction and the advancement of women. She talked at educational conferences, assisted the radical *député* Ferdinand Buisson, a supporter of voting rights for women, in founding a lay society for child assistance, and worked with Maria Deraismes, one of the founders of the first feminist group in France, Société pour la Revendication des Droits de Femme, an organization from which she received a modest study grant. By the time of the 1900 Paris World Exhibition, Louise Napias was a central figure in more than one

French educational and left-wing political group.[3] Writing under the pseudonym Blanche Galien, she contributed several articles on health and medical topics to the feminist newspaper *La Fronde* in 1898 and 1899.[4] Founded by Marguerite Durant, this publication, run entirely by women, was devoted to factual reporting on national and international concerns in much the same style as other newspapers; in addition it strove to further the advancement of women, especially in such areas as women's access to higher education. With its title *La Fronde*, a name with connotations of the seventeenth-century anti-establishment revolts, and its editorial staff of leading French female intellectuals, including Clémence Royer (translator of Darwin's *Origin of Species*) and astronomer Dorothea Klumpke, as well as Louise Napias, the paper presented a strong stand against the notion that women were creatures with domestic interests only, removed from the world's serious work.[5]

Louise Napias also had a role in the organization of the Fourteenth Arrondissement Université Populaire, where she met Augustin Chaboseau in 1902. Chaboseau (1868–1946), descendant of a French aristocratic family, had studied widely in a great many fields, including medicine and spiritualism, but above all he was a champion of human rights. Active in a range of socialist politics, he advocated public higher education and the advancement of women; the passing of labor laws, especially those affecting women and children, and the organization of a wide network of employment bureaus were also major interests. He and Louise Napias were married on 17 December 1902; they had two children, Jean and Claudienne.

From about 1904 Louise Chaboseau-Napias contributed a number of articles to socialist journals. "Les partis socialistes: All[e]magne: . . ." appeared in 1904, and two years later she brought out her work on the franchise rights of women, a study that remained of interest for social historians for decades.[6] For a brief period about 1907 she served as secretary of a Paris section of a women's commission, part of a drive to set up a women's section of the Workers' International.[7]

Throughout the years she also practiced as a pharmacist, from 1911 in the newly reorganized Salpêtière-Pitié Hospital in the Thirteenth Arrondissement; she continued to work at a dispensary in the Thirteenth Arrondissement until she retired. A widow for the last six years of her life, she died on 15 May 1952, at the age of seventy-four, at Villemomble, a town to the east of Paris, in the department of Seine-Saint Denis.

A HUNGARIAN ALPINIST

Hermine Tauscher-Geduly[8] (d. 1923), whose 1887 geographical article was listed in the Royal Society's *Catalogue*, was one of the early women mountaineers who took up the challenge of the high peaks of the Alps and Dolomites during the 1870s and 1880s. One of seven children of an eminent Bratislava theologian and pastor, Louis Geduly, and his wife, Anna Svehla, she married a leading Bratislava surgeon, Bela Tauscher, and with him climbed regularly, season after season. The Tauscher-Gedulys became well known and admired figures among climbers in Europe and further afield.

The first Hungarian woman on Mont Blanc (1881), Hermine made more than fifty climbs over the course of her mountaineering career, most of them on peaks in the thirteen-thousand-foot range in the eastern Alps. She described her excursions in a series of lively articles that not only gave details of the routes taken and the obstacles encountered, but also offered glimpses of the personalities of the climbers and their guides, and of the rivalry at times between parties vying for the honor of being the first to reach a summit (see bibliography).

One of her early climbs was on the 13,313-foot Piz Bernina, on the Swiss-Italian border. In fact, she is often credited with being the first woman to reach the summit. However, as she herself makes clear in her detailed 1881 report (see bibliography), the expedition of August 1880 in which she took part reached the summit by a particular route, that of the west face. The first woman to stand on the summit of Piz Bernina was the English mountaineer Lucy Walker (1836–1916), who made the climb in 1869, a decade earlier.[9]

Hermine Tauscher-Geduly found her climbing deeply satisfying; in expressing her feelings about the mountains she borrowed the words of English climber Edward Whymper: "Meine Alpenfahrten haben mir zwei der besten Dinge gegeben die Mensch besitzen kann, Gesundheit und Freunde" (My alpine journeys have given me two of the best things man can have, health and friends).[10] She was prominent in social work in Bratislava, particularly among the city's poor, and served as president of the Bratislava Women's Association. Hermine Tauscher-Geduly was one of the founders of the Hungarian Red Cross. She died in 1923.

TWO MEDICAL WOMEN FROM THE BALKANS

After the settlements reached at the Congress of Berlin in 1878, at the end of the Russo-Turkish War, Bulgaria became an independent state after five centuries of Turkish rule; the province of Macedonia, the birthplace of ZAKHARINA DIMITROVA[11] (1873–1940), remained under Turkish administration, a matter that seriously affected Macedonian women aiming for higher education and professional careers. In independent Bulgaria the new government, very aware of the country's urgent need to improve health conditions, placed no restrictions on women's access to medical education and careers.[12] State scholarships, open to women as well as men, were made available for training abroad, particularly during the periods 1885–1900 and 1913–1918.[13] Medical schools in French provincial cities, such as Montpellier, Lyons, Bordeaux, and Nancy, were preferred destinations for aspiring female physicians, these cities being seen by Orthodox Christian Bulgarians as relatively free from the "big-city evils" of metropolitan centers like Paris.

Zakharina Dimitrova came from an old Bulgarian family, several of whose members, including her paternal grandfather, Micho Vel'o, were prominent in the country's struggles against the Turks for political and religious freedom. She was born in the town of Ressen, in the southwest of the present-day Republic of Macedonia, on 26 November 1873. By the early 1890s she had completed her studies at a high school for girls in Thessalonica, Greece, and in 1895, at the age of twenty-two, began medical training.

Supported by a Bulgarian Orthodox archbishop scholarship, she enrolled in the School of Medicine at the University of Nancy. Her dissertation research, pioneering work for which she is remembered, concerned the microstructure of the pineal gland, the small nugget of tissue in the roof of the brain that releases the hormone melatonin, the regulator of circadian rhythms in mammals. Under the guidance of anatomy professor Adolphe Nicolas, Dimitrova carried out a painstaking histological study of the fine structure of the gland and was the first to provide an accurate description of the pinealocytes, the organ's main cells. She pointed out the characteristic vacuoles in the nuclei of the pinealocytes and related her finding to a possible endocrine function. This highly acclaimed work was awarded the medical school's gold medal and was published, in 1901, as a beautifully illustrated sixty-page paper in the recently established Belgian neurology journal *Névraxe*;[14] it was quickly recognized as groundbreaking research by specialists in neurosecretion. Dimitrova was awarded her doctorate on 27 February 1900.

She carried out two other studies during her years in Nicolas's research team. One, on the development of the lung structure in sheep, appeared in the *Comptes Rendus* of the Paris Biology Society in 1897; the other described an abnormal human form that lacked a heart and was reported the same year (see bibliography).

Dimitrova returned briefly to Macedonia in 1900, but, unable to either continue research or open a practice there, she very soon moved to Bulgaria. Over the course of a career of almost three decades she worked in hospitals and schools in the towns of Sliven and Pazardzhik—in Sliven from 1901 to 1910, and after her marriage to military pharmacist Major Paniot Dimitrov, in Pazardzhik, or Tatar-Pazardzhik, as it was then known. As school doctor and district doctor for Pazardzhik, she was a well-known and well-liked figure in the area, not only for her medical work but for the generous financial assistance she and her husband gave to charitable organizations. During her later years she shared a practice, but in 1930 poor health caused her to retire.

Like many Bulgarian women doctors, Dimitrova participated in public health work for many years. Through lectures in schools and factories the women doctors vigorously and persistently addressed the pressing need for such reforms as the introduction of up-to-date health standards and practices, the control of epidemics, and the founding of tuberculosis sanatoria. The last ten years of her life, from her retirement in 1930 until 1940, came just after a time of considerable territorial adjustment and upheaval in the Balkans; Dimitrova and her husband were especially active in assisting refugees and finding educational opportunities for the children among them. She died in her sixty-seventh year, on 14 April 1940, when armed conflict in Europe was breaking out once again.

ELENA DENSUŞIANU,[15] later PUŞCARIU (1875–1968), was the first woman professor in a medical faculty in Romania and one of the first in Europe.[16] An energetic and productive scientist, author of a notable number of studies on pathological conditions of the eye and surrounding tissue, she was widely recognized among west European ophthalmologists.

The daughter of Aron Densuşianu (1837–1910) and his wife, Elena, née Circa, she was born on 3 March 1875 in the ancient city of Făgăraş, Braşov County, a mountainous region about 150 miles north of Bucharest. Aron Densuşianu was professor of history and philosophy at the University of Iaşi. Situated in the city of Iaşi in northeast Romania, near the border with Bessarabia, and now known as the Alexandru Ioan Cuza University, the University of Iaşi is the oldest institution of higher education in Romania. Elena's brother Ovid (1873–1938) became professor of philosophy and Romanian languages and literature at the University of Bucharest; widely known as an author and publisher of periodicals, he founded the Bucharest School of Linguistics.

Elena attended the United Principalities high school in Iaşi and then enrolled in the faculty of medicine at the University of Iaşi. She received her doctorate in medicine and surgery in 1898, at the age of twenty-three.[17] Moving to Paris for further experience and training in pathological anatomy, she worked with André Victor Cornil, a specialist in this area, and his assistant Pierre Ernest Pinoy. She also got to know a number of other distinguished scientists, including the two Pasteur Institute researchers, bacteriologist Ilya Metchnikov, and biologist Emil Roux, remembered for his development of the serum treatment for diphtheria control.

When she returned to Romania she was appointed *Dozent* at the University of Bucharest and junior physician in the laboratory of pathological anatomy. Some years later she went back to her alma mater, to a position as *Dozent* in the ophthalmology department of the medical faculty and work in the St. Spiridon Hospital in Iaşi. In 1920 she became head of the university's ophthalmology clinic and the first female professor in a medical faculty in Romania; possibly she was also the first woman in Europe to hold a professorship in ophthalmology, a field that by then had a notable number of women practitioners.[18]

Elena Densuşianu's research on pathological conditions included early histological work on subjects such as tumors of the mammary gland, an area in which she was guided by her future husband, Iaşi bacteriologist and histologist Emil Puşcariu (1859–1926). However, diseases of the eye soon became her major interest, especially after she returned to the University of Iaşi. Her publications included reports on a number of pioneering studies (such as the ocular effect of syphilis), which led the way to changes and improvements in preventive and therapeutic procedures; some of her reports presented results of large-scale statistical investigations worked out in her Bucharest and Iaşi clinics. Among those with whom she collaborated were several distinguished scientists, such as Bucharest bacteriologist and histologist Victor Babes.[19]

After retiring she continued to work at the University of Iaşi ophthalmology clinic and kept up the technical editing work she had done for many years. Her last publication, brought out by the Paris firm of Doin in 1963, appeared just two years before her ninetieth birthday.[20] Elena Puşcariu held memberships in ophthalmic associations and societies in Britain, France, and Italy. For a time she was president of the Iaşi Society of Physicians and Naturalists. She died in 1966, when she was about ninety-one, in Timişoara, a city in the lowland region to the west of the Transylvanian Alps. In 2006, forty years after her death, the ophthalmic clinic of the St. Spiridon Hospital in Iaşi was named the Professor Dr. Elena Puşcariu Ophthalmological Clinic, commemorating the 130th anniversary of her birth.

FIVE CONTRIBUTORS OF UNVERIFIED NATIONALITY (PROBABLY EUROPEANS)

Very little information has been found about the following women who each authored or coauthored one paper listed in the Royal Society's *Catalogue*: Julia Braddon, A. E. Crétier, A. Coleman, E. Naville, and Helen M. O'Neil.

JULIA BRADDON contributed a well-illustrated observational note to the Paris journal *Astronomie* in 1885 (see bibliography). Possibly one of the many northern Europeans who moved to Italy as long-term residents in the late nineteenth century, Julia Braddon wrote from the Mediterranean coast city of San Remo, near the French-Italian border. Her article described the extraordinary sight of six very large water spouts over a calm sea with a clear horizon; each spout lasted for ten minutes, the whole phenomenon taking place in the space of half an hour.

Mlle A. E. CRÉTIER[21] was a student and coworker of Swiss botanist and phycologist Robert Chodat, director of the University of Geneva's renowned Botanical Laboratory. Crétier collaborated with Chodat in investigations of the fruiting bodies of the lower blue-green algae, one of the subjects that had been of major interest to Chodat over many years. Their short joint report, presented by Chodat to the Geneva Société de Physique et d'Histoire Naturelle, was published in the society's *Archives* in 1900 (see bibliography). An honorary member of the Institut National Genevois, Crétier would appear to have lived in Geneva until at least the 1920s.

Mlle A. COLEMAN coauthored an 1896 note with Paris-based Romanian research physiologist Mariette Pompilian on physiological processes in the muscles of cold-blooded animals.[22] Their report was published in the *Mémoires* of the Paris Société de Biologie (see bibliography).

Mme E. NAVILLE carried out histological studies on aspects of eye development in the University of Paris medical faculty's histological laboratory. Her report appeared in the *Mémoires* of the Paris Société de Biologie in 1896 (see bibliography).

HELEN M. O'NEIL did morphological research in the Anatomische Institut in the southern German university town of Freiburg im Breisgau in the late 1890s. Although not officially open to women students until 1900, a few foreign women were auditing courses and carrying out graduate-level research at the University of Freiburg im Breisgau by the mid-1890s; three had been awarded doctorates by 1898,[23] but there is no clear indication that Helen O'Neil intended to proceed toward a degree.

Working under the guidance of anatomists Ernst Gaup (1865–1916) and his colleague Robert Wiedersheim (1848–1923), both of the University of Freiburg medical faculty, she studied the sheath surrounding the cranium and spinal cord in amphibians, part of the programs of morphological research of both her mentors.[24] Her results were published in a sixteen-page report in the journal *Morphologische Arbeiten* in 1898 (see bibliography).

THE FAR EAST: A TECHNOLOGIST, A MEDICAL WOMAN, AND AN EDUCATOR

The lives and work of Davy, from India, of Jin, from China, and of Tsuda, from Japan, all reflect the influence of Western penetration into regions of the Far East, a penetration that intensified throughout the eighteenth and nineteenth centuries. In India, British military and administrative control became dominant throughout the subcontinent; in China and Japan, American Christian missions played a role, perhaps most significantly in changing the lives of women.

ZENUT (ZIENUT) DAVY[25] was the illegitimate daughter of Major William Davy (d. 1784) of the Bengal Native Infantry. She was most likely born in the mid-1770s, before her father married his English wife. Major Davy had been commissioned into the Bengal Army of the British East India Company in 1767. Well versed in Persian and knowledgeable in the language of government and culture during the period of the Mogul Empire in India, he worked for a time as Persian secretary to Warren Hastings, Governor of Bengal; he translated not only official papers but also several of the great Mogul literary works. In 1784, following a year as Persian assistant to Major William Palmer at Lucknow, he resigned his posts, probably because of ill health, and left India; he died at sea on his way home.

Although his widow and five legitimate children are known to have returned to England, no information has been discovered about the upbringing of Zenut, to whom her father left £500 in his will. She evidently had a reasonably good education, was fluent in English, and spoke Bengali. She was not well off and for some of her life dwelt in rural regions, but at least by the 1840s she had an address in Calcutta (now Kolkata), the cultural center of Bengal.

About 1840 Miss Davy was asked by Dr. Henry Spry of the British East India Company, secretary of the Agricultural and Horticultural Society of India, to arrange for a trial manufacture of cloth from the fibers extracted from pineapple plant leaves.[26] After unsuccessful attempts to persuade the principal weavers in Dacca (now Dhaka) to take on the experiment, she found a weaver in a country district to carry out the work.[27] Assisted by a grant of one hundred rupees from the Agricultural and Horticultural Society, she also arranged for the production of a sample of cloth made from the fibers extracted from the leaves of the Moorgavie plant, *Sanseviera zeylanica*.[28] The society sent samples of both materials to the Asiatic Society in London but found that most British manufacturers favored flax fiber as more likely to be of use. Nevertheless, a patent was taken out, and the suggestion made that the fibers of both the pineapple plant leaves and the Moorgavie leaves be used in combination with more conventional materials, an idea that was in fact followed up.[29]

Possibilities for the cloth were a matter of considerable interest to the members of the Agricultural and Horticultural Society, and a good deal of correspondence with Zenut Davy took place. It may be noted that the Dacca (Dhaka) muslin industry, which stretches back to the time of the Egyptian pharaohs, has by no means disappeared in the twenty-first century.[30]

The story of Zenut Davy's life is regrettably absent from readily available records, and the extent of her contribution to the development of cloth manufacture from fibers other than cotton in colonial India was undoubtedly minor; nevertheless, she deserves a short footnote in the annals of the Calcutta (Kolkata)-based Agricultural and Horticultural Society.

Jin Yamei[31] (Y. May King, 1864–1934), the first Chinese woman to be awarded a university degree, was born in Ningpo County, Zhejiang Province, about one hundred miles south of Shanghai, where her father served as a Christian missionary. Orphaned at the age of two, she lived for a time with a relative, but about 1868 she was adopted by friends of her parents, the American Protestant medical missionary Divie Bethune McCartee (1820–1900) and his wife and fellow missionary, Juana (née Knight). McCartee went to China in 1844, at the request of the Board of Foreign Missions of the Presbyterian Church. He started a mission in Ningpo, one of the five ports in China then open to foreign trade, and by 1845 had organized the first Protestant church on Chinese soil. For three years, starting in 1862, he served as United States vice-consul to Japan, but by 1865 he was back in Ningpo and had resumed his missionary work.

Jin Yamei had an eventful childhood in the McCartee family. About four years after her adoption, McCartee was transferred to Shanghai, where he soon gave up his missionary duties and joined the Shanghai consular staff as interpreter and assessor to the Shanghai Mixed Court.[32] For about five years during the 1870s she lived in Tokyo, where Divie McCartee was secretary and interpreter for a Chinese government commission and also lectured in law and science at recently organized Tokyo Imperial University.[33]

Jin Yamei received a good education, and became fluent in Japanese as well as English. In 1880, when she was sixteen, the family returned to the United States and she was immediately enrolled at the Women's Medical College of the New York Infirmary for Women and Children. After graduating with high honors in 1885, she spent about a year gaining further experience in a succession of women's hospitals in the eastern United States, but by 1887 had moved to China as physician to the Amoy mission.

From Amoy she sent her one contribution to Western medical literature, an article in the *New York Medical Journal* (see bibliography) on the advantages of combining photography with microscopy, a relatively new technique in scientific work. After listing the necessary apparatus—microscope, light, condensing lens, and camera with plate-holder—she explained in detail the many steps and precautions required before good results could be expected.

Her stay in Amoy was brief, cut short by a severe attack of cholera. To recuperate she went to Kobe, Japan, where her foster parents were then stationed, and when recovered she stayed on in Japan for a year or two and worked in American Southern Baptist missions. About 1894 she married de Silva, a musician and educator of Portuguese background. They moved to Hawaii for a time and a son was born in 1896.[34] But the marriage failed, ending in divorce in 1904.

From then on Jin Yamei gave her energies to promoting women's rights and particularly the provision of medical training for women. Until about 1915 she made frequent visits to the United States and kept in contact with American women's rights organizations. Settling eventually in Beijing, she worked to establish hospitals, schools, orphanages, and laboratories for medical research. Perhaps her most notable achievement in the area of medical education was her contribution to the setting up of the school of nursing in Tianjin, the first such school in China. Jin Yamei died in Beijing on 3 April 1934, when she was about seventy.

UME (UMEKO) TSUDA[35] (1864–1929), a pioneer of higher education for women in early twentieth-century Japan, is remembered as the founder of Tsuda College in Tokyo. From its beginning in 1900 as a small private college called the Women's English School, this institution has developed into a broad-based liberal arts college that includes programs leading to doctoral degrees in a variety of fields.

Higher education for women, which would bring about the possibility of women leading independent lives, was then seen by most middle- and upper-class Japanese as outside the traditional social norms and, as such, as something to be opposed; the highest role for a Japanese woman was that of obedient wife and wise mother—all other matters were men's business. Tsuda's goal was to quietly bring higher education within the grasp of young, aspiring middle-and upper-class women without discouraging the claims of the traditional woman's sphere. Her task was a delicate one; almost the only avenue then open to her was to offer advanced teacher training, teaching being an occupation considered within the bounds of womanly behavior.

The second daughter of Tsuda Sen and his wife, Tsuda Hatsuko, Tsuda Ume was born on 31 December 1864 in the city of Edo (renamed Tokyo a year later). Tsuda Sen, a low-ranking samurai, was an ambitious man intent on establishing a prominent position for himself in middle-class society; a convert to Christianity and an admirer of Western technology, he was something of a leader in modern agricultural methods and was also active in agriculture education. In 1871, when a reformist Japanese government was investigating the sources of Western success and power, Tsuda Sen volunteered Ume to join the student section of a mission to the United States and Europe. Known as the Iwakura Mission, this group included many of Japan's political leaders, a number of male students sent to study in specified technical fields, and five girls to learn about the domestic roles of Western women. An overseas education was then becoming a status symbol, and a daughter's education might well bring significant advantage to her family. Ume, aged six, was the youngest member of the group. She had little preparation for this ten-year experiment, being a substitute for her older sister when the latter balked at the prospect.

The Iwakura Mission left Yokohama for San Francisco on the *America* on 23 December 1871, and by late February 1872 the five girls reached Washington, D.C. After several changes of possible long-term homes, Ume Tsuda was settled with Charles and Adeline Lanman in the Washington suburb of Georgetown. Lanman, a librarian by profession, was then a secretary at the Japanese legation; he and his wife were childless, and they received Ume with warmth and affection, making her fully part of their family.

Her education was very much that of a middle-class American girl. At Lucy Stephenson's Georgetown Collegiate Institute she progressed quickly, completing the program offered by 1874, when she was about ten years old; she then went on to the Archer Institute, a small private secondary school, where she did well in a fairly broad range of subjects, including science and mathematics in addition to English, French, Latin, and music. Her need to feel more securely settled in her new home led Ume, when still a child of seven, to ask to be baptized Christian. Although Episcopalians and regular churchgoers, the Lanmans decided that the appropriate ceremony for Ume would be nonsectarian, and the rite was duly performed in the Old Swedes Church in Bridgeport, Pennsylvania, in the summer of 1873.

Ume gradually became more and more a typical middle-class American child, and the socially conservative ideas about women's education and roles in life that she imbibed and would carry back with her to Japan were those of America's pioneers of higher education for women. In 1882, when she was eighteen, she returned to her native land, where she found herself something of an outsider, unable anymore to speak Japanese. After her eleven-year stay in the United States at Japanese government expense, she expected to have a role in the work of improving the provision of higher education for women, part of the process of building a new, modern Japan. But she had come back too late to catch Japan's Westernization wave of the 1860s and 1870s. The rise of popular political dissent in the 1880s triggered a conservative reaction that led to repression of all change, including the improvement of education for women. Ume Tsuda's inquiries to the Ministry of Education went unanswered, although the male students sent on the 1871 Iwakura Mission promptly received government positions on their return. Meanwhile, education for Japanese women remained about seventy years behind that of most Western countries.

Thanks to the good offices of Ito Hirobumi, one of the high officials who had taken part in the Iwakura Mission, in 1884 Ume Tsuda was appointed to the position of English teacher in Tokyo Peeresses' School, a school for upper-class girls under the direct management of the Imperial Household Ministry. She also became an imperially approved civil servant (*sonin*), a position one rank above that normally open to women, and she was paid a salary sufficient for her to rent a house in central Tokyo.

With four imperial princesses among the school's elite students, formalities were formidable and time-consuming; a wide range of subjects was taught at a shallow level, loyalty and filial piety were stressed, and the traditional family system, with its emphasis on the age-old theme of being "good wives and wise mothers," remained prominent in the overall instructional program. Ume had no success in getting Western ideas across to students.

In 1886 she was elevated to civil servant of intermediate rank with a salary of 500 yen a year, an unprecedented honor for a female government employee. But two years later, realizing that if she was ever going to reach her goal of making a substantive improvement in women's education, she needed further directed training herself.

She left Japan in 1889 on a two-year paid leave of absence from the Peeresses' School. Her request for leave was carefully camouflaged in conservative rhetoric that stressed the idea that well-educated mothers are better equipped to form the characters of the young people who would ensure Japan's future success. Her wish to serve the rising generation of Japanese women was wholly genuine, but her only way to introduce what was fundamentally a radical reform without alarming the citizenry was to stress the link between higher education and future anticipated sex roles. Her foremost concern in 1889, however, was most likely her own career.

With help from Mary Wistar Morris, a Philadelphia Quaker who had known her during her days at the Archer Institute, Ume enrolled as a special student at Bryn Mawr College. She made good progress in a range of subjects and concentrated in particular on biology. A skillful laboratory worker, she assisted Thomas Hunt Morgan, a scientist later widely known for his genetics research, in a study of the earliest stages in the development of the eggs of frogs. Morgan published the results in an 1894 paper that listed Ume Tsuda as coauthor (see bibliography). When offered a fellowship to stay on and work for a degree, Ume, then in her late twenties, declined; her work in Tokyo took precedence over her interest in further academic research.

While at Bryn Mawr she became active in various social-reform groups and charitable organizations, where she gained experience in public speaking and fund-raising; she also established contacts with women's networks in the United States and those then forming in Japan. Martha Carey Thomas, Bryn Mawr's forceful president, was for Ume an invaluable model of the successful leader of a very good college for women. Although Carey Thomas did not share the persistent Japanese stress on the need to include instruction in domestic duties in the higher-education curriculum for women, she became one of Ume's primary examples for the work of organizing the daily operation of a small college. Ume particularly admired Carey Thomas's insistence on high academic standards, equal to those of the best colleges for men, and later made that a key feature in her own college.[36]

Among Ume's important early goals was the creation of a scholarship fund of $8000 to enable a few selected Japanese girls to study in the United States. As part of her efforts to raise money for this fund, she collaborated with Alice Bacon, an American teacher friend who had spent some time in Tokyo, in writing the classic study *Japanese Girls and Women*. A detailed exposition of the lives of Japanese women of all classes, the book was published under Bacon's name only, so that Ume Tsuda might avoid the full force of the anticipated hostile reviews in Japan.[37] It was well received in the United States, where it generated much interest.

This publication and her fund-raising talks throughout the eastern United States were markedly successful, and a committee was set up to campaign for and administer the fund. Mary Morris and Martha Carey Thomas led this Philadelphia Committee, most of whose members were affluent wives of notably successful businessmen.

Given a one-year extension of her leave from the Peeresses' School, Ume spent a term at a Teachers' College in Oswego, northern New York, studying pedagogical methods, and she also visited two manual-training institutions, both of which impressed her. By 1892 she was back in Japan, where she immediately started to promote her scholarship program for four years of study at specified colleges in the United States. Stringent conditions were established by the Tokyo extension of the Philadelphia

Committee, but the first scholarships were awarded in 1893. Of major importance for Ume Tsuda personally was the fact that this fund demonstrated publicly, in the United States and Japan, her connection with higher education for women.

She resumed her position at the Peeresses' School, and was chosen to represent Japan at the 1898 Convention of Women's Clubs in Denver, Colorado. From there she went on to visit a number of women's colleges in England, particularly those in Oxford, Cambridge, and London; Dorothea Beale, principal of elite Cheltenham Ladies College in Gloucester, became an inspiration. Ume also took the precaution of seeking a blessing on her future work from the archbishop of York in order to justify in religious terms her aggressive, unwomanly ambitions.

Arguing that delay in advancing women's education would be a disgrace to the country and a drag on Japan's national progress, she finally opened her own school, the Women's English School, in Tokyo in 1900. Resigning from the Peeresses' School, and giving up her official rank, title, and court connection, she became a commoner, free to do what she wanted. A growing amount of reliable moral support came from social-reform movements in Japan, but great care was still required, and the orthodox role of the domestic sphere could not be ignored; financial support came mainly from the United States, via Mary Morris and the Philadelphia Committee.

The Women's English School started with four part-time teachers, including Alice Bacon, and fifteen students. It offered a three-year course of preparation for the government licensing examination for public-school teachers; since higher education and specialty courses were then still a male preserve in Japan, the Women's English School may rightly claim to have been the first of its kind. Umeko Tsuda (the name she took in 1902) adopted much of the curriculum of the leading men's colleges in Japan, as Carey Thomas had followed the Harvard curriculum. English language and literature were her special focus, but she developed a broad and thorough humanities-based education. The school prospered, enrollments rapidly increased, and a number of well-respected male instructors from Tokyo Imperial University and the Higher Normal School were brought in as part-time lecturers. Official recognition as a vocational college came from the government's Education Department in 1904, and after 1905 graduates were no longer required to pass the teachers' licensing examination.

Increasingly teacher training lost its central position in the college; more and more of the students were the daughters of government officials seeking higher education as a means to social mobility, or ambitious young women who went on to become vocal and active in questioning the traditional roles of their fellow countrywomen. Umeko kept a tight hand on her students' extracurricular activities, but in many ways she sympathized with those who were impatient with the slow rate of change. These activists were to a considerable extent her intellectual heirs, and graduates of her college were some of those who, from the 1920s on, made significant gains for Japan's women.

After two decades of constant hard work and worry about keeping the college running, Umeko's health broke down. As well as teaching at her college she had earned money by performing part-time work at other schools, compiling textbooks, and editing magazines; she also kept up a huge correspondence. With no institutional funding from the Japanese government, she had to rely entirely on financial support from the United States, which, understandably, was somewhat erratic at times. In 1917 respiratory problems hospitalized her for many weeks, and by 1920 she was forced to give up most of her teaching and administrative work.

She died of heart failure, at the age of sixty-four, on 16 August 1929, in her summer cottage in Kamakura, Kanagawa, an ancient town about thirty miles south of Tokyo, where she had lived since retiring. Four years after her death the Women's English School was renamed Tsuda Women's College (Tsuda Eigaku Juku); after World War II it became Tsuda College and about that time was given university status.

Twice transplanted, Umeko Tsuda tried to fit into each of the two societies that had molded her character. But the late nineteenth-century middle-class American values that she absorbed during her early years remained dominant in her outlook throughout her forty-five-year career in Japan. This meant that her life was a lonely one; to share thoughts and ideas she had to turn to her American friends. By the mid-1920s Umeko Tsuda's brand of late nineteenth-century feminism was failing to address the issues of the day. Nevertheless her limited effort, against considerable odds, to unlock the gateway to higher education for Japanese women was not in vain. Tsuda College is a legacy of which she can be proud. Her ashes lie buried on its grounds.

A MEXICAN SKYWATCHER

Ana Maria Fernández (fl. 1890s), a sky-phenomena observer in Mexico, coauthored a short note in the Central Meteorological Observatory's monthly *Boletin*. Her coauthor, meteorologist Vicente Fernández, was most likely a relative.

Their report (see bibliography) described strange sky effects seen on 14 November 1899. Between noon and one o'clock on that date drifts of delicate white fibers were glimpsed beneath the sun's rays at what appeared to be about the height of cirrus clouds; also observed were small, white, wool-like masses to which were linked clouds of particles looking like tiny insects shining in the sun. The authors compared these effects with similar unusual phenomena observed elsewhere. They suggested a possible connection between the observed daytime phenomena and the appearance of a fireball, or perhaps the shooting-star events common in mid-November. No biographical or other information about Ana Maria Fernández has been uncovered.[38]

NOTES

1. "Louise Rosalie Napias," http://fr.wikipedia.org/wiki/Louise_Rosalie_Napias, July 2014 (there are some errors in this article); Christian Rebisse, "Pierre-Augustin Chaboseau (1868–1946) Un Servidor Desconocido," http://www.biblioteca.amorc6.org/personajes/doc_download/513-pierre-augustin-chaboseau-1868-1946-un-servidor-desconocido (July 2014); Ancestry.co.uk, http://search.ancestry.co.uk/cgi-bin/sse.dll?gl=34&rank=1&sbo=t&gsbco=Sweden&gsln=Napias (July 2014).

2. For Blanche Edwards-Pilliet, see Mary R. S. Creese, *Ladies in the Laboratory II: West European Women in Science, 1800–1900* (Lanham, Md.: Scarecrow Press, 2004), 58–62.

3. Charles Sowerwine, *Sisters or Citizens? Women and Socialism in France since 1876* (Cambridge: Cambridge University Press, 1982), 119, 127.

4. Blanche Galien, "Ce que nous mangeons et ce que nous buvons," *La Fronde* no. 92, 11 March 1898; "A propos du tetanus," *La Fronde* no. 119, 6 April 1898; "A l'Institut Pasteur," *La Fronde* no. 197, 23 June 1898; "Conseils aux mères," *La Fronde* no. 263, 28 August 1898; "Une élection à l'Académie des Sciences," *La Fronde* no. 420, 1 February 1899; "Le microbe du cancer," *La Fronde* no. 489, 11 April 1899; "La Peste. Le retour du docteur Calmette," *La Fronde* no. 663, 2 October 1899; "L'incident du concours de l'internat en médecine," *La Fronde* no. 675, 14 October 1899.

5. See Mary Louise Roberts, "Copie subversive: Le journalism féministe en France à la fin du siècle dernier," *CLIO: Histoire, femmes et sociétés* no. 6 (Autumn 1997): 230–49. For Klumpke and Royer, see Creese, *Ladies in the Laboratory II*, 79–81 and 85–87.

6. Louise Chaboseau-Napias, "Les partis socialistes: All[e]magne: l'organisation des femmes socialistes," *Le mouvement socialiste: Revue de critique sociale, littéraire et artistique* (Paris), 15 December 1904. See also "Le droit de suffrage pour les femmes," *La Revue Socialiste* 44, no. 260 (August 1906): 145–66.

7. Sowerwine, *Sisters or Citizens?*, 119.

8. "Tauscherné Geduly, Hermina," http://www.fsz.bme.hu/mtsz/mhk/csarnok/t/tauscher.htm (July 2014); Geduly, Lajos Gábor, http://zope.lutheran.hu/honlapok/protestans/felvidek/pozsony/lelkeszek/geduly (July 2014). Google translations available for both.

9. Lucy Walker (1836–1916) was the oldest child of Frank Walker, a Liverpool lead merchant, and his wife, Jane McNeile. Lucy started her climbing when she was about twenty-two, after her doctor had prescribed walking as a cure for rheumatism. At that time many women were making walking expeditions below the snowline and some had crossed high passes, but few had reached summits; Lucy was the first to climb regularly in the Alps. A big, strong woman, equal to any climber in pace, endurance, and determination, she climbed with her father and younger brother Horace, both Alpine Club members, along with the Oberland guide Melchior Anderegg. Clad in her usual white-print dress, and on a diet of sponge cake and champagne, she had climbed four major peaks, including Mont Blanc, by 1862. Two years after her 1869 ascent of Piz Bernina she reached the summit of the Matterhorn, climbing with her father and Anderegg. She was the first woman to accomplish this feat, which was especially notable since it came only six years after English climber Edward Whymper had become the first to reach the summit; the accident that killed four of his party on the descent had given the peak a certain notoriety. After her father's death in 1872, Lucy Walker climbed with her brother and Anderegg. She continued to add new peaks to her list until 1879; she was then forty-three, and her doctor advised her to give up long expeditions, although she continued to walk in the Alps. Celebrated as the pioneer of women climbers, in 1909 she joined the

newly formed Ladies Alpine Club and was its second president (1913–1915). She made ninety-eight expeditions but never published any accounts.

At her home in Liverpool she led the life of a cultured Victorian lady, an expert needlewoman, fluent in several languages, and a charming hostess. She died in Liverpool on 10 September 1916, after a long illness. See Janet Adam Smith, "Walker, Lucy (1836–1916)," in *Oxford Dictionary of National Biography* (Oxford: Oxford University Press, 2004), 56:873–74.

10. Tauscher, "Wettlauf über den Wolken," *Osterreichischen Alpenzeitung* (1881): 242.

11. D. Paskalev, D. Radoinova, and B. Galunska, "Doktor Zaharina Dimitrova (1873–1940): Pioner v oblasti issledovania mikrostruktury epefiza" [Dr. Zaharina Dimitrova (1873–1940): Pioneer in research on the microstructure of the pineal gland], *Zhurnal Nefrologiia, Istoriia Meditsiny* [Journal of Nephrology, History of Medicine section] 15, no. 2 (2011): 115–18. English trans. of abstract available online: http://journal.nephrolog.ru/toc.php?id=24 (July 2014); Georgeta Nazarska, "Bulgarian Women Medical Doctors in the Social Modernization of the Bulgarian National State (1878–1944)," *Historical Social Research* 33, no. 2 (2008): 232–46; Simone Gilgenkrantz, "Les premières étudiantes étrangères en médecine à Nancy (1894–1914)," http://www.professeurs-medecine-nancy.fr/premieres_etudiantes_etrangeres.htm (July 2014); P. Petkov and A. K'nev, "130 godini ot rozhdenieto na d-r Zakharina Dimitrova: Izvestna b sveta, nepoznata b Bulgariia" [130th anniversary of the birth of Dr Zakharina Dimitrova: Famous worldwide, little known in Bulgaria], *Asklepios* 17 (2004): 184–86.

12. Bulgaria's liberal policy on women's medical education and careers stands in notable contrast to that in many Western countries and tsarist Russia at the time. It must be acknowledged, however, that when they came to practice, Bulgarian women doctors tended to occupy low posts on the professional ladder, as did their contemporaries elsewhere when they entered the workforce.

13. Nazarska, "Bulgarian Women Medical Doctors," 234.

14. Zaharina Dimitrova, "Recherches sur la structure de la glande pinéale chez quelques mammifères," *Névraxe* 2, fasc. 3 (1901): 261–321.

15. Şerban N. Ionescu, *Who Was Who in Twentieth Century Romania*, East European Monographs, Boulder (New York: Columbia University Press, 1994); *Enciclopedia României*, http://enciclopediaromaniei.ro/wiki/Elena_Densuşianu-Puşcariu (July 2014); Richard Constantinescu, "O personalitate a medicinei româneşti: Elena Densusianu-Puscariu," ROST 42–43 (August–September 2006), http://www.rostonline.org/rost/aug-sep2006/densusianu.shtml (July 2014); Richard Constantinescu, "Priviri anatomorpatologice," *Viata Medicala* 27, 17 January 2012, http://www.viata-medicala.ro/Priviri-anatomopatologice.html*articleID_4710-dArt.html (July 2014); *Bibliografia românească modernă, Autori-Litera P*, http://www.biblacad.ro/bnr/brmautori.php?aut=p&page=2320&limit=20 (July 2014).

16. Elena Densuşianu-Puşcariu is sometimes referred to as Romania's first woman to qualify as a physician, but readily available evidence suggests that that honor probably belongs to Maria Cuţarida-Crătunescu (1857–1919). The daughter of Basile Cuţarida, a property owner in Bucharest, Maria was born on 10 February 1857 in the city of Călăraşi in southeast Romania, near the Bulgarian border. After completing her high school course at the Central School in Bucharest, she began studies in the medical faculty of the University of Zurich, starting in the winter semester of 1877. In 1880 she transferred to the University of Montpellier, where she earned a bachelor's degree and carried out a period of hospital work. Specializing in gynecology with Sigismond Jaccoud, professor of clinical medicine at several Paris hospitals, she went on to doctoral research on early indications of uterine cancer. The University of Paris faculty of medicine awarded her a degree in 1884. See Marie V. Coutzarida, "De l'hydrorriiée et de sa valeur séméiologique dans le cancer du corps de l'utérus," doctoral thesis, medical faculty, University of Paris, 1884, no. 241 (Paris: A. Parent, 1884), 96 pp.

She returned to Romania the same year, passed the Romanian state qualifying examinations *magna cum laude*, and opened a private clinic as well as joining the staff of the gynecological section of the Bucharest Philanthropy Hospital (Spitalul Filantropia), where she quickly rose to a leading position. A militant feminist with the health of women and children her major concern, she played an important role in the founding, in 1897, of the Maternal Society (Societatea Materna). Until the First World War she kept abreast of medical advances in western Europe, sometimes attending international congresses. Maria Cuţarida-Crătunescu died on 16 November 1919, at the age of sixty-two. See the matriculation record, Zurich University, http://www.matrikel.uzh.ch/active/static/25081.htm (July 2014); *Enciclopedia României*, http://enciclopediaromaniei.ro/wiki/Maria_Cuţarida-Crătunescu (July 2014); BIU Santé, faculté de médecine de Paris, http://www2.biusante.parisdescartes.fr/livanc/?cote=TPAR1884x241 (July 2014).

17. Elena Densuşianu, "Consideraţiuni asupra Neoplasmelor Sînului cu deosebire din punct de vedere al histologiei patologice," doctoral dissertation for a degree in medicine and surgery, medical faculty, University of Iaşi, October 1898 (Tipografia Naţională, 1898). See *Bibliografia românească modernă*, note 62.

18. The work of another early woman ophthalmologist, the Russian Raissa Shlykova, later Kerschbaumer (1854–1928), is perhaps worth recalling here. Specializing in diseases of the eye, Raissa Shlykova graduated from

the University of Bern in 1876. She was the first woman to practice independently in Austria-Hungary, and opened her own ophthalmic clinic in Salzburg in 1890. Shlykova earned an enviable reputation for successful cataract removal. In 1896 she returned to Russia, where for eleven years she provided sorely needed ophthalmic care to communities in the Caucasus region and along the Trans-Siberian Railway. Then followed a few more years in Vienna before she immigrated to the United States in 1911. See Creese, *Ladies in the Laboratory II*, 171–73.

19. Elena Puşcariu's publications include the following: Elena Densuşianu and Ernest Pinoy, "Action du cantharidate de potasse sur la cellule nerveuse," *Société de Biologie. Comptes rendus* 53, no. 5 (1901): 101–2; Elena Densuşianu and Victor Babes, "Sur la néphrite hématogène des pyramides," *Archives de médecine expérimentale et d'anatomie pathologique*, Paris (July 1902): 502–14; Elena Puşcariu, "Un caz de tromboză a venei nasale superioare a retinei de natură syphilitică," *Buletinul Medicilor* şi *Naturalişti din Iaşi* (1909); "Câteva instrucţiuni pentru campania contra conjunctivitei granuloase," Institutul de Arte Grafice, *Speranţa* (1913); "Le traitement de la syphilis oculaire," *Revista Ştiinţelor Medicale*, no. 7–8 (July–August 1914); "Operaţiuni şi metode noi în chirurgia ochiului. Trepanaţia sclero-corneană in glaucom. Extracţia intra-capsulară a cataractei. Avansament şi lambou conjunctival in operaţia cataractei," *Revista Ştiinţelor Medicale*, no. 7–8 (July–August 1914); "Le trachome (La conjonctivite granuleuse) au point de vue clinique et avec des considérations sur son état en Roumanie," (tip. Modernă Cultura, Societate Colectivă, 1914); with Eugénie Lazaresco, "Deux observations de 'xeroderma pigmentosum' avec épithéliomas épibulbaires," (n.p., 1930); "Le plasmome de la conjonctive . . . ," *Annales d'Oculistique* (December 1930): 1017–26; "La kératite pustuliforme profonde (Fuchs). À propos de l'observation d'un cas chez une hérédosyphilitique . . . ," *Annales d'Oculistique* (April 1931): 243–55; N. Hortolomei, and Dumitru Lazaresco, "Paraglangliome intercarotidien," *Lyon-chirurgical* (May–June 1931): 298–316; *Vingt syndromes oculaires hérédo-familiaux et congénitaux*, (Paris: Doin, 1963). See also bibliography.

20. See note 19 above.

21. *Bulletin de l'Institut National Genevois* (Membres honoraires, section des sciences et mathématiques) 45, no. 1 (1922): 339. See also the Russian botanist Olga Malinesco, discussed in chapter 3, who worked with Chodat on blue-green algae.

22. For Mariette Pompilian, who made her career in Paris in laboratory research in physiological and neurological processes, see Creese, *Ladies in the Laboratory II*, 75–76, 248.

23. These three were Constance Gelderblom, from Holland, who graduated in 1895, and two Americans, Elizabeth Bickford (1895) and Julia Platt (1898). See E. Th. Nauck, *Das Frauenstudium an der Universität Freiburg im Breisgau* (Freiburg im Breisgau: Eberhardt Albert Universitätsbuchhandlung, 1953), 62; James C. Albisetti, *Schooling German Girls and Women* (Princeton, N.J.: Princeton University Press, 1988). For Platt, see Mary R. S. Creese, *Ladies in the Laboratory? American and British Women in Science, 1800–1900* (Lanham, Md.: Scarecrow Press, 1998), 101–3.

24. For a few general remarks about his first women students, see Robert Wiedersheim, *Lebenserinnerungen* (Tübingen: J. E. B. Mohr, 1919), 111–14. Gaup and Wiedersheim also sponsored and directed the research of Julia Platt, whose work on the development of the cartilaginous skull of the aquatic salamander *Necturus* was the subject of her 1898 doctoral thesis.

25. BBC-Your Paintings: Tilly Kettle, Major William Davy (d. 1784), Bengal Army, Persian Secretary to the Governor-General, in Persian Dress, http://www.bbc.co.uk/arts/yourpaintings/paintings/major-william-davy-d-1784-bengal-army-persian-secretary-t182741 (July 2014); BBC-Your Paintings: Tilly Kettle, Major William Davy (d. 1784), Bengal Native Infantry, http://www.bbc.co.uk/arts/yourpaintings/paintings/major-william-davy-d-1784-bengal-native-infantry-182477 (July 2014).

26. Henry Harpur Spry (1804–1842), a native of Truro, Cornwall, was an East India Company surgeon appointed to the company's Bengal staff in 1827. Having a strong interest in the new science of statistics, he brought out a two-volume work that listed animals, plants, manufactures, and other data about India, which was designed to educate the British public about the subcontinent's agricultural and manufacturing capabilities. *Modern India: With Illustrations of the Resources and Capabilities of Hindustan* (London, 1837) is now considered to be a rambling and even sensational work. Spry became secretary of the Agricultural and Horticultural Society of India and also held the post of secretary to the statistics department of the Asiatic Society of Bengal. He died at Fort William, Calcutta (Kolkata), in 1842. See Katherine Prior, "Spry, Henry Harpur (1804–1842)," in *Oxford Dictionary of National Biography* (Oxford: Oxford University Press, 2004), 52:3.

27. The weavers of Dacca (Dhaka) were famous for centuries for the production of the very thin cotton cloth known as muslin. The finest muslins, materials of delicate, gossamer quality, were made from a cotton grown at certain locations on the banks of the Bramaputra River and its branches. Much prized by emperors and the nobility, by the eighteenth century this fabric was highly valued in Europe. Only the competition of cheaper, machine-made substitutes eventually caused its use to decline. See Abdul Karim, "Muslin," in *Banglapedia, National Encyclopedia of Bangladesh* (Dhaka: Asiatic Society of Bangladesh, 2003) 7:179–81.

28. Known as snake plant, mother-in-law's tongue, or bowstring hemp in the West, Sanseviera is popular as an ornamental.

29. The fiber extracted from pineapple plant leaves is now used in combination with silk or polyester to make the fabric known as *piña*, a very fine cloth produced mainly in the Philippines.

30. Efforts have been made at the Computer-Aided Design Centre at Krishnanagar in West Bengal to revive the muslin textile industry and produce cloth that incorporates indigenous but varied designs for the popular Dhakai saris. See K.R.N. Swamy, "Reviving the Once-Famous Muslin Industry," *Chandigarh* (India) *Tribune*, 19 October 2002, http://www.tribuneindia.com/2002/20021019/windows/main4.htm (July 2014).

31. Kai Wing Chow, "Jin Yamei," trans. Stephanie Po-yin Chung, in *Biographical Dictionary of Chinese Women*, ed. Lily Xiao Hong Lee, and Agnes D. Stefanowska, assisted by Sue Wiles, vol. 1, *The Qing Period, 1644–1911*, ed. Clara Wing-chung Ho (Armonk, N.Y.: M. E. Sharpe, 1998), 94–96; Mark Rowley, "Y. May King (Jin, Yanei), 1864–1934," in Cabinet of Art and Medicine, http://www.artandmedicine.com/biblio/authors/america/King.html (July 2014); Carole Ann Duff, "Christianity, Science and Society: Two Nineteenth-Century American Medical Missionaries in the Far East," master's thesis, University of Texas, Dallas, 1977 (Dallas: University of Texas, 1977); Divie Bethune McCartee and Robert Elliott Speer, *A Missionary Pioneer in the Far East: A Memorial to Divie Bethune McCartee*, ed. Robert E. Speer (New York: Fleming H. Revell, 1922).

32. The Shanghai Mixed Court (1863–1880) was a British colonial institution established in 1863. Originally intended to deal with criminal cases that arose between British and Chinese nationals, it soon expanded to try both civil and criminal disputes between Chinese operating in the foreign settlement. See Kelly Anne Hammond, "The Shanghai Mixed Court, 1863–1880," master's thesis, Simon Fraser University, 2007, online at http://summit.sfu.ca/item/8387 (July 2014).

33. The Tokyo Imperial University was then emphasizing the introduction of Western expertise in science and technology, as well as starting a program of legal studies. See also the sketch of Ume Tsuda, below.

34. Sources differ on whether Jin Yamei's child was a boy or a girl.

35. Barbara Rose, *Tsuda Umeko and Women's Education in Japan* (New Haven, Conn.: Yale University Press, 1992); Yoshiko Furuki, *The White Plum: A Biography of Ume Tsuda* (New York and Tokyo: Weatherhill, 1991); "Bryn Mawr Celebrates Centennial of Tsuda College," http://www.brynmawr.edu/alumnae/bulletin/tsuda.htm (July 2014); Tsuda College—Profile, http://www.tsuda.ac.jp/en/about/ (July 2014); "Umeko Tsuda—1899," St. Hilda's College, University of Oxford, http://www.st-hildas.ox.ac.uk/umeko-tsuda (July 2014); Amy D. Yamashiro and Ethel Ogane, "Tsuda Ume: Pioneering Education for Women and ELT," *Language Teacher* 22.5, May 1998. See also Japan Association of Language Teachers, Kyoto, http://jalt-publications.org/tlt/articles/2302-tsuda-ume-pioneering-education-women-and-elt (Sept. 2014).

36. Although the first American schools to offer advanced education for girls and women in the early years of the nineteenth century had to justify their existence in terms of making women more capable of bringing up the next generation, this argument was no longer necessary in the 1880s. By then much of American society had accepted the idea that women should not be denied the right to a college-level education. For an account of the opening of the first American female seminary (the Troy Female Seminary in northern New York), see Margaret W. Rossiter, *Women Scientists in America: Struggles and Strategies to 1940* (Baltimore: Johns Hopkins University Press, 1982), 4–8. Tsuda's acceptance of the earlier model of the first American female seminaries during her impressionable teenage years may have helped her to more willingly take on the Japanese restrictions as being part of a necessary early stage in the introduction of higher education for women.

37. Alice Mabel Bacon, *Japanese Girls and Women* (Boston and New York: Houghton Mifflin, 1891; rev. ed., 1902). Ume Tsuda is credited in the preface with more than usual assistance.

38. The observed white fibers were most likely masses of giant webs produced by social spiders (species *Anelosimus eximinus*), which live in colonies sometimes composed of thousands of individuals. The webs can stretch from the ground up to tree canopies or sixty-five-foot-high buildings; they can be detached from their moorings by strong winds and moved considerable distances, newly hatched young being shaken out en route. The right light conditions make visible such events, which have been observed now from Brazil to northern Texas. See Rachel Nuwer, "It's Raining Spiders in Brazil," Smithsonian.com, 18 Feb. 2013, http://www.smithsonianmag.com/science-nature/its-raining-spiders-in-brazil-19885877/?no-ist (July 2014). My thanks to Dr. Anna Creese for translating the Fernández article from the Spanish.

Bibliography of Papers by Women of Imperial Russia in Scientific Periodicals, 1800–1900

Entries are from the *Catalogue of Scientific Papers, 1800–1900*, 19 volumes, compiled by the Royal Society, London (Cambridge: Cambridge University Press, 1867–1925). The list has been supplemented by additional entries, enclosed in braces {}. Most of the latter were collected from periodicals not examined by the Royal Society indexers: minor journals, regional journals of short run, general-interest magazines, and journals in fields not covered in the *Catalogue* (such as some clinical-medicine journals). For consistency, only entries derived from the *Catalogue* are included in comparisons of productivity. The *Catalogue*'s style of presentation and its nomenclature conventions have been retained for the most part. A key to abbreviations follows the bibliography.

POLAND

Botany

Balicka-Iwanowska, Gabrielle

Balicka-Iwanowska, (Mme.) Gabrielle

Contribution à l'étude anatomique et systématique du genre Iris et des genres voisins. *Arch. Sci. Phys. Nat.*, **28** (1892) 413–35; **29** (1893) 185–200; 225–41

La structure des cyanophycées. *Arch. Sci. Phys. Nat.*, **33** (1895) 206–8

Die Morphologie des Thelygonum Cynocrambe. *Flora*, **83** (1897) 357–66

Contribution a l'étude du sac embryonnaire chez certains gamopétales. *Flora*, **86** (1899) 47–71

Chodat, Robert; Iwanowska, (Mme.) G.

Hybrides de Montbretia potsii et de Crocosmia aurea. *Arch. Sci. Phys. Nat.*, **24** (1890) 665–66

Sur la structure anatomique des feuilles des Iridées. *Arch. Sci. Phys. Nat.*, **26** (1891) 496–501; **27** (1892) 128–31

La feuille des Iridées, essai d'anatomie systématiques. *J. Bot., Paris*, **6** (1892) 220–32; 253–67

[Caractères des Trémandracées.] *Arch. Sci. Phys. Nat.*, **29** (1893) 325–26

Goldflus, Mathilde

Goldflus, (Mlle.) Mathilde

Sur la structure et les fonctions de l'assise épithéliale et des antipodes chez les composées. *J. Bot., Paris*, **12** (1898) 374–84; **13** (1899) 9–17; 49–59; 87–96

Chodat, Robert; Goldflus, (Mlle.) Mathilde

Sur certaines particularités de l'ovule des composées. *Arch. Sci. Phys. Nat.*, **5** (1898) 390–92

Chodat, Robert; Goldflus, Mathilde; Preda, Agiluf

{Note sur la culture des Cyanophycées et sur le développement d'oscillatoriées coccogènes. *Univ. de Genève, Lab. bot.*, série 4, fasc. 6 (1897)}

Maliniak, Marie
Maliniak, (Mlle.) Marie
Recherches sur la formation des matières protéiques à l'obscurité dans les végétaux supérieurs. *Rev. Gén. Bot.*, **12** (1900) 337–43

Entomology

Pavlova, Mariia Ivanovna
Pavlov, M.
Ueber ampullenartige Blutcirculationsorgane im Kopfe verschiedener Orthopteren. *Zool. Anz.*, **18** (1895) 7–13; 92
Zum Bau des Eingeweidenervensystems der Insecten. *Zool. Anz.*, **18** (1895) 85–87
{Zur Frage der Metamorphose bei der Familie der Mantidae. *Varshava Lab. Zool. Raboty* (1896) 155–82}

Medicine

Kleindienst, Natal'ia
Kleindienst, Nathalie
Ueber Pental als Anaestheticum. [1892] *Deutsche Ztschr. Chirurg.*, **35** (1893) 333–50
Poplawska, Stanislawa
Poplawska, (Mme.) Stanislawa
Zur Aetiologie der Panophthalmie nach Verletzung durch Fremdkörper. *Fortschr. Med.*, **8** (1890) 489–92
Zur Aetiologie der Entzündung des Auges nach Verletzung durch Fremdkörper. [1890] *Arch. Augenheilk.*, **22** (1891) 337–54; *Arch. Ophthalm.*, **21** (1892) 22–32

Neurology and Physiology

Gruzewska, Z.
Gruzewska, (Mlle.) Z.
Crystallisation de l'albumine du sang. *Paris, Ac. Sci. C. R.*, **128** (1899) 1535–37
Joteyko, Joséphine
Joteyko, (Mlle.) J.
Action toxique curarisante de la neurine. *Paris, Soc. Biol. Mém.*, **49** *(C. R.)* (1897) 341–43
La fatigue et la réparation du muscle lavé de sang. *Paris, Soc. Biol. Mém.*, **50** *(C. R.)* (1898) 420–21
La méthode graphique appliquée à l'étude de la fatigue. *Rev. Sci.*, **10** (1898) 486–94; 516–22
Recherches expérimentales sur la fatigue des centres nerveux par l'excitation électrique. *Paris, Soc. Biol. Mém.*, **51** *(C. R.)* (1899) 384–85
Recherches expérimentales sur la fatigue des organes terminaux. *Paris, Soc. Biol. Mém.*, **51** *(C. R.)* (1899) 386–87
L'effort nerveux et la fatigue. Recherches ergographiques et dynamométriques. [1899] *Arch. de Biol.*, **16** (1900) 479–535
Le quotient de la fatigue H/N. *Paris, Ac. Sci. C. R.*, **130** (1900) 527–29; 756
Le travail des centres nerveux spinaux. *Paris, Ac. Sci. C. R.*, **130** (1900) 667–69
Effets du travail de certains groupes musculaires sur d'autres groupes qui ne font aucun travail. *Paris, Ac. Sci. C. R.*, **131** (1900) 917–19
Recherches expérimentales sur la résistance des centres nerveux médullaires à la fatigue. *Inst. Solvay Trav.*, **3**, fasc. 2 (1900) 55 pp.
Joteyko, (Mlle.) Joséphine; Richet, Charles
Réparation de la fatigue musculaire par la respiration élémentaire du muscle. *Paris, Soc. Biol. Mém.*, **48** *(C. R.)* (1896) 146–47

Joteyko, (Mlle.) Joséphine; Stefanowska, (Mlle.) Micheline

Anesthésie générale et anesthésie locale du nerf moteur. *Paris, Ac. Sci. C. R.*, **128** (1899) 1606–8

Nusbaum-Hilarowicz, Rozalia

Nusbaum, Rozalia

[Lombroso's hypothesis in the light of recent anatomical researches.] *Kosmos (Lwow)*, **22** (1898) 42–64

Przedniewicz, Eugénie Antoine de

Przedniewicz, (Mlle.)

Le phospore et l'alimentation. Les applications de l'acide phosphorique. *Congr. Int. Méd. C. R.* **6** (1900) 186–205

Stefanowska, Michalina

Stefanowska, (Mlle.) Micheline

La disposition histologique du pigment dans les yeux des arthropodes sous l'influence de la lumière directe et de l'obscurité complète. [1890] *Rec. Zool. Suisse*, **5** (1892) 151–200

{Action des alcaloïdes et de diverses substances médicamenteuses sur les coeurs lymphatiques de la grenuille: expériences faites aux laboratoires de physiologie de l'Université libre de Bruxelles et de l'Université de Varsovie. *Soc. Sci. Méd. Nat. Bruxelles. Ann.*, **5**, fasc. 3 (1896) 425–55}

Les appendices terminaux des dendrites cérébraux et leurs différentes états physiologiques. [Introduction et notes par M. le docteur Paul Heger.] *Inst. Solvay Trav.*, [**1**], fasc. 3 (1897) 57 pp.; {*Soc. Sci.* Méd. *Nat. Bruxelles. Ann.*, **6**, no. 1 (1897) 351–407}

{Sur les appendices des dendrites. *Soc. Sci. Méd. Nat. Bruxelles. Bull.*, **55** (1897) 76–81}

Sur le mode d'articulation entre les neurones cérébraux. *Paris, Soc. Biol. Mém.*, **49** *(C. R.)* (1897) 969–70

{Développement des cellules corticales chez la jeune souris. *Soc. Sci.* Méd. Nat. Bruxelles. *Bull.*, **56** (1898) 28–31}

Évolution des cellules nerveuses corticales chez la souris après la naissance. *Inst. Solvay Trav.*, **2**, fasc. 2 (1898) 44 pp.

Localisation des altérations cérébrales produites par l'éther. *Inst. Solvay Trav.*, **3**, fasc. 3 (1900) 25–81

Sur le mode de formation des varicosités dans les prolongements des cellules nerveuses. *Inst. Solvay Trav.*, **3**, fasc. 3 (1900) 83–100

Joteyko, (Mlle.) Joséphine; Stefanowska, (Mlle.) Micheline

Anesthésie générale et anesthésie locale du nerf moteur. *Paris, Ac. Sci. C. R.*, **128** (1899) 1606–8

Szczawinska, Wanda

Szczawinska, (Mlle.) Wanda

{Contribution à l'étude des yeux de quelques Crustacés et recherches expérimentales sur les mouvements du pigment granuleux et des cellules pigmentaires soux l'influence de la lumière et de l'obscurité dans les yeux des Crustacés et des Arachnides. *Arch. de Biol.*, **10** (1890) 523–32; 559–66}

{Recherches récentes sur la structure histologique de la cellule nerveuse. *Année psychologique,* **2**, issue 1, no. 2 (1895) 510–20}

Sur la structure réticulaire des cellules nerveuses centrales. *Paris, Ac. Sci. C. R.*, **123** (1896) 379–80

Recherches sur le système nerveux des Sélaciens. *Arch. de Biol.*, **15** (1898) 463–509

Tomaszewicz-Dobrska, Anna

Tomaszewicz, A.

Die Wirkung des Chlorals und der Trichloressigsäure. *Pflüger, Arch. Physiol.*, **9** (1874) 35–43

Zielinska, Mariia

Zielinska, Marie

Beiträge zur Kenntniss der normalen und strumösen Schilddrüse des Menschen und des Hundes. *Virchow, Arch.*, **136** (1894) 170–94

Chemistry

Bushmakina, S. V.

Wagner, E. E.; Bušmakina, S. V.

[Sur l'oxydation du safrole et de l'isosafrole par le permanganate de potassium.] [1891] *Varsovie Soc. Nat. Trav. (C. R., Phys. Chim.),* no. 2 (1891–1892) 1–2

Haberkant, Wanda

Kehrmann, J. A. L. Friedrich; Haberkant, (Mlle.) Wanda

[Derivées de l'acide naphtopicrique.] *Arch. Sci. Phys. Nat.*, **4** (1897) 93–94

Zur Kenntniss der Naphthopikrinsäure. [1897] *Berlin, Chem. Ges. Ber.*, **31** (1898) 2420–24

Idz'kovskaia, Mathilde

Idzkovskij, (Mlle.) M.; Wagner, Egor E.

{[Sur l'oxydation des corps alicycliques. Ch. 1. Oxydation de l'acétyle-trimethylène.] *Varsovie Soc. Nat. Trav. (C. R., Phys. Chim.)*, mém. 2, fasc. 2 (1897)}

{[Contributions to the knowledge of the reaction of oxidation of alicyclic compounds. Oxidation of acetyltrimethylene.] *Paris Soc. Chim. Bull.*, **20** (1898) 850–51; *Russ. Phys.-Chem. Soc. J.*, **30** (Chem) (1898) 259–69}

Kehrmann, Friedrich; Idzkovskij, (Mlle.) M.

Nitroquinone. *Arch. Sci. Phys. Nat.*, **7** (1899) 591; *Berlin, Chem. Ges. Ber.*, **32** (1899) 1065–72; 4082

Mazurowska, Mariia (pseud. Orlowsky, Marjan)

Orlowsky, Marjan

Ueber die Einwirkung des Sulfurylhydroxylchlorids auf Alkohole. *Berlin, Chem. Ges. Ber.*, **8** (1875) 332–35

Mazurowska, Marja

Ueber Schwefelsäureäther. [1875] *J. Prakt. Chem.*, **13** (1876) 158–75

Polonowska, Natalia

Polonowska, Natalie

Ueber den sogenannten Carbacetessigäther. *Berlin, Chem. Ges. Ber.*, **19** (1886) 2402–4

Ueber die Oxime des Benzils. *Berlin, Chem. Ges. Ber.*, **21** (1888) 488–91

Goldschmidt, Heinrich; Polonowska, Natalie

Ueber das Diphenyloxäthylamin. *Berlin, Chem. Ges. Ber.*, **20** (1887) 492–95

Ueber das Anisamin. *Berlin, Chem. Ges. Ber.*, **20** (1887) 2407–13

Srzedzińska, R.

Jaworowski, Adam; Srzedzińska, R.

[Contributions to the science of the investigation of alkaloids using aldehydes and ketones.] [1894] *Kosmos (Lwow)*, **20** (1895) 446–47

Geology

Czarnowska, Jósefa

Czarnowska, Jósefa

[Notes on the geological history and the present condition of the Kara Bugaz according to the memoire of N. Andrusow.] *Kosmos (Lwow)*, **25** (1900) 88–107

RUSSIA

Bacteriology

Mechnikova, Ol'ga (née Belokopytova)

Metchnikoff, (Mme.) Olga

Contribution à l'étude de la vaccination charbonneuse. *Ann. Inst. Pasteur*, **5** (1891) 145–57

Raskin, Mariia
Raskin, Marie
Zur Züchtung der pathogenen Mikroorganismen auf aus Milch bereiteten festen und durchsichtigen Nährböden. [1887–1888] *St. Petersb. Med. Wschr.*, **12** (1887) 357–60; *Biol. Centrbl.*, **8** (1889) 462–70
Klinisch-experimentelle Untersuchungen über Secundärinfection bei Scharlach. *Centrbl. Bakt.*, **5** (1889) 433–44; 465–79
Ueber die Aetiologie des Scharlachs. *Centrbl. Bakt.*, **5** (1889) 683
{Micrococcus pyogenes tenuis in einem Falle von Pyämie. *Centrbl. Bakt.*, **5** (1889) 714}
Rechtsamer, Mariia
Rechtsamer, (Mme.) M.
Ueber die feinen Spirillen in Dejektionen Cholerakranker. *Centrbl. Bakt.*, **15** (1894) 795–98
Schultz, Nadezhda Karlova
Schultz, (Mme.) Nadežda Karlova
{Ueber die Bereitung des flussigen Blutserums. *Centrbl. Bakt.*, **5** (1889) 713}
Zur Frage von der Bereitung einiger Nährsubstrate. *Centrbl. Bakt.*, **10** (1891) 52–64
De l'action des antiseptiques sur le Bacillus pestis hominis et de la désinfection d'effets et de locaux contaminés par la peste bubonique. *Centrbl. Bakt. (Abt. 1)*, **23** (1898) 594–98; *St. Pétersb. Arch. Sci. Biol.*, **6** (1898) 397–426
Afanasjev, Michael Ivanovich; Schulz, (Frau)
Ueber die Aetiologie der Actinomycosis. *Centrbl. Bakt.*, **5** (1889) 683–84
Solomon, Vera
Solomon, Vera
Experimentelle Untersuchungen über Rabies. 1) Neue Methoden der experimentellen Diagnose der Rabies. 2) Wirkung der Galle auf das Virus der Rabies. *Centrbl. Bakt. (Abt. 1)*, **28** (1900) 70–79
Tsiklinskaia, Praskoviia Vasilievna
Tsiklinski, (Mlle.) P.
Recherches sur la virulence de la bactéridie. *Ann. Inst. Pasteur*, **6** (1892) 465–77
Sur les microbes thermophiles. *Ann. Micrógr.*, **10** (1898) 286–88
Sur les Mucédinées thermophiles. *Ann. Inst. Pasteur*, **13** (1899) 500–505
Sur les microbes thermophiles des sources thermales. *Ann. Inst. Pasteur*, **13** (1899) 788–95
Zaïdmann, Rosalie
Rhodet, A. J.; Zaïdmann, (Mlle.)
Injections intra-spléniques de Bacilles d'Eberth et coli. *Paris, Soc. Biol. Mém.*, **52** *(C. R.)* (1900) 1007–9
Ziber-Shumova, Nadezhda Olimpievna
Sieber-Šumov, (Mme.) Nadežda Olimpievna
[Recherches sur les streptococcus pathogènes.] *St. Pétersb. Arch. Sci. Biol.*, **1** (1892) 264–97
Contribution à l'étude des poissons venimeux. Sur le bacillus Piscicidus agilis, microbe pathogène pour les poissons. *Pharm. Ztschr. Russland*, **34** (1895) 193–99; 209–15; 225–32; 241–48; *St. Pétersb. Arch. Sci. Biol.*, **3** (1895) 226–56
Les sérums thérapeutiques anticocciques. *Pharm. Ztschr. Russland*, **35** (1896) 641–643; 657–61; 673–77; 689–93; *St. Pétersb. Arch. Sci. Biol.*, **4** (1896) 415–28
Entgegnung [an Prof. Dr. Oscar Wyss: Ueber eine Fischseuche durch Bacterium vulgare]. *Ztschr. Hyg.*, **28** (1898) 159–61
Nencki, M.; Sieber-Šumov, (Mme.) N. O.; Šumov-Simanovskij, (Mme.) E. O.
Die Entgiftung der Toxine durch die Verdauungssäfte. *Centrbl. Bakt. (Abt. 1)*, **23** (1898) 840–47; 880–87
Nencki, M.; Sieber-Šumov, (Mme.) N. O.; Wyžnikiewicz, W.
Recherches sur la peste bovine. *Centrbl. Bakt. (Abt. 1)*, **23** (1898) 529–38; *St. Pétersb. Arch. Sci. Biol.*, **6** (1898) 374–96; **7** (1899) 303–336; {Ueber die Rinderpest. *Berlin. Klin. Wochenschr.*, **24** (1897)}

Biochemistry

Ekunina, Mariia

Ekunina, Marie

Ueber die Ursache der sauren Reaction der thierischen Gewebe nach dem Tode. *J. Prakt. Chem.*, **21** (1880) 478–88

Manasseina, Mariia Mikhailovna (née Korkunova)

Manassein, Marija Michajlovna

{K ucheniiu ob alkogol'nom brozhenii. [Toward a theory of alcoholic fermentation.] *St. Petersb., Med. Vestnik*, **8** (1871) 57–59}

{[Toward a theory of alcoholic fermentation.] *Voenno-med. Zhurnal*, **3**, no. 8 (1871) 155–75}

Zur Frage von der alkoholischen Gährung ohne lebende Hefezellen. *Berlin, Chem. Ges. Ber.*, **30** (1897) 3061–62

Zur Frage von der alkoholischen Gährung ohne lebende Hefezellen und über die Gährung im Allgemeinen. [1898] *Physiologiste Russe*, **1** (1898–1899) 88–94

Shumova-Simanovskaia, (Mme.) Ekaterina Olimpievna

Šumov-Simanovskij, (Mme.) E. O.

[Sur le suc stomacal et la pepsine chez les chiens.] *St. Pétersb. Arch. Sci. Biol.*, **2** (1893) 462–93

Simanovskij, N. P.; Šumov-Simanovskij, (Mme.) E. O.

Ueber den Einfluss des Alcohols und des Morphiums auf die physiologische Oxydation. *Pflüger, Arch. Physiol.*, **33** (1884) 251–64

Schoumow-Simanowsky, E. O.

{Ueber den Magensaft und des Pepsin bei Hunden. *Arch. Exper. Path.*, **33** (1894) 336–52}

Pavlov, Ivan Petrovic; Sumov-Simanovskij, (Mme.) E. O.

Innervation der Magendrüsen beim Hunde. [1889] *Centrbl. Physiol.*, **3** (1890) 113–14

{[The innervations of the gastric glands in the dog.] *Vrach*, no. 15 (1889) 352–53; no. 41 (1890) 929–31}

Beiträge zur Physiologie der Absonderungen. Die Innervation der Magendrüsen beim Hunde. [1890] *Arch. Anat. Physiol. (Physiol. Abt.)* (1895) 53–69

Nencki, M.; Šumov-Simanovskij, (Mme.) E. O.

Études sur le chlore et les halogèns dans l'organisme animal. *St. Pétersb. Arch. Sci. Biol.*, **3** (1895) 191–211

Nencki, M.; Sieber-Šumov, (Mme.) N. O.; Šumov-Simanovskij, (Mme.) E. O.

Die Entgiftung der Toxine durch die Verdauungssäfte. *Centrbl. Bakt. (Abt. 1)*, **23** (1898) 840–47; 880–87

Ziber-Shumova, Nadezhda Olimpievna

Sieber-Šumov, (Mme.) Nadežda Olimpievna

Ueber die antiseptische Wirkung der Säuren. *J. Prakt. Chem.*, **19** (1879) 433–44

Ueber die angebliche Umwandlung des Eiweisses in Fett beim Reifen des Roquefort-Käses. *J. Prakt. Chem.*, **21** (1880) 203–221

Beiträge zur Kenntniss der chemischen Zusammensetzung der Schimmelpilze. *J. Prakt. Chem.*, **23** (1881) 412–21

Ueber die Pigmente der Chorioidea und der Haare. *Arch. Exper. Path.*, **20** (1886) 362–67

Ueber die Umikoff'sche Reaction in der Frauenmilch. *Ztschr. Physiol. Chem.*, **30** (1900) 101–12

Nencki, M.; Sieber-Šumov, (Mme.) N. O.

Untersuchungen über die physiologische Oxydation. *J. Prakt. Chem.*, **26** (1882) 1–41

Ueber das Vorkommen von Milchsäure im Harn bei Krankheiten und die Oxydationen in den Geweben der Leukämischen. *J. Prakt. Chem.*, **26** (1882) 41–47

Ueber eine neue Methode die physiologische Oxydation zu messen und über den Einfluss der Gifte und Krankenheit auf dieselbe. *Pflüger, Arch. Physiol.*, **31** (1883) 319–49

Untersuchungen über den Blutfarbstoff. *Arch. Exper. Path.*, **18** (1884) 401–22; *Berlin, Chem. Ges. Ber.*, **17** (1884) 2267–76; **18** (1885) 392–99

Ueber das Hämin. *Arch. Exper. Path.*, **20** (1886) 325–32
Venöse Hämoglobinkrystalle. *Berlin, Chem. Ges. Ber.*, **19** (1886) 128–30; 410
Weitere Beiträge zur Kenntniss der thierischen Melanine. [1887] *Arch. Exper. Path.*, **24** (1888) 17–26
Ueber das Hämatoporphyrin. [1888] *Arch. Exper. Path.*, **24** (1888) 430–446; *Mhefte. Chem.* (1888) 115–32; *Wien, Ak. Sber.*, **97** (Abt. 2b) (1889) 80–97
Zur Kenntniss der bei der Eiweissgährung auftretenden Gase. [1889] *Mhefte. Chem.* (1889) 526–31; *Wien, Ak. Sber.*, **98** (Abt. 2b) (1890) 417–22
Ueber die Bildung der Paramilchsäure durch Gährung des Zuckers. [1889] *Mhefte. Chem.* (1889) 532–40; *Wien, Ak. Sber.*, **98** (Abt. 2b) (1890) 423–31
[Sur la composition chimique du goudron de pin et sur ses propriétés désinfectantes.] *St. Pétersb. Arch. Sci. Biol.*, **2** (1893) 358–419; {*Arch. Exper. Path.*, **33** (1894) 1–45}

Sieber-Šumov, (Mme.) Nadežda Olimpieva; Šubenko, —
[Sur la formation de méthylmercaptan par fusion de l'albumine avec la potasse caustique.] *St. Pétersb. Arch. Sci. Biol.*, **1** (1892) 314–21

Macfadyen, Allan; Nencki, M.; Sieber-Šumov, (Mme.) N. O.
Untersuchungen über die chemischen Vorgänge im menschlichen Dünndarm. *Arch. Exper. Path.*, **28** (1891) 311–50; *J. Anat. Physiol.*, **25** (1891) 390–427

Botany

Fedchenko, Ol'ga Alexandrovna (née Armfel'd)
Fedčenko, Olga A.
[Beitrag zur Flora des Gouvernments Archangelsk.] *Allg. Bot. Ztschr.* (1898) 91–92; 112–13; *Moscou Soc. Nat. Bull.*, **11** (1898) 469–83

Fedčenko, Olga A.; Fedčenko, Boris A.
Beitrag zur Flora des südlichen Altai. *Engler, Bot. Jbüch.*, **25** (1898) 483–94
[Ranunculaceae des russischen Turkestan.] *Kazan Soc. Nat. Trans.*, **33**, no. 3 (1899) 85 pp. (*Res.* 82–85); *Engler, Bot. Jbüch.*, **27** (1900) 341–90

Karsakova, Natal'ia Vladimirovna
Karsakov, (Mlle.) N.
Quelques remarques sur le genre Myriotrichia. *J. Bot., Paris*, **6** (1892) 433–44
Sur deux Floridées nouvelles pour la flore des Canaries. *Ann. Sci. Nat. (Bot.)*, **4** (1896) 281–91

Malinesco, Olga
Chodat, Robert; Malinesco, (Mme.) O.
La structure cellulaire des Cyanophycées. [1892] *Arch. Sci. Phys. Nat.*, **29** (1893) 108–10
Sur le polymorphisme de certaines Protococcoïdées. *Arch. Sci. Phys. Nat.*, **29** (1893) 431–32

Schipilova, Ekaterina (Catherine)
Schipiloff, (Mlle.) Catherine
Sur l'irrégularité florale chez les Légumineuses. *Arch. Sci. Phys. Nat.*, **28** (1892) 166–68

Sokolova, Ekaterina Mikhailovna
Sokolov, (Mlle.) C.
Naissance de l'endosperme dans le sac embryonnaire de quelques gymnospermes. *Moscou Soc. Nat. Bull.*, **4** (1891) 446–97
Ueber das Wachsthum der Wurzelhaare und Rhizoiden. *Moscou Soc. Nat. Bull.*, **11** (1898) 167–277

Entomology

Freîmut, Elizavet Karlovna
Freîmut, Elizavet Karlovna
[Pompholyx dimorpha, *n. sp.* New wingless form in the family of saw-flies (Tenthredinidae); and note on other new species of that family.] *Moscou, Soc. Sci. Bull.*, **8**, no. 1 (1870) col. 213–24

Poletaeva, (Mme) Ol'ga Gustavovna
Poletaev, Olga
{Peterburgskaiia Odonata. [Odonata of St. Petersburg district.] *St. Petersb., Russ. Entomol. Soc. [Trudy]*, **11** (1880) 97–119}
{Neskol'ko slov ob organakh dukhaniia lichinok strekoz. [A few words on the respiratory organs of the dragonfly.] *St. Petersb., Russ. Entomol. Soc. [Trudy]*, **11** (1880) 182–89}
Quelques mots sur les organes respiratoires des larves des odonates. [1879] *St. Petersb., Russ. Entomol. Soc. [Horae]*, **15** (1880) 436–52
{Zamtka o prichinakh raspravki kryl'ev naskomago pri perekhod ego b sovershennoe sostoianie. [Remarks on the smoothing out of the wings of flying insects in the change to the adult stage.] *St. Petersb., Russ. Entomol. Soc. [Trudy]*, **13** (1881–1882) 19–30}
{Dobavleniia k spesku Odonata C.-Peterburgckoi gubernii. [Additions to the list of Odonata for the St. Petersburg district.] *St. Petersb., Russ. Entomol. Soc. [Horae]*, **19** (1885) v}
{Soobshchenie o serdtsie nasiekomykh. [Communication on the hearts of insects.] *St. Petersb., Russ. Entomol. Soc. [Horae]*, **19** (1885) xvi–xviii}
Du coeur des insectes. *Zool. Anz.*, **9** (1886) 13–15

Tikhomirova, Olga
Tichomirov, (Mme.) Olga
[Contribution to the history of the development of Pulex serraticeps, *Gerv.*] *Moscou, Soc. Sci. Bull.*, **67**, no. 3 (1890) 33–36
Sur l'histoire du développement de Chrysopa perla. (L'origine du mésoderme des cellules vitellines.) *Congr. Int. Zool. [C. R.]*, part 1 (1892) 112–19
[Contribution to the biology of the Pseudoscorpions of the environs of Moscow.] *Moscou, Soc. Sci. Bull.*, **86** (1894) 12–14
Élevage des vers à soie en Russie au moyen du Scorsonère (Scorzonera hispanica). *France Soc. Acclim. Bull.*, **43** (1896) 471–79

Medicine

Hasse, Sof'ia (later Erismann)
Stoff, Olga; Hasse, Sophie
Einige Notizen über die Circulations-Verhältnisse der Milz. *Centrbl. Med. Wiss.*, **10** (1872) 753–56

Hornstein, Sof'ia
Hornstein, Sophie
Ein Fall von halbseitigem Riesenwuchs. *Virchow, Arch.*, **133** (1893) 440–63

Kovalevskaia, Ol'ga
Kovalevskij, Olga
Relations de la chimotaxie et de la leucocytose avec l'action antiphlogistique de diverses substances. *Ann. Microgr.*, **8** (1896) 185–226

Lubnitzky, Sof'ia
Lubnitzky, Sophie
Die Zusammensetzung des Thrombus in Arterienwunde in den ersten fünfe Tagen. *Arch. Exper. Path.*, **19** (1885) 185–208

Sandberg, Dina
Sandberg, Dina
Die Abnahme der Lungenschwindsucht in England während der drei letzten Decennien nach Beruf und Geschlecht. Eine statistische Studie. *Ztschr. Hyg.*, **9** (1890) 369–82

Schlee, Mariia
Schlee, Marie
Ueber die Dehnung der Bauchwand während der Schwangerschaft. *Ztschr. Geburtshlf. Gynäk.*, **13** (1886) 1–15

Simonovich, Rosa (Rosalie)
Simonowitsch, Rosa

{Ueber Hyoscyamin und dessen Bedeutung für die Augenheilkunde. *Arch. Augen– und Ohrenheilkunde*, **4** (1874) 1–44; English trans. *Arch. Ophthalm.*, **4** (1874) 157–201}

Stoff, Olga von (later Siebold)

Stoff, Olga; Hasse, Sophie

Einige Notizen über die Circulations-Verhältnisse der Milz. *Centrbl. Med. Wiss.*, **10** (1872) 753–56

Suslova, Nadezhda Prokof'evna (later Erismann, then Golubev)

Suslowa, Nadeshda

{Izmenenie kozhnykh oshushchenii pod vliianiem elektricheskogo razdrazheniia. [Changes in skin sensations under the influence of electrical stimulation.] *St. Petersb., Med. Vestnik*, nos. 21, 22 (1862)}

Veränderungen der Hautgefühle unter dem Einflusse electrischer Reizung. *Ztschr. rationelle Med.*, **17** (1863) 155–60.

Die Modification der Hautempfindungen unter dem Einfluss electrischer Reizungen. *St. Petersb. Med. Ztschr.*, **5** (1863) 317–18.

Beiträge zur Physiologie der Lymphherzen. *Ztschr. rationelle Med.*, **31** (1868) 224–33

Suslova-Erismann, N.

{O vospitanii detei v pervye gody zhizni. [On the upbringing of children in their early years.] *Arch. sudebnoi med. obshchestvennoi gigieny*, **6**, no. 4 (1870) 21–31}

Tienes-Gomberg, Malka Sof'ia

Gomberg, Sophie

Ueber die Belebungsversuche beim Chloroformtode. *Deutsche Ztschr. Chirurg.*, **39** (1894) 474–99

Vaschkevich, Tat'iana

Vaškevič, Tatiana

Ueber grosszellige Heerde in den Milzfollikeln bei Diphtheritis und anderen Affectionen. *Virchow, Arch.*, **159** (1900) 137–51

Anatomy, Neurology, and Physiology

Amitin, Sarah

Amitin, Sarah

Ueber den Tonus der Blutgefässe bei Einwirkung der Wärme und der Kälte. *Ztschr. Biol.*, **35** (1897) 13–42

Bégoune, Anna (later Roux)

Bégoune, Anna

Ueber die Gefässversorgung der Kröpfe mit besonderer Berücksichtigung der Struma cystica. *Deutsche Ztschr. Chirurg.*, **20** (1884) 258–92

Berlinerblau, Stefaniia (Fanny)

Berlinerblau, Fanny

Ueber den direkten Uebergang zwischen Arterien und Venen. *Arch. Anat. Physiol.* (1875) 177–88

Berlin, Fanny

{Three cases of complete prolapsus uteri operated upon according to the method of Leon Le Fort. *Amer. J. Obst.*, **14** (1881) 866–70.}

Beschasnova, Pélagia

Bečasnov, (Mme.) Pélagie

Abhängigkeit der Pulsfrequenz des Froschherzens von seinem Inhalte. *Arch. Anat. Physiol. (Physiol. Abt.)* (1898) 531; *Schweiz. Natf. Ges. Verh.* (1898) 123–24; *J. Physiol.*, **23**, suppl. (1898–1899) [12]

Besser, Lidiia Egon de

Besser, (Mlle.) Lydie Egon de

De l'action mécanique des muscles des doigts et du poignet. *Lausanne, Soc. Vaud. Bull.*, **34** (1898) 299–347

De la rétraction des muscles après la section de leur tendon. *Lausanne, Soc. Vaud. Bull.*, **34** (1898) 348–66

Bugnion Édouard; [Besser, L. Egon (Mlle.)]
Sur la rétraction des muscles après la section de leur tendon. [Résultats des expériences de Mlle. L. Egon Besser.] *Arch. Sci. Phys. Nat.*, ser. 3, **34** (1895) 469–73

Divine, Iuliia
Divine, Julia
Ueber die Athmung des Krötenherzens. *Arch. Anat. Physiol. (Physiol. Abt.)* (1898) 533–34; *Schweiz. Natf. Ges. Verh.* (1898) 124–25; *J. Physiol.*, **23** (1898–99) [12]–[13]
Kronecker, Hugo; Divine, (Mlle.)
[Respiration of the heart of the tortoise.] *Nature*, **58** (1898) 481

Eliaschova, Ida
Eliaschoff, Ida
Ueber die Wirkung des Cantharidins auf die Nieren. *Virchow, Arch.*, **94** (1883) 323–42

Familiant, Victoria
Familiant, Victoria
Beiträge zur Vergleichung der Hirnfurchen bei den Carnivoren und den Primaten. Im Anschlusse an die Untersuchen eines Löwen-Gehirness. *Bern Mitth.*, Heft. 2–3 (1885) 49–81

Frenkel (ńee Rabinowitsch), Sof'ia
Frenkel, Sophie
Klinische Untersuchungen über die Wirkung von Coffein, Morphium, Atropin, Secale cornutum und Digitalis auf den arteriellen Blutdruck, angestellt mittelst des v. Basch'schen Sphygmomanometers. *Deutsch. Arch. Klin. Med.*, **46** (1890) 542–82

Gitiss, Anna
Gitiss, Anna
Beiträge zur vergleichenden Histologie der peripheren Ganglien. *Bern Mitth.* (1887) 24–39

Handler (Gandler), Sof'ia
Handler, Sophie
Ueber die Reduction des Hämoglobins im Herzen. *Ztschr. Biol.*, **26** (1890) 233–58

Kleimann, Anna
Kleimann, A.; Simonowitsch, R.
Experimentelle Untersuchungen über den Brechact. II. *Pflüger, Arch. Physiol.*, **5** (1872) 280–82

Konev, Elena
Koneff, Helene
Beiträge zur Kenntniss der Nervenzellen in den peripheren Ganglien. *Bern Mitth.* (1886) 13–44
Flesch, Maximilian H. J.; Koneff, Helene
Bemerkung über die Struktur der Ganglienzellen. *Neurol. Centrbl.*, **5** (1886) 145–47

Kossovich, Barbara
Kossowitsch, Barbara
Untersuchen über den Bau des Rückenmarks und der Medulla oblongata eines Mikrocephalen. *Virchow, Arch.*, **128** (1892) 497–527

Kotlarevskaia, Anna
Kotlarewsky, Anna
Physiologische und mikrochemische Beiträge zur Kenntniss der Nervenzellen in den peripheren Ganglien. [1886] *Bern Mitth.* (1887) 3–23

Kovalenskaia - von Kompaneiskaia, Ekaterina Ivanovna von
Kovalenskij, Katerina
Beiträge zur vergleichenden mikroskopischen Anatomie der Hirnrinde des Menschen und einiger Säugethiere. *Bern Mitth.* (1886) 59–90

Leonova, Ol'ga Vassilievna von (later von Leonova von Lange)
Leonova, O.
Ein Fall von Anencephalie. Ueber den feineren Bau des Rückenmarkes eines Anencephalus. *Arch. Anat. Physiol. (Anat. Abt.)* (1890) 403–22
Ueber das Verhalten der Neuroblasten des Occipitallappens bei Anophthalmie und Bulbusatrophie, und seine Beziehungen zum Sehact. *Arch. Anat. Physiol. (Anat. Abt.)* (1893) 308–18

Zur pathologischen Entwickelung des Centralnervensystems. (Ein Fall von Anencephalie combinirt mit totaler Amyelie.) *Neurol. Centrbl.*, **12** (1893) 218–27; 263–67

Contribution à l'étude de l'évolution pathologique du système nerveux. Anencéphalie totale combinée avec une amyélie et un rhachischisis totaux chez un embryon humain. *Moscou Soc. Nat. Bull.*, **7** (1894) 191–98

Die Sinnesorgane und die Ganglien bei Anencephalie und Amyelie. *Deutsch. Natf. Verh.* (1894) Th. 2, Hälfte 2, 176–77

Beiträge zur Kenntniss der secundären Veränderungen der primären optischen Centren und Bahnen in Fällen von congenitaler Anophthalmie und Bulbusatrophie bei neugeborenen Kindern. [1895] *Arch. Psychiatr.*, **28** (1896) 53–96; 642

Einige Bemerkungen zu [ihrer] im *Archiv f. Psychiatrie* Bd. 28, H. 1, erschienenen Abhandlung: Beiträge zur Kenntniss der secundären Veränderungen der primären optischen Centren und Bahnen in Fällen von congenitaler Anophthalmie und Bulbusatrophie bei neugeborenen Kindern. *Moscou Soc. Nat. Bull.*, **10** (1897) 570–74

Levin, Ida

Levin, I.

[On the nerve-endings in the cornea and their correlation with the nerve-endings in the tail of the tadpole.] [1884] *St. Pétersb. Ac. Sci. Mém.*, **49** (1885) 1–11

Lomakina, Nadezhda (Nadine)

Lomakina, Nadine

Ueber die nervösen Verbindung auf den Herzen der Hunde und Pferde. *Arch. Anat. Physiol. (Physiol. Abt.)* (1898) 531–32; *J. Physiol.*, **23** (1898–1899) [11]–[12]; *Schweiz. Natf. Ges. Verh.* (1898) 125–26

Ueber Verlauf und Bedeutung der Herznerven. *Ztschr. Biol.*, **39** (1900) 377–429

Ludkevich, Anna

Ludkewitch, Anna

L'articulation de l'épaule. Étude d'arthologie comparée. *Lausanne, Soc. Vaud. Bull.*, **35** (1899) 297–371

Bugnion, Édouard; [Ludkewitch, Anna]

[L'articulation de l'épaule. Étude d'arthologie comparée, par Anna Ludkewitch. Introduction et appendice.] *Lausanne, Soc. Vaud. Bull.*, **35** (1899) 297–382

Manasseina, Mariia Mikhailovna (née Korkunova)

Manassein, Marija Michajlovna

De l'antagonisme qui existe entre chaque effort de l'attention et des innervations motrices. *Congr. Med. Int., Atti*, **2**, Fisiol. (1894) 48–56

Suppléance d'un hémisphère cérébral par l'autre. *Congr. Med. Int. Atti*, **2**, Fisiol. (1894) 162–67; *Arch. Ital. Biol.*, **21** (1894) 326–32

Quelques observations expérimentales sur l'influence de l'insomnie absolue. *Congr. Med. Int. Atti*, **2**, Fisiol. (1894) 174–177; {*Arch. Ital. Biol.*, **21** (1894) 322–25}

Markova, Klavdia

Claparède, Ed.; Markova, (Mlle.) K.

Nouveau procédé pour étudier la perception des forms par le toucher. *Arch. Sci. Phys. Nat.*, **10** (1900) 84–86.

Miropolskaia, Mariia

Miropolsky, Marie

Du passage dans le sang du foetus des substances solides contenues dans le sang de la mère. *Arch. de Physiol.*, **6** (1885) 101–8

{[On the passage into fetal blood of solids in the maternal blood.] *Vrach*, **5**, no. 46 (1886) 776–79; *London Med. Rec.*, n.s., **13**, no. 126 (1885) 112 (abst.)}

Ott, Nadiia

Ott, Nadine

Ueber peri- und endoneurale Wucerungen in den Nervenstämmen einiger Thierspecies. *Virchow, Arch.*, **136** (1894) 69–82

Popova, Nadezhda
Popov, Nadine
Ueber die Bildung von Serumalbumin im Darmkanale. *Ztschr. Biol.*, **25** (1889) 427–52
Kronecker, Hugo; Popov, Nadine
Ueber die Bildung von Serumalbumin im Darmcanale. *Arch. Anat. Physiol. (Physiol. Abt.)* (1887) 345–47; *Bern Mitth.* (1887) xvi–xvii
Reinstein-Mogilova, Anna
Reinstein-Mogilowa, Anna
Ueber die Betheiligung der Zellschichte des Chorion an der Bildung der Serotina und Reflexa. *Virchow, Arch.*, **124** (1891) 522–45
Schachova, Seraphima
Schachova, S.
Ueber intercellulares Knochenwachsthum. *Centrbl. Med. Wiss.*, **11** (1873) 900–902
Schilina, Liudmila
Schilina, Ludmilla
Vergleich von Ludwig's Kymographen mit Hürthle's Tonographen. *Arch. Anat. Physiol. (Physiol. Abt.)* (1898) 526–30; *Schweiz. Natf. Ges. Verh.* (1898) 126; *J. Physiol.*, **23** (1898–1899) [13]; *Ztschr. Biol.*, **38** (1899) 433–86
Schipilova, Ekaterina (Catherine)
Schipiloff, (Mlle.) Catherine
Ueber die Entstehungsweise der Muskelstarre. Vorläufige Mitteilung. *Centrbl. Med. Wiss.*, **20** (1882) 291–94
Ueber den Einfluss der Nerven auf die Erweiterung der Pupille bei Fröschen. *Pflüger, Arch. Physiol.*, **38** (1886) 219–73
{Recherches sur la nature et les causes de la rigidité cadavérique. *Rev. Méd. Suisse rom.*, **9**, no. 10 (1889) 556–80}
Recherches sur les ferments digestifs. *Arch. Sci. Phys. Nat.*, **22** (1889) 185–213; 314–52; **23** (1890) 256–64; **26** (1891) 462–68
Recherches sur l'influence de la sensibilité générale sur quelques fonctions de l'organisme. *Arch. Sci. Phys. Nat.*, **24** (1890) 149–75; 266–84; **25** (1891) 518–41; 680–94; **26** (1891) 157–81
Recherches relatives à l'influence de la chaleur sur les muscles. *Arch. Sci. Phys. Nat.*, **26** (1891) 572–75
Recherches sur les fonctions de la VIIIme paire. *Arch. Sci. Phys. Nat.*, **28** (1892) 109–11
Sur un nouveau procédé d'épuration de l'eau. [1892] *Arch. Sci. Phys. Nat.*, **29** (1893) 99–100
Recherches sur les nerfs de la VIIIme paire cranienne et sur les fonctions du cerveau et de la moelle chez les grenouilles, avec un aperçu comparatif des fonctions du système nerveux central dans la classe des Batraciens. [Communication préliminaire.] *Arch. Sci. Phys. Nat.*, **32** (1894) 174–82
Recherches sur les nerfs de la VIIIe paire cranienne et sur les fonctions du cerveau et de la moelle chez les grenouilles, avec un aperçu comparatif des fonctions du système nerveux central dans la classe des Batraciens. [1895] *Genève Soc. Phys. Mém.*, **32** (1894–1897) no. 6, 115 pp.
Schipiloff, Catherine; Danilevskii, Alexandr Yakovlevich
Ueber die Natur der anisotropen Substanzen des quergestreiften Muskels und ihre räumliche Verteilung im Muskelbündel. *Ztschr. Physiol. Chem.*, **5** (1881) 349–65
Schipiloff, (Mlle.) Catherine; Espine, Ad. d'
[Expériences sur la valeur du filtre Schipiloff.] *Arch. Sci. Phys. Nat.*, **4** (1897) 185–86
Schultz, Nadezhda Karlova
Schultz, Nadezhda
{Ueber die Vernarbung von Arterien nach Unterbindungen und Verwundungen. *Deutsche Ztschr. Chirurg.*, **9** (1878) 84–121}
Simonovich, Rosa (Rosalie)
Kleimann, A.; Simonowitsch, R.
Experimentelle Untersuchungen über den Brechact. II. *Pflüger, Arch. Physiol.*, **5** (1872) 280–82

Sollogub, Liudmila
Sollogub, Liudmila
[On the endings of Remak's fibres in the muscles of the frog.] *St. Pétersb. Ac. Sci. Mém.*, **52** (1886) 142–51
Vassilîeva-Kleimann, Mariia
Wassilieff-Kleimann, Marie
Ueber Resorption körniger Substanzen von Seiten der Darmfollikel. *Arch. Exper. Path.*, **27** (1890) 191–201

Zoology and Natural History

Balasheva, Mariia Dmitrievna
Balaševa, M. D.
(De l'influence du milieu extérieur et principalement celle des dimensions du bassin d'eau sur quelques Mollusques.) *New Russ. Soc. Nat. Mem.*, **12**, no. 1 (1887) 21–28
(De l'influence du milieu extérieur et principalement celle de la température de l'eau et de l'air sur les Planorbis Vertia.) *New Russ. Soc. Nat. Mem.*, **13**, no. 1 (1888) 161–84
[On the influence of the external environment in general and plants in particular on Mollusca belonging to the genera Limnaeus and Planorbis.] *Moscou, Soc. Sci. Bull.*, **67**, no. 6 (1891) 15–18
[On the influence of the external environment in general and plants in particular on some molluscs of Russian waters.] *Moscou, Soc. Sci. Bull.*, **86** (1894) 19–21
Halperin, Rebecca
Halperin, Rebecca
Die abnorme Krümmung der Wirbelsäule bei congenitaler Spaltbildung der Leibeswand. *Arch. Wiss. Prakt. Thierheilk.*, **15** (1889) 48–65
Mechnikova, Ludmila Vassilievna (née Fedorovich)
Mechnikov, Il'ya Il'ich; Mechnikov, Ludmilla
[Further information about Siphonophora and Medusae.] *Moscou, Soc. Sci. Bull.*, **8**, no. 1 (1870) col. 295–370
Mechnikova, Ol'ga (née Belokopytova)
Metschnikoff, Olga
[On the pelvic and scapular arch of cartilaginous fishes. Preliminary communication.] *New Russ. Soc. Nat. Mem.*, **6**, no. 1 (1879) 2 pp.
Notiz über den Beckenbogen des Scaphirhynchus hermanni. *Carus, Zool. Anz.*, **3** (1880) 21–23
Zur Morphologie des Becken- und Schulterbogens der Knorpelfische. [1879] *Ztschr. Wiss. Zool.*, **33** (1880) 423–38; **34** (1880) 366
Pereiaslavtseva, Sof'ia Mikhailovna
Perejaslavcev, Sofja
[On the development of Rotifera.] *New Russ. Soc. Nat. Mem.*, **9**, no. 1 (1884) 19 pp.; *Nature*, **32** (1885) 579–80
Sur le développement des Turbellariés. *Zool. Anz.*, **8** (1885) 269–71
[Protozoen des Schwarzen Meers.] *New Russ. Soc. Nat. Mem.*, **10**, no. 2 (1886) 79–114; *Arch. Slaves Biol.*, **4** (1887) 116
Études sur le développement des Amphipodes. [Première partie. Le développement de Gammarus pœcilurus, *Rthk.* Troisème partie. Le développement de Caprella ferox, *Chrnw.*] *Moscou Soc. Nat. Bull.*, **2** (1888) 183–219; 582–97
Monographie des Turbellariés de la Mer Noire. *New Russ. Soc. Nat. Mem.*, **17**, no. 3 (1892) xx+303 pp.
Mémoire sur l'organisation de la Nerilla antennata, *O. Schmidt. Ann. Sci. Nat. (Zool.)*, **1** (1896) 277–345
Les premiers stades du développement des Pédipalpes. *Paris, Ac. Sci. C. R.*, **125** (1897) 319–21
Les derniers stades du développement des Pédipalpes. *Paris, Ac. Sci. C. R.*, **125** (1897) 377–80

Rossiîskaia-Kozhevnikova, Mariia Aleksandrovna
Rossiîskij-Koževnikov, Marie
Études sur le développement des Amphipodes. Deuxième partie. Le développement d'Orchestia littorea, *Spence Bate.* [Quatrième partie. Développement de la Sunamphitoë valida, *Czerniavski*, et de l'Amphitoë picta, *Rathke.*] *Moscou Soc. Nat. Bull.*, **2** (1888) 561–81; **4** (1891) 82–103
Sur la formation des organes génitaux chez les Amphipodes. (Note préliminaire.) [1893] *Zool. Anz.*, **16** (1894) 33–35
Les organes embryonnaires du Sphæroma serratum, *Fabr.* (Note préliminaire.) *Zool. Anz.*, **18** (1895) 151–53
Étude sur le développement embryonnaire du Gammarus pulex (Gammarus neglectus, *var.* kossinensis). *Moscou Soc. Nat. Bull.*, **10** (1897) 53–64

Schirmann, Daria
Schirman, Daria
Ueber die Rückbildung der Dickdarmzotten des Meerschweinchens. *Würzb. Phys. Med. Verh.*, **32** (1899) [1]–[9]

Wagner, Ekaterina
Wagner, (Mme.) Catherine
Études sur le développement des Amphipodes. Cinquième partie. Développement de la Melita palmata. *Moscou Soc. Nat. Bull.*, **5** (1892) 401–9

ASTRONOMY

Bobrinskaia, Nadezhda A. (Comtesse)
Bobrinskoj, (Comtesse) Nadejda
Détermination de l'orbite de la comète 1890 VI. [1893] *St. Pétersb. Ac. Sci. Bull.*, **36** (1894) 227–46
Étude sur l'amas stellaire C. G. 4294 = M.92. *St. Pétersb. Ac. Sci. Bull.*, **3** (1895) 163–72
Ephéméride de la planète (147) Protogeneia. *St. Pétersb. Ac. Sci. Bull.*, **7**, no. 1 (1897) 1–2

Maksimova, Evgeniia Aleksandrovna
Maksimov, Evgenija
Ephémeride de la planète (209) Dido. 1895, août 6.5–septembre 11.5. *St. Pétersb. Ac. Sci. Bull.*, **2** (1895) 239–41
Elemente und Ephemeride des Planeten (209) Dido. *Astr. Nachr.*, **141** (1896) 403–4
Bahnbestimmung des Planeten (209) Dido. *St. Pétersb. Ac. Sci. Bull.*, **12** (1900) 331–54

Zhilova, Mariia Vasil'evna
Šilov, Marie
Grössenbestimmung der Sterne im Sternhaufen 20 Vulpeculæ. *St. Pétersb. Ac. Sci. Bull.*, **2** (1895) 243–51
Positionen von 1041 Sternen des Sternhaufens 5 Messier, aus photographischen Aufnahmen abgeleitet. *St. Pétersb. Ac. Sci. Bull.*, **8** (1898) 253–312
Angenäherte absolute Bahn des Planeten (147) Protogeneia. *St. Pétersb. Ac. Sci. Bull.*, **13** (1900) 345–50
Planet (147) Protogeneia. *Astr. Nachr.*, **154** (1901) 161–62

CHEMISTRY

Balandina, V. A.
Balandina, V. A.
{[Results of a chemical investigation of the waters of lake Plotbishche, Yenisei gubernia.] *Russ. Phys.-Chem. Soc. J.*, **32** (1900) 194–95}

Bogdanovskaia-Popova, Vera Evstaf'evna
Bogdanovskij, Vera E.

Ueber Dibenzylketon und dibenzylcarbinol. *Berlin, Chem. Ges. Ber.*, **25** (1892) 1271–77

[Sur les réactions de condensation et de reduction de la dibenzylcétone.] *Russ. Phys.-Chem. Soc. J.*, **24** (Chem.) (1892) 355–65; *Paris Soc. Chim. Bull.*, **10** (1893) 129–30

[Transformation par oxydation des cétones en oxacides.] *Russ. Phys.-Chem. Soc. J.*, **26** (Chem.) (1894) 161–62; *Paris Soc. Chim. Bull.*, **14** (1895) 1285–86

Bogoslovskaia, N. V.

Bogoslovskij, (Mme.) N.

[Action du triméthylène sur le benzène en présence du clorure d'aluminium.] *Russ. Phys.-Chem. Soc. J.*, **26** (Chem., part 2) (1894) 6; *Paris Soc. Chim. Bull.*, **14** (1895) 1294

Davidova, Ol'ga Aleksandrovna

Davidoff, (Mlle.) Olga

Société Chimique Russe, Correspondance Russe. *Paris Soc. Chim. Bull.*, **39** (1883) 507–10; 595–98; **40** (1883) 22–28; 71–75; 183–95; {**41** (1884) 253–61; 309–20; 340–41; 387–94; 547–554; **42** (1884) 315–31; **43** (1885) 108–29; 246–52; 380–95}

[Ein Fall der Bildung von Bernsteinsäureäthylester.] *Berlin, Chem. Ges. Ber.*, **19** (1886) 406–8; *Russ. Phys.-Chem. Soc. J.*, **18** (Chem.) (1886) 172–74

[Ueber die Einwirkung von Silberoxyd auf Methan und Aethylen.] *Berlin, Chem. Ges. Ber.*, **21** (*Ref.*) (1888) 442–43; *Russ. Phys.-Chem. Soc. J.*, **20** (Chem.) (1888) 362–63

Fomina-Zhukovskaia, Evdokiia Aleksandrovna

Fomin, Evdokia Aleksandrovna (Mme. Fomin-Žukovskij)

Recherches sur quelques combinaisons du groupe de l'euxanthone. *Arch. Sci. Phys. Nat.*, **21** (1889) 429–40

Ueber α-und β-Phenonaphtoxanthon und über Methylxanthon. *Liebig's Ann.*, **257** (1890) 87–95

Fomin, Evdokia Aleksandrovna; Jakovkin, Aleksandr A[leksandrovich]

Épuration des eaux de fabrique par le procédé Moser et Plata. *Russ. Phys.-Chem. Soc. J.*, **27** (Chem.) (1895) 309–13; *Paris Soc. Chim. Bull.*, **16** (1896) 645–46

Jebenko, Anna

Konowaloff, M. I.

{[Nitration of methane and triethylmethane.] *Russ. Phys.-Chem. Soc. J.*, **31** (Chem.) (1899) 1027–29 [Jebenko, Anna, along with Kotzina, Sophie, are listed as coauthors in subheadings]}

Kaufmann, L. E. (née Solovelchik)

Kaufmann, (Mme.) L. E.

[Influence des bromures métalliques sur la vitesse de bromuration des carbures aromatiques.] *Russ. Phys.-Chem. Soc. J.*, **32** (Chem.) (1900) 270; *Paris Soc. Chim. Bull.*, **26** (1901) 43

Kikina, Zenaida Viktorovna

Kikin, Zenaida V.

[Sur les nitro- et aminoflavindulines.] *Paris Soc. Chim. Bull.*, **24** (1900) 900–903; *Russ. Phys.-Chem. Soc. J.*, **32** (Chem.) (1900) 170–86

Kehrmann, Friedrich; Kikin, Zenaida V.

Ueber Nitro- und Amino-Flavinduline. *Berlin, Chem. Ges. Ber.*, **32** (1899) 2633–37

Konowaloff, M. I.

{Ueber die nitrirende Wirkung der Salpetersäure auf gestättigten Charakter besitzende Kohlenwasserstoffe und deren Derivative. *Berlin, Chem. Ges. Ber.*, **29** (1896) 2199–2205 [Kikina, Zenaida V., along with Nikin, A., and Tschitschkin, A., are listed as coauthors in a subheading]}

Kotzina, Sof'ia

Konowaloff, M. I.

{[Nitration of methane and triethylmethane.] *Russ. Phys.-Chem. Soc. J.*, **31** (Chem.) (1899) 1027–29 [Kotzina, Sophie, and Jebenko, Anna, are listed as coauthors in subheadings]}

Lazareva, Pelagia E.

Lazarev, (Miss) P.

[Mutual displacement of haloids.] *Russ. Phys.-Chem. Soc. J.*, **22** (Chem.) (1890) 383–85; *Chem. Soc. J.*, **62** (Abs.) (1892) 1270–71

[Dispacement of hydrogen by haloids in aromatic hydrocarbons.] *Russ. Phys.-Chem. Soc. J.*, **22** (Chem.) (1890) 385–88

Lermontova, Iuliia Vsevolodovna

Lermontoff, Julie

Ueber die Zusammensetzung des Diphenins. *Berlin, Chem. Ges. Ber.*, **5** (1872) 230–36

Ueber das Verhalten des Methylenjodids gegen einige Amine. *Berlin, Chem. Ges. Ber.*, **7** (1874) 1252–57

Ueber das Sulfocyanmethylen. *Berlin, Chem. Ges. Ber.*, **7** (1874) 1282–84

Ueber die Darstellung von Trimethylenbromid. *Liebig's Ann.*, **182** (1876) 358–62

{Société Chimique Russe, Correspondance Russe. *Paris Soc. Chim. Bull.*, **27** (1877) 260–66; 366–73 (which includes Julie Lermontoff, Sur la formation du bromure de triméthylène, 368–69); 445–57; 501–5; 553–55; **28** (1877) 105–12; 151–55; 346–51; 559–64; **29** (1878) 210–15; 368–77; 535–41; **30** (1878) 20–28 (which includes Julie Lermontoff, Action de l'iodure de butyle tertiaire sur l'isobutylène en présence de l'oxyde de calcium, 21–22); 533–38; **31** (1879) 67–72; 197–200; 487–97}

Sur l'action de l'iodure de butyle tertiaire sur l'isobutylène en présence d'oxydes métalliques. [1878] *Liebig's Ann.*, **196** (1879) 116–22; *St. Pétersb. Ac. Sci. Bull.*, **25** (1879) col. 203–9

Lukanina-Rykachevna-Paevskaia, Adelaida Nikolaevna

Lukanin, (Frau) Adelheid von

Ueber die Einwirkung des Chlorsuccinyls auf Benzoin. [1872] *St. Pétersb. Ac. Sci. Bull.*, **18** (1873) col. 72–75; {*Russ. Phys.-Chem. Soc. J.*, **4** (1872) 129–33}

Markova, E. A. (née Bulatova)

Gustavson, Gavriil Gavriilovič; Bulatov, (Mlle.) H.

Ueber die Bildung von Ketopentamethylen aus Vinyltrimethylenbromid. *J. Prakt. Chem.*, **56** (1897) 93–95

Miropolskaia, Mariia A.

Miropolskij, (Mme.) M. A.

[Sur la présence du mercure dans l'acide sulfurique commercial.] *Russ. Phys.-Chem. Soc. J.*, **16** (Chem.) (1884) 689–90; *Paris Soc. Chim. Bull.*, **43** (1885) 383–84

Popper, O. M.

Gustavson, Gavriil Gavriilovič; Popper, (Frl.) O.

Beitrag zur Constitution von Pentaerythrit. *J. Prakt. Chem.*, **56** (1897) 95–96

Ueber Dimethyl-1,1-trimethylen. *J. Prakt. Chem.*, **58** (1898) 458–61

[Sur la structure du diméthlytriméthylène.] *Russ. Phys.-Chem. Soc. J.*, **30** (Chem., part 2) (1898) 214–15; *Paris Soc. Chim. Bull.*, **22** (1899) 586

Rikhter-Rzhevskaia, N. P.

Richter, (Mme.) N. P.

[Sur la vitesse d'hydration de l'anhydride acétique.] *Russ. Phys.-Chem. Soc. J.*, **32** (Chem.) (1900) 349; *Paris Soc. Chim. Bull.*, **26** (1901) 165

Rodzîanko, Aleksandra Nikolaevna

Rodzîanko, A. N.

[On mono- and di-nitroparaazobenzoic acids.] *Russ. Phys.-Chem. Soc. J.*, **20** (Chem.) (1888) 18–29; *Chem. Soc. J.*, **56** (Abs.) (1889) 141–42

[On some properties of humin and humic acids.] *Russ. Phys.-Chem. Soc. J.*, **22** (Chem.) (1890) 208–29; *Chem. Soc. J.*, **62** (Abs.) (1892) 1373

Rudinskaia, (Mme.)

Rudinskij, (Mme.); Bogomoletz, Ivan [Vasilievich]

[Action of ammonia on parabamic acid.] *Russ. Phys.-Chem. Soc. J.*, **17** (Chem.) (1885) 278–80; *Chem. Soc. J.*, **50** (Abs.) (1886) 141–42; *Paris Soc. Chim. Bull.*, **45** (1886) 250

Samelson, Minna

Friedheim, Carl; Samelson, Minna

Ueber Permanganmolybdate. I. Abhandlung. *Ztschr. Anorg. Chem.*, **24** (1900) 65–107

Volkova, Anna Fedorovna

Volkow, Anna Fedorovna

[Ueber die isomeren Toluosulfosäuren.] *Russ. Chem. Soc. J.*, **2** (1870) 161–75; *Ztschr. Chem.*, **6** (1870) 321–27

[Ueber die Säuren, welche durch Vertretung des Wasserstoffe in den Amiden der Toluolsulfosäuren durch saure Radicale entstehen.] *Russ. Chem. Soc. J.*, **2** (1870) 243–352; *Ztschr. Chem.*, **6** (1870) 577–81

[Effect of *para*-toluenesulphonyl chloride on amides of the type N(RH2).] *Russ. Chem. Soc. J.*, **3** (1871) 239–42

[Ueber neue Amidsäuren.] *Russ. Chem. Soc. J.*, **3** (1871) 242–46; **4** (1872) 7–15; *Ztschr. Chem.*, **7** (1871) 422–23

Ueber die Einwirkung des β-toluolsulfosäure-Chlorides auf Säure-amide. *Ztschr. Chem.*, **7** (1871) 421–22

Ueber die Einwirkung von Phosphorpentachlorid auf einige Aciamide. *Berlin, Chem. Ges. Ber.*, **5** (1872) 139–43

Iso-Crotylether. *Berlin, Chem. Ges. Ber.*, **6** (1873) 196

Ziber-Shumova, Nadezhda Olimpievna

Sieber, Nadina; Nencki, Marcellus

Ueber eine neue Synthese des Glykocyamins. *J. Prakt. Chem.*, **17** (1878) 477–80

Ueber die Verbindungen der ein- und zweibasischen Fettsäuren mit Phenolen. *J. Prakt. Chem.*, **23** (1881) 147–56; 537–46

Ueber die Zersetzung des Traubenzuckers und der Harnsäure durch Alkalien bei der Bruttemperatur. *J. Prakt. Chem.*, **24** (1881) 498–506

Ueber zwei neue Derivate des Sulfoharnstoffs. [1881] *J. Prakt. Chem.*, **25** (1882) 72–81

Ueber das Uroroseïn, einen neuen Harnfarbstoff. *J. Prakt. Chem.*, **26** (1882) 333–36

Sieber-Šumov, (Mme.) Nadežda Olimpievna; Smirnov, A.

Ueber das Verhalten der drei isomeren Nitrobenzaldehyde im Thierkörper. *Mhefte. Chem.* (1887) 88–93; *Wien, Ak. Sber.*, **95**, Ab. 2 (1887) 309–14

Geology

Goloubtzova, Vera

Goloubtzoff, (Mlle.) Vera

Étude sur la lave basaltique de Tounka (Sibérie). *Brux. Soc. Belge Géol. Bull.* (Proc.-Verb.) (1887) 109–10

Pavlova, Mariia Vasil'evna (née Gortynskaia, later Illych-Shishatskaia)

Pavlov, Marija V.

Les ammonites du groupe Olcostephanus versicolor. *Moscou Soc. Nat. Bull.*, **62**, part 2 (1887) 27–43

Études sur l'histoire paléontologique des ongulés en Amérique et en Europe. *Moscou Soc. Nat. Bull.*, n.s., **1** (1887) 343–73; **2** (1889) 135–82; **3** (1890) 653–716; **6** (1893) 137–221; **13** (1900) 268–328

Notice sur l'Hipparion crassum du Roussillon. *Moscou Soc. Nat. Bull.*, **5** (1892) 161–64

Qu'est-ce que c'est que l'Hipparion. *Moscou Soc. Nat. Bull.*, **5** (1892) 410–14

Note sur un nouveau crâne d'Amynodon. *Moscou Soc. Nat. Bull.*, **7** (1894) 37–42

[On Russian mastodons and their relation to the mastodons of other countries.] [1894] *Moscou Soc. Nat. Bull.*, **8** (1895) 146–54; *St. Pétersb. Ac. Sci. Mém.*, **1** (1895) no. 3, 43 pp.

Nouveaux mammifères tertiaires trouvés en Russie. *Moscou Soc. Nat. Bull.*, **10** (1897) 174–86

[On the mammoth found near Jaroslavl in the summer of 1896.] *Russ. Geol. Min. Annu.*, **2**, Sec. 1 (1897–1898) 43–50; *Moscou Soc. Nat. Bull.*, **11** (Prot.) (1898) 3–5

[On the development of the teeth of the more ancient ruminants.] *Moscou Soc. Nat. Bull.*, **12** (Prot.) (1899) 27–31

Solomka-Sotiriadis, Evgeniia Viktorovna

Solomka-Sotiriadis, (Mlle.) Evgenija

Vorläufige Mittheilung über die Mikrostruktur der Stromatoporen. *Neues Jbuch. Min.*, Bd. 2 (1885) 168–71

[Stromatopora des Devonischen System Russlands.] *St. Petersb. Min. Ges. Verh.*, **23** (1887) 1–48; *St. Pétersb., Com. Géol. Bull.*, **7**, suppl. (1889) 47

Die Jura- und Kreidekorallen der Krim. *St. Petersb. Min. Ges. Verh.*, **24** (1888) 67–231

Tsvetaeva, Mariia Kuz'minichna

Cvétaev, Marie

Cephalopodes de la section supérieure du calcaire carbonifère de la Russie centrale. [1888] *St. Pétersb., Com. Géol. Mém.*, **5** (1888–1890) no. 3, 58 pp. (Rés. 41–58)

Nautiloidea et Ammonoidea de la section inférieure du calcaire carbonifère de la Russie centrale. [1898] *St. Pétersb., Com. Géol. Mém.* (1888–1898) no. 4, 46 pp. (Rés. 35–46)

Mathematics

Kovalevskaia, Sof'ia Vasil'evna (née Korvin-Krukovskaia)

Kovalevskij, (Mme.) Sofja Vasiljevna

Zur Theorie der partiellen Differentialgleichungen. [1874] *Crelle, J. Math.*, **80** (1875) 1–32

Sur la propagation de la lumière dans un milieu cristallisé. *Paris, Ac. Sci. C. R.*, **98** (1884) 356–57; *Stockh., Öfvers.*, no. 2 (1884) 119–21

Ueber die Reduction einer bestimmten Klasse Abel'scher Integrale 3ten Ranges auf elliptische Integrale. *Acta Math.*, **4** (1884) 393–414

Ueber die Brechung des Lichtes in cristallinischen Mitteln. *Acta Math.*, **6** (1885) 249–304

Zusätze und Bemerkungen zu Laplace's Untersuchung über die Gestalt der Saturnsringe. *Astr. Nachr.*, **111** (1885) 37–48

Sur le problème de la rotation d'un corps solide autour d'un point fixe. *Acta Math.*, **12** (1889) 177–232

Sur une propriété du système d'équations différentielles qui définit la rotation d'un corps solide autour d'un point fixe. *Acta Math.*, **14** (1890–1891) 81–93

Sur un théorème de M. Bruns. *Acta Math.*, **15** (1891) 45–52

Mémoire sur un cas particulier du problème de la rotation d'un corps pesant autour d'un point fixe, où l'intégration s'effectue à l'aide de fonctions ultraelliptiques du temps. *Paris, Mém. Savants Étrang.*, **31** (1894) no. 1, 62 pp.

Shiff, Vera Iosifovna (Frau)

Schiff, Věra Os.

Ueber Symmetricaxen der centrischen Curven vierter Ordnung. [Sur les axes de symétrie des courbes à centre du quatrième ordre.] [1892] *Fortschr. Math.* (1892) 600; *Kharkov Math. Soc. Commun.*, **3** (1893) 163–72

Archaeology

Chebysheva, V. M.

Čebyševa, V. M.

[Excavations of tumuli in the government of Smolensk, Dorogobuzh district, in the summer of 1879.] [1879] *Moscou, Soc. Sci. Bull.*, **49**, no. 1 (1886) 67–70

[On excavations of tumuli in the Smolensk district.] [1881] *Moscou, Soc. Sci. Bull.*, **49**, no. 3 (1886) 14–24

RUSSIA/ITALY

Anatomy, Neurology, and Physiology

Bakunina, Sof'ia Mikhailovna

Bakunin, Sofia

Recherches sur la respiration et la sécrétion renale de l'embryon. *Congr. Med. Int., Atti*, **3** (Pediatr.) (1894) 56–60

Sur l'activité sécrétrice des épithéliums de Wolff et des épithéliums rénaux dans les premiers jours de développement embryonnaire. *Arch. Ital. Biol.*, **23** (1895) 350–54

Sur l'évolution des fonctions embryonnaires. Recherches expérimentales. *Arch. Ital. Biol.*, **23** (1895) 420–23

Zoology and Natural History

Vedenisov, Tatiana

Vedenissova, Tatiana

Di alcuni Crostacei raccolti nel Paese dei Somali dall'ing. L. Bricchetti-Robecchi. *[Italia], Soc. Ent. Bull.*, **26** (1894) 408–24

Chemistry

Bakunin, Maria Michailovna (Marussia)

Bakunin, M.

{Sugli acidi fenilnitrocinnamici e sui loro isomeri stereometrici. *Gazz. Chim. Ital.*, **25** (1895) 137–89}

{Stereoisomeri degli acidi fenilnitrocinnamici. *Gazz. Chim. Ital.*, **27** (1897) 37–48}

{Stereoisomero dell' acido fenilcinnamico. *Gazz. Chim. Ital.*, **27** (1897) 48–55}

{Sul' acido ossifenilcinnamico. *Napoli, Rend.*, ser. 3, **6** (1900) 209–15}

Oglialoro-Todari, Agostino; Bakunin, Marussia

Sugli acidi meta e paranitro-fenilcinnamici e sopra alcuni loro derivati. *Napoli, Rend.*, **29** (1890) 104–14

RUSSIA?

Anatomy and Physiology

Eliasheva, P.

Eliašev, (Mme.) P.

Contribution à l'étude des matières extractives non dialysables des urines. *Paris, Soc. Biol. Mém.*, **43** (1891) 71–80

Krantz, Cath. (Catherina?)

Krantz, Cath.

La tortue caspienne dans les environs de Moscou. *France Soc. Acclim. Bull.*, **38**, Sem. 1 (1891) 871–74

Les races de harengs dans la Baltique. *France Soc. Acclim. Bull.*, **39**, Sem. 2 (1892) 369–73

Sur la propagation des poissons d'eau douce. *France Soc. Acclim. Bull.*, **40**, Sem. 2 (1893) 74–78

Essai de croisement entre le cerf wapiti et le cerf commun. *France Soc. Acclim. Bull.*, **40**, Sem. 2 (1893) 97–100

Pokrychkina, Tatiana

Capitan, Louis; Pokrychkine, (Mlle.) Tatiane

Les changements de form du coeur sous l'influence de la course étudié par la phonendoscopie. *Paris, Soc. Biol. Mém.*, **49** *(C. R.)* (1897) 642–44

APPENDIX

Bulgaria: Anatomy and Physiology

Dimitrova, Zakharina

Dimitrov, (Mlle.) Z.

Observation d'un monstre humain acardiaque (mylacéphale, I. Geoffroy Saint-Hilaire). *Bibliogr. Anat.*, **5** (1897) 206–12

Nicolas, Adolphe; Dimitrov, (Mlle.) Z.
Note sur le developpement de l'arbre bronchique chez le mouton. *Paris, Soc. Biol. Mém.*, **49** *(C. R.)* (1897) 1019–21

China: Medicine

Jin, Yamei (King, Y. May)
King, Y. May
The photo-micrography of histological subjects. *N. Y. Med. J.*, **46** (1887) 7–11

France: Biochemistry

Napias, Louise Rosalie
Napias, (Mlle.) L.
Action de la bacteridie charbonneuse sur les hydrates de carbone. *Ann. Inst. Pasteur*, **14** (1900) 232–48
Pottevin, Henri; Napias, (Mlle.) L.
Sur la "sucrase" de la levure. *Paris, Soc. Biol. Mém.*, **50** *(C. R.)* (1898) 237–38

Hungary: Geography

Tauscher-Geduly, Hermine
Tauscher-Geduly, Hermine
{Ein Wettlauf über den Wolken. Erste Ersteigung des Piz Bernina (4052M) vom Tschierva-Gletscher. *Österr. Alpenzeitung*, **3** (1881) 225–42}
{Ersteigung der Dent Blanche. *Schweiz. Alpenclub Jbuch.*, **18** (1883) 119–43}
{Ersteigung der Trafoier Eiswand vom Norden. *Deutsch. Alpenver. Ztschr.*, **15** (1884) 87–97}
{Auf den Ortler über die Hinteren Wandln. *Deutsch. Alpenver. Ztschr.*, **16** (1885) 355–65}
{Traversirung des Fluchthorns. *Deutsch. Alpenver. Ztschr.*, **17** (1886) 273–80}
Der Grosslitzner (Silvretta-Gruppe). *Deutsch. Alpenver. Ztschr.*, **18** (1887) 254–65

India: Technology

Davy, Zenut
Davy, (Miss) Zenut
Correspondence relative to the manufacture of cloth from the fibres of the Pine-apple leaf and from the Moorva or Moorgavie Plant (Sanseviera zeylanica). *India Agr. Soc. J.*, **3**, part 1, no. 1 (1844) 46–47; 54
Manufacture of cloth from the thread of the Pine-apple and Moorva (Sanseviera zeylanica). *India Agr. Soc. J.*, **3**, part 2, no. 2 (1844) 89–93

Japan: Zoology

Tsuda, Ume (Umeko)
Morgan, T. H.; Tsuda, Umé
The orientation of the frog's egg. *Quart. J. Micr. Sci.*, **35** (1894) 373–405

Mexico: Entomology

Fernández, Ana Maria
Fernández, Vincente; Fernández, Ana Maria
Telarañas. Papalotes de San Juan. *Méx. Obs. Bol.* (1899) 110

Romania: Anatomy, Neurology, and Physiology

Densuşianu-Puşcariu, (Mme.) Elena

Densusianu, (Mlle.)

Dégénérescences et régéneration des terminaisons nerveuses motrices à la suite de la section des nerfs périphériques. [With discussion.] *Congr. Int. Méd. C. R.*, **3** (Anat. Path.) (1900) 208–16

Densuşianu, Elena

{La réparation des plaies aseptiques des muscles. *Soc. Anat. Bull. Mém.* (July 1899) 684–85}

{Histologie des arthrites croniques et spécialement de l'arthrite blennorrhagique. *Arch. Méd. Expér.*, Paris, **13** (1900) 79–92}

Country Undetermined: Botany

Crétier, A. E.

Chodat, Robert; Crétier, (Mlle.)

Sur le noyau des algues vertes inférieures. *Arch. Sci. Phys. Nat.*, **10** (1900) 387–88

Country Undetermined: Medicine

Naville, E.

Naville, (Mme.) E.

Sur le développment des follicules clos dans la conjonctive oculaire. *Paris, Soc. Biol. Mém.*, **48** *(C. R.)* (1896) 451–54

Country Undetermined: Zoology

O'Neil, Helen M.

O'Neil, Helen M.

Hirn- und Rückenmarks-Hüllen bei Amphibien. *Morphol. Arb.*, **8** (1898) 48–64

Country Undetermined: Physiology

Coleman, A.

Coleman, (Mlle.) A.; Pompilian, (Mlle.) M.

Influence de la température sur la contraction musculaire des animaux à sang froid: grenouille, écrevisse. *Paris, Soc. Biol. Mém.*, **48** *(C. R.)* (1896) 696–98

Country Undetermined: Meteorology

Braddon, Julia

Braddon, Julia

Six trombes marines observées dans l'espace d'une demi-heure. *Astronomie* (1885) 110–11

ABBREVIATIONS KEY

Abbreviations follow the Royal Society *Catalogue* usage. For those periodicals not included in the *Catalogue* but added to the bibliography a similar style of abbreviation has been adopted.

Acta Math.	*Acta Mathematica.* Stockholm.
Allg. Bot. Ztschr.	*Allgemeine Botanische Zeitschrift für Systematik, Floristik, Pflanzen geographie, etc.* Karlsruhe.

Amer. J. Obst.	*American Journal of Obstetrics and Diseases of Women and Children*. New York.
Anatomische Abteilung	*Archiv für Anatomie und Entwickelungsgeschichte*. Leipzig. *(Physiol. Abt.) Physiologische Abteilung. Archiv für Physiologie.* Leipzig.
Ann. Inst. Pasteur	*Annales de l'Intitut Pasteur. (Journal de Microbiologie.)* Paris.
Ann. Microgr.	*Annales de Micrographie spécialement consacrées à la Bactériologie, aux Protophytes et aux Protozoaires*. Paris.
Ann. Sci. Nat.	*Annales des Sciences Naturelles (Bot. Botanique)*. Paris. *(Zool.) Zoologie et Paléontologie, comprenant l'Anatomie, la Physiologie, la Classification et l'Histoire Naturelle des Animaux*. Paris.
Année psychologique	*L'Année psychologique: revue de psychologie cognative.* Paris.
Arch. Anat. Physiol.	*Archiv für Anatomie und Physiologie*. Leipzig. *(Anat. Abt.)*
Arch. Augen- und Ohrenheilkunde	*Archiv für Augen- und Ohrenheilkunde.* Karlsruhe.
Arch. Augenheilk.	*Archiv für Augenheilkunde. In deutscher und englischer Sprache.* Wiesbaden.
Arch. de Biol.	*Archives de Biologie*. Gand. Leipzig. Paris. Liège.
Arch. de Physiol.	*Archives de physiologie normale et pathologique.* Paris.
Arch. Exper. Path.	*Archiv für experimentelle Pathologie und Pharmakologie*. Leipzig.
Arch. Ital. Biol.	*Archives Italiennes de Biologie. Revues, Résumés, Reproductions des Travaux Scientifiques Italiens.* Turin.
Arch. Méd. Expér.	*Archives de médecine expérimentale et d'anatomie pathologique*. Paris.
Arch. of Ophthalm.	*Archives of Ophthalmology*, edited and published in English and German. New York. Before 1879, *Archives of Opthalmology and Otology*.
Arch. Psychiatr.	*Archiv für Psychiatrie und Nervenkrankheiten.* Berlin.
Arch. Sci. Phys. Nat.	*Bibliothèque Universelle. Archives des Sciences Physiques et Naturelles.* Genève.
Arch. Slaves Biol.	*Archives Slaves de Biologie*. Paris.
Arch. sudebnoi med. Obshchestvennoi	*Archiv sudebnoi meditsiny i obshchestvennoi gigieny gigieny.* St. Petersburg. (Archives of legal medicine and public health.)
Arch. Wiss. Prakt. Thierheilk.	*Archiv für Wissenschaftliche und Praktische Thierheilkunde*. Berlin.
Astr. Nachr.	*Astronomische Nachrichten begrundet von H. C. Schumacher. Unter Mitwirkung des Vorstandes Astronomischen Gesellschaft herausgegeben von* Kiel.
Astronomie	*L'Astronomie*. Paris.
Berlin, Chem. Ges. Ber.	*Berichte der Deutschen Chemischen Gesellschaft.* Berlin.
Berlin. Klin. Wochenschr.	*Berliner klinische Wochenschrift. Organ für praktische* Ärzte; *mit Berücksichtigung der preussischen Medicinal-Verwaltung und Medicinal-Gesetzgebung nach amtlichen Mittheilungen.* Berlin.
Bern Mitth.	*Mittheilungen der Naturforschenden Gesellschaft in Bern*. Bern.
Bibliogr. Anat.	*Bibliographie Anatomique. Revue des Travaux en langue française. Anatomie. Histologie. Embryologie. Anthropologie*. Paris. Nancy.
Biol. Centrbl.	*Biologisches Centralblatt*. Erlangen. Leipzig.
Brux. Soc. Belge Géol. Bull.	*Bulletin de la Société Belge de Géologie, de Paléontologie et d'Hydrolgie (Bruxelles)*. Bruxelles.
Carus, Zool. Anz.	See *Zool. Anz.*
Centrbl. Bakt.	*Centralblatt für Bacteriologie und Parsitenkunde.* *(Abt. 1) Erste Abteilung. [Medizinisch-hygienische Bakteriologie und Tierische Parasitenkunde.]* Jena.

	(Abt. 2) Zweite Abteilung. [Allgemeine, landwirtwchaftlich-technologische Bakteriologie, Gärungsphysiologie und Planzenpathologie.] Jena.
Centrbl. Med. Wiss.	*Centralblatt für die Medicinischen Wissenschaften.* Berlin.
Centrbl. Physiol.	*Centralblatt für Physiologie.* Leipzig. Wien.
Chem. Soc. J.	*Journal of the Chemical Society.* London.
Congr. Int. Méd. C. R.	*Comptes-Rendus du XII. Congrès International de Médecine, Moscou, 7(19)–14(26) août 1897.* Moscou.
Congr. Int. Zool. [C. R.]	*Congrès International de Zoologie.* Moscou. 1892.
Congr. Med. Int., Atti	*Atti dell' XI. Congresso Medico Internazionale. Roma.* 29 marzo–5 aprile 1894. Torino. Roma.
Crelle, J. Math.	*Journal für die Reine und Angewandte Mathematik; Fortsetzung des von A. L. Crelle und C. W. Borchardt herausgegebenen Journals.* Berlin.
Deutsch. Alpenver. Ztschr.	*Zeitschrift des Deutschen und Oesterreichischen Alpenvereins.* Salzburg. München. Wien. Berlin. Graz.
Deutsch. Arch. Klin. Med.	*Deutsches Archiv für Klinische Medicin.* Leipzig.
Deutsch. Natf. Verh.	*Verhandlungen der Gesellschaft Deutscher Naturforscher und Aerzte.* Leipzig.
Deutsche Ztschr. Chirurg.	*Deutsche Zeitschrift für Chirurgie.* Leipzig.
Engler, Bot. Jbüch.	*Botanische Jahrbücher für Systematik, Pflanzengeschichte und Pflanzengeographie herausgegeben von A. Engler.* Leipzig.
Flora	*Flora oder Allgemeine Botanische Zeitung [früher] herausgegeben von der Königl. Bayer. Botanischen Gesellschaft in Regensburg.* Regensburg. Marburg.
Fortschr. Math.	*Jahrbuch* über *die Fortschritte der Mathematik begründet von Carl Ohrtmann.* Berlin.
Fortschr. Med.	*Forschritte der Medicin.* Berlin.
France Soc. Acclim. Bull.	*Bulletin de la Société Nationale d'Acclimatation de France. [(Revue des Sciences Naturelles Appliquées.)]* Paris. *Revue des Sciences Naturelles Appliquées. Bulletin bimensuel de la Société Nationale d'Acclimatation de France.* Paris.
Gazz. Chim. Ital.	*[La] Gazzetta Chimica Italiana.* Palermo. Roma.
Genève Soc. Phys. Mém.	*Mémoires de la Société de Physique et d'Histoire Naturelle de Genève.* Genève.
India Agr. Soc. J.	*Journal of the Agricultural and Horticultural Society of India.* Calcutta.
Inst. Solvay Trav.	*Institut Solvay. Travaux de Laboratoire publiés par Paul Heger.* Bruxelles.
[Italia], Soc. Ent. Bull.	*Bullettino della Società Entomologica Italiana.* Firenze.
J. Anat. Physiol.	*The Journal of Anatomy and Physiology, normal and pathological [human and comparative].* London. Cambridge. Edinburgh.
J. Bot., Paris	*Journal de Botanique.* Paris.
J. Physiol.	*The Journal of Physiology.* Cambridge. London.
J. Prakt. Chem.	*Journal für praktische Chemie.* Leipzig.
Kazan Soc. Nat. Trans.	*Trudy Obshchestva Estestvoispytateleii pri Imperatorskom' Kazanskom' Universitet'.* [Transactions of the Society of Naturalists of the Imperial University of Kazan.] Kazan.
Kharkov Math. Soc. Commun.	*Soobshchenia Kharkovskago Matematicheskago Obshchestva. Communications de la Société Mathématique de Kharkov.* Kharkov.
Kosmos (Lwow)	*Kosmos. Czasopismo polskiego Tow. Przyroników imienia Kopernika.* [Cosmos. The journal of the Polish Society of Naturalists founded in honour of Copernikus.] Lwow. [Lemberg.]
Lausanne, Soc. Vaud. Bull.	*Bulletin de la Société Vaudoise des Sciences naturelles.* Lausanne.

Liebig's Ann.	*Justus Liebig's Annalen der Chemie.* Leipzig. Heidelberg.
London Med. Rec.	*The London Medical Record.* London.
Méx. Obs. Bol.	*Ministerio de Fomento de la República Mexicana. Boletin Mensual del Observatorio Meteorológico [-Magnetico] Central de México.* México.
Mhefte. Chem.	*Monatshefte für Chemie und verwandte Theile anderer Wissenschaften. Gesammelte Abhandlungen aus den Sitzungsberichten der kaiserlichen Akademie der Wissenschaften.* Wien.
Morphol. Arb.	*Morphologische Arbeiten.* Jena. Superceded by: *Ztschr. Morphol. u. Anthrop.*, q.v.
Moscou Soc. Nat. Bull.	*Bulletin de la Société Impériale des Naturalistes de Moscou.* Moscou.
Moscou, Soc. Sci. Bull.	*Izvestiia Imperatorskago Obschestva Liubitelei Estestvoznaniia, Antropologii i Ethnografii, sostoiashchago pri Imperatorskom' Moskovskom' Universitete.* [Bulletin of the Imperial Society of Lovers of Natural Science, Anthropology, and Ethnography, in connection with the Imperial University of Moscow.] Moscow.
Napoli, Rend.	*Rendiconto dell' Accademia della Scienze Fisiche e Matematiche. (Sezione della Società Reale di Napoli.)* Napoli.
Nature	*Nature. A weekly illustrated Journal of Science.* London. New York.
Neues Jbuch. Min.	*Neues Jahrbuch für Mineralogie, Geologie und Palaeontologie.* Stuttgart.
Neurol. Centrbl.	*Neurologisches Centralblatt.* Leipzig.
New Russ. Soc. Nat. Mém.	*Zapiski Novorossiiskago Obshchestva Estestvoispi'tatelei. Mémoires de la Société des Naturalistes de la Nouvelle-Russie.* Odessa.
N. Y. Med. J.	*The New York Medical Journal.* New York.
Österr. Alpenzeitung	*Österreichische Alpenzeitung.* Wien.
Paris, Ac. Sci. C. R.	*Comptes Rendus hebdomadaires des Séances de l'Académie des Sciences.* Paris.
Paris, Mém. Savants Étrang.	*Mémoires présentés par divers Savants à l'Académie des Sciences de l'Institut de France.* Paris.
Paris, Soc. Biol. Mém.	*Comptes Rendus [hebdomadaires] des Séances et Mémoires de la Société de Biologie.* Paris
Paris Soc. Chim. Bull.	*Bulletin de la Société Chimique de Paris.* Paris.
Pflüger, Arch. Physiol.	*Archiv für die gesammte Physiologie des Menschen und der Thiere. Herausgegeben von Dr. E. F. W. Pflüger.* Bonn.
Pharm. Ztschr. Russland	*Pharmaceutische Zeitschrift für Russland.* St. Petersburg.
Physiologiste Russe	*Le Physiologiste Russe.* Moscou.
Quart. J. Micr. Sci.	*Quarterly Journal of Microscopical Science.* London.
Rec. Zool. Suisse	*Recueil Zoologique Suisse, comprenant l'Embryologie, l'Anatomie, et l'Histologie comparées, la Physiologie, l'Éthologie, la Classification des Animaux vivants et fossils.* Genève. Bâle. Continued as *Rev. Suisse Zool.*, q.v.
Rev. Gén. Bot.	*Revue générale de Botanique.* Paris.
Rev. Méd. Suisse rom.	*Revue médicale de la Suisse romande.* Lausanne.
Rev. Sci.	*[La] Revue Scientifique [de la France et de l'Étranger. Revue des Cours Scientifiques.]* Paris.
Rev. Suisse Zool.	*Revue Suisse de Zoologie [et] Annales [de la Société Zoologique Suisse et] du Musée d'Histoire Naturelle de Genève.* Genève. Continuation of *Rec. Zool. Suisse.*, q.v.
Russ. Chem. Soc. J.	*Zhurnal' Russkago Khimicheskago Obshchestva.* St. Petersburg.
Russ. Geol. Min. Annu.	*Ezhegodnik' po Geologii i Mineralogii Rossii. Annuaire Géologique e Minéralogique de la Russie.* Varshava. Varsovie. Novaia-Aleksandriia. Novo-Alexandria.

Russ. Phys.-Chem. Soc. J.	*Zhurnal' Russkago Fiziko-Khimicheskago Obshchestva pri Imperatorskom' S.-Peterburgskom' Universitet.* [Journal of the Russian Physico-Chemical Society of the Imperial University of St. Petersburg. S. Peterburg'.] [St. Petersburg.] There are two series of this publication, one for the physical and one for the chemical section of the society; they have the same vol. numbers and dates.
Schweiz. Alpenclub Jbuch.	*Jahrbuch des Schweizer Alpenclub.* Bern.
Schweiz. Natf. Ges. Verh.	*Verhandlungen der Schweizerischen Naturforschenden Gesellschaft. Actes de la Société Helvétique des Sciences Naturelles. [Atti della Società Elvetica delle Scienze Naturali.]* [Variable location.]
Soc. Anat. Bull. Mém.	*Bulletins et mémoires de la Société Anatomique de Paris.* Paris.
Soc. Sci. Méd. Nat. Bruxelles. Ann.	*Annales de la Société royale des Sciences Médicales et Naturelles de Bruxelles.* Bruxelles.
Soc. Sci. Méd. Nat. Bruxelles. Bull.	*Société royale des Sciences Médicales et Naturelles de Bruxelles. Bulletin des séances.* Bruxelles.
Stockh., Öfvers.	Öfversigt *af Kongl. Vetenskaps-Akademiens Förhandlingar.* Stockholm.
St. Pétersb. Ac. Sci. Bull.	*Bulletin de l'Académie Impériale des Sciences de St. Pétersbourg.* St. Pétersbourg. *Izvestiia Imperatorskoi Akademii Nauk'. Bulletin de l'Académie Impériale des Sciences de St. Pétersbourg.* Peterburg'. St. Pétersbourg.
St. Pétersb. Ac. Sci. Mém.	*Zapiski Imperatorskoi Akademii Nauk'.* [Series I–VI] *S.-Peterburg'. Mémoires de l'Académie Impériale des Sciences de Saint-Pétersbourg.* [Series VII] St.-Pétersbourg. *Zapiski Imperatorskoi Akademii Nauk' po Fiziko-Matematicheskomu Otdieleniiu. Mémoires de l'Académie Impériale des Sciences de Saint-Pétersbourg. Classe des Sciences physiques et mathématiques.* [Series VIII] St. Pétersbourg.
St. Pétersb. Arch. Sci. Biol.	*Arkhiv' Biologicheskikh Nauk' izdavaemyi Imperatorskim' Institutom' Eksperimental'noi Meditsinu v' S.-Peterburg'. Archives des Sciences Biologique publiées par l'Institut Impérial de Médecine Expérimentale à St.-Pétersbourg.* St. Pétersbourg.
St. Pétersb., Com. Géol. Bull.	*Izvestiia Geologicheskago Komiteta.* S.-Peterburg'. *Bulletins du Comité Géologique, & Supplements.* St.-Pétersbourg.
St. Pétersb. Com. Geol. Mém.	*Trudy Geologicheskago Komiteta. Mémoirs du Comité Géologique.* S.-Petersburg' St. Pétersbourg.
St. Petersb., Med. Vestnik	*Meditsinskii Vestnik.* [Medical bulletin.] St. Petersburg.
St. Petersb. Med. Wschr.	*St. Petersburger Medicinische Wochenschrift.* St. Petersburg.
St. Petersb. Med. Ztschr.	*St. Petersburger Medicinische Zeitschrift.* St. Petersburg.
St. Petersb. Min. Ges. Verh.	*Zapiski Imperatorskago S.-Peterburgskago Mineralogicheskago Obschchestva.* Sanktpeterburg'. *Verhandlungen der Russisch-Kaiserlichen Mineralogischen Gesellschaft zu St. Petersburg.* St. Petersburg.
St. Petersb., Russ. Entomol. Soc.	*Russkago Entomologicheskago Obshchestva, Trudy, Horae.* S.-Peterburg'. [Russian Entomological Society.] St. Petersburg.
Univ. de Genève, Lab. bot.	*Laboratoire de botanique, Université de Genève.* Genève.
Varshava Lab. Zool. Raboty	*Raboty iz Laboratorii Zoologicheskago kabineta Imperatorskago Varshavskogo Universiteta.* [Contributions from the Zoological Laboratory, Imperial University of Warsaw.] Varshava.
Varsovie Soc. Nat. Trav.	*Trudy Obshchestva Estestvoispytatelei pri Imperatorskom' Varshavskom' Universitete. Travaux de la Société des Naturalistes de l'Université Impériale de Varsovie. (C. R., Phys. Chim.) Protokoly Otdelenia Fiziki i Khemii.* Varshava. *Comptes Rendus de la Section de Physique et de Chimie.* Varsovie.

Virchow, Arch.	*Archiv für Pathologische Anatomie und Physiologie und für Klinische Medicin. Herausgegeben von Rudolf Virchow*. Berlin.
Voenno-med. Zhurnal	*Voenno-medisinskii zhurnal.* [Military medical journal.] St. Petersburg. Moscow.
Vrach	*Vrach.* [Physician.] St. Petersburg.
Wien, Ak. Sber.	*Sitzungsberichte der Kaiserlichen Akademie der Wissenschaften. Mathematisch-Naturwissenschaftliche Classe*. Wien.
Würzb. Phys. Med. Verh.	*Verhandlungen der Physikalisch-Medicinischen Gesellschaft zu Würzburg*. Würzburg.
Zool. Anz.	*Zoologischer Anzeiger herausgegeben von Prof. J Victor Carus in Leipzig. [Zugleich Organ der Deutschen Zoologischen Gesellschaft.]* Leipzig.
Ztschr. Anorg. Chem.	*Zeitschrift für anorganische Chemie*. Hamburg. Leipzig.
Ztschr. Biol.	*Zeitschrift für Biologie*. München. Leipzig.
Ztschr. Chem.	*Zeitschrift für Chemie.* Göttingen.
Ztschr. Geburtshlf. Gynäk.	*Zeitschrift für Geburtshülfe und Gynäkologie. Unter Mitwirkung der Gesellschaft für Geburtshülfe und Gynäkologie [in Berlin].* Stuttgart.
Ztschr. Hyg.	*Zeitschrift für Hygiene [und Infectionskrankenheiten].* Leipzig.
Ztschr. Morphol. u. Anthrop.	*Zeitschrift für Morphologie und Anthropologie.* Stuttgart. Continuation of *Morphol. Arb.*, q.v.
Ztschr. Physiol. Chem.	*[Hoppe-Seyler's] Zeitschrift für Physiologische Chemie*. Stuttgart.
Ztschr. rationelle Med.	*Zeitschrift für rationelle Medicin. Herausgegeben von Dr. J. Henle und Dr. C. v. Pfeufer.* Zurich. Heidelberg. Leipzig.
Ztschr. Wiss. Zool.	*Zeitschrift für wissenschaftliche Zoologie*. Leipzig.

Selected Bibliography

Alpert, Carl. "Berlin, Fanny." In *The Universal Jewish Encyclopedia*, edited by Isaac Landman, 10 vols., 2:216. New York: Universal Jewish Encyclopedia, 1940.

Anon. "E. A. Fomina-Zhukovskaia." *Journal of the Russian Physical-Chemical Society* 26 (Chem.) (1894): 432–34.

———. "Evgenia Solomka-Sotiriadis [*sic*]." *Ezhegodnik po Geologii i Mineralogii Rossii* 3, sect. 1 (1898–1899): 137.

Annual Report of the Registrar-General of Births, Deaths, and Marriages in England. London: Her Majesty's Stationary Office, 1859–1887.

Apotheker, Jan, and Livia Simon Sarkadi, eds. *European Women in Chemistry.* Weinheim: Wiley-VCH Verlag & Co. KGaA, 2011.

Audin, Michele. *Remembering Sofia Kovalevskaia.* London: Springer, 2011. Translation from the French, *Souvenirs sur Sofia Kovalevskaya.* N.p.: Calvage et Mounet, 2008.

Bachmann, Barbara, and Elke Bradenahl. "Medizinstudium von Frauen in Bern, 1871–1914." Doctoral dissertation, Bern University, 1990.

Baza osób polskich—Polnische Personendatenbank [Database of Polish people]. http://baza-nazwisk.de/erw-suche.php?data=59376&hits=3560&ds=2171&title=Haberkant%20Wanda (Aug. 2014).

BBC—Your Paintings: Tilly Kettle, Major William Davy (d. 1784), Bengal Native Infantry. http://www.bbc.co.uk/arts/yourpaintings/paintings/major-william-davy-d-1784-bengal-native-infantry-182477 (Aug. 2014).

Bessudnova, Z. A. "Pavlova, Mariia Vasil'evna (1854–1938)." In Starodubtseva, J. A., et al., *Pavlovskaia Geologicheskaia Shkola*, 21–26. Moscow: Nauka, 2004.

Bogdanov, Anatolii Petrovich. *Materialy dlia istorii nauchnoi i prikladnoi deiatel'nosti v Rossii po zoologii i soprikasaiushchimsia s neiu otrasliam znaniia, preimushchestvenno za poslednee tridtsatipiatiletie, 1850–1887g. Izvestiia Imperatorskago obshchestva liubitelei estestvoznaniia, antropologii i etnografii* [Proceedings of the Imperial Society of Natural History, Anthropology, and Ethnography]. 4 vols. Moscow, 1888–1892.

Bologa, Alexandru S. "Development of Marine Biological Institutions around the Black Sea." In *Ocean Sciences bridging the Millennia: A Spectrum of Historical Accounts.* Paper presented at the Sixth International Congress on the History of Oceanography, Quingd'ao, People's Republic of China, 15–20 Aug. 1998. Paris: UNESCO, 2004.

Bonner, Thomas Neville. "Rendezvous in Zurich: Seven Who Made a Revolution in Women's Medical Education, 1864–1874." *Journal of the History of Medicine* 44, no. 1 (Jan. 1989): 7–27.

Borisiak, A. A., and V. V. Menner. "Maria Vasilevna Pavlova." *Vestnik Rossiskii Academii Nauk* 6 (1939): 78–80.

Brinkschulte, Eva, ed. *Weibliche* Ärzte*: Die Durchsetzung des Berufsbildes in Deutschland.* Berlin: Edition Heintrich, 1993.

Briquet, John. *Biographies des botanists à Genève de 1500 à 1931.* Edited by François Cavilliet. Geneva: Albert Kundig, 1940.

Bruziewicz-Miklaszewska, Barbara. "Żeby ślad pozostal i pamięć: Medicial [*sic*] Women's International Association—MWIA—1919–1999." *Gazeta Lekarska* no. 4 (99), s. 51 (1999): 53. http://www.oil.org.pl/xml/nil/gazeta/numery/n1999/n199904/n19990424 (Aug. 2014).

Carr, E. H. *Michael Bakunin.* New York: Octagon Books, 1975.

Cassino, Samuel Edson, comp. *Naturalists' Universal Directory.* 19th ed. Salem, Mass.: S. E. Cassino, 1905.

Catalogue des Étudiants *de l'Université de Lausanne, 1895–1896.* Lausanne: Couchoud, 1895.

Chasin, Judith. "Fanny Berlin." In *Jewish Women: A Comprehensive Historical Encyclopedia*. Jewish Women's Archive. Last modified 1 March 2009. http://jwa.org/encyclopedia/article/berlin-fanny (Aug. 2014).

Ciardi, Marco, and Miriam Focaccia. "Maria Bakunin (1873–1960)." In *European Women in Chemistry*, edited by Jan Apotheker and Livia Simon Sarkadi, 51–54. Weinheim: Wiley-VCH, 2011.

Constantinescu, Richard. "O personalitate a medicine româneşti: Elena Densusianu-Puscariu." ROST 42–43, August–September 2006. http://www.rostonline.org/rost/aug-sep2006/densusianu.shtml (Aug. 2014).

———. "Priviri anatomopatologice." *Viata Medicala* 27, 17 January 2012. http://www.viata-medicala.ro/Priviri-anatomopatologice.html*articleID_4710-dArt.html (Aug. 2014).

Cornish-Bowden, Athel. "The Origins of Enzymology." *Biochemist* 19, no. 2 (1999): 36–38.

Creese, Mary R. S. *Ladies in the Laboratory? American and British Women in Science, 1800–1900.* Lanham, Md.: Scarecrow Press, 1998.

———. "Early Women Chemists in Russia: Anna Volkova, Iuliia Lermontova, and Nadezhda Ziber-Shumova." *Bulletin for the History of Chemistry* 21 (1998): 19–24.

———. *Ladies in the Laboratory II: West European Women in Science, 1800–1900*. Lanham, Md.: Scarecrow Press, 2004.

———. "Fossil Hunters, a Cave Explorer and a Rock Analyst." In *The Role of Women in the History of Geology*, edited by C. V. Burek and B. Higgs, 39–49. London: Geological Society Special Publication no. 281, 2007.

Das Frauenstudium an der schweizer Hochschulen, ed. Schweizerischer Verband der Akademikerinnen. Zurich: Rascher, 1928.

Dettelbacher, Werner. "Dr. med. Marie Siebold—eine der ersten Ärztinnen in Serbien, in der Türkei und in Ägypten." *Münchener Medizinische Wochenschrift* 45, no. 115 (1973): 2051–53.

Duff, Carole Ann. "Christianity, Science and Society: Two Nineteenth-Century American Medical Missionaries in the Far East." MA thesis, University of Texas–Dallas, 1977. Dallas: University of Texas, 1977.

"Early Women at the University of Bern." In Scandola, Pietro, and Franziska Rogger, *The History of the University*. Bern University Archives. http://www.unibe.ch/eng/pdf/history.pdf (Aug. 2014).

Elina, Olga Yu. "Private Botanical Gardens in Russia: Between Noble Culture and Scientific Professionalization (1760s–1919)." In *The Global and the Local: The History of Science and Cultural Integration of Europe*, edited by M. Kokowski, 579–85. Proceedings of the Second ICESHS, Cracow, Poland, 6–9 Sept. 2006.

Enciclopedia României. http://enciclopediaromaniei.ro/wiki/Elena_Densuşianu-Puşcariu (Aug. 2014).

Enke, Ulrike. "Geburtshelferin der englischen Königen." *Hessisches* Ärzteblatt 8 (2009): 525–26.

Eyler, John M. "Farr, William (1807–1883), Statistician and Epidemiologist." In *Oxford Dictionary of National Biography*, 19:99–101. Oxford: Oxford University Press, 2004.

Fabbris, Faber Renato. "Biography of Renato Caccioppoli." In *The MacTutor History of Mathematics Archive*. School of Mathematics and Statistics, University of St. Andrews, Scotland. http://www-history.mcs.st-andrews.ac.uk/Biographies/Caccioppoli.html (Aug. 2014).

Fedchenko, Olga. "*Eremurus*: Kritische Übersicht der Gattung." *Mémoires de l'Académie Impériale des Sciences de St. Pétersburg s. 8. Classe physico-mathématique* 23, no. 8 (1909). Reprinted in Plant Monographs Reprints, ed. J. Cramer and H. K. Swann. Lehre, Niedersachsen: J. Cramer, 1968.

Feliksiak, Stanislaw, ed. *Słownik Biologów Polskich.* Warsaw: Polish Academy of Science, 1987.

Filar, Zbigniew. *Anna Tomaszewicz Dobrska: Karta z dziejów polskich lekarek.* Polskiego Towarzystwo Historii Medycyny. Życiorysy no. 4 (1959).

Furst, Lilian R., ed. *Women Healers and Physicians: Climbing a Long Hill*. Lexington: University Press of Kentucky, 1997.

Furuki, Yoshiko. *The White Plum: A Biography of Ume Tsuda: Pioneer in the Higher Education of Japanese Women.* New York and Tokyo: Weatherhill, 1991.

Gilgenkrantz, Simone. "Les premières étudiantes étrangères en médecine à Nancy (1894–1914)." http://www.professeurs-medecine-nancy.fr/premieres_etudiantes_etrangeres.htm (Aug. 2014).

Grinstein, Louise S., and Paul J. Campbell, eds. *Women of Mathematics: A Biobibliographic Sourcebook.* Westport, Conn.: Greenwood Press, 1987.

Gustavson, G. "Vera Evstaf'evna Bogdanovskaia." *Journal of the Russian Physical-Chemical Society* 29 (Chem.) (1897): 147–51.

Hammond, Kelly Anne. "The Shanghai Mixed Court, 1863–1880." MA thesis, Simon Fraser University, 2007.

Ho, Clara Wing-chung, ed. *Biographical Dictionary of Chinese Women*. Vol. 1, *The Qing Period, 1644–1911*. Armonk, N.Y.: M. E. Sharpe, 1998.

Jahresverzeichnis der Schweizerischen Universitätsschriften, 1897–1898; *Catalogue des* Écrits *Académiques Suisse, 1897–1898*. Basel: Schweighauserische Buchdruckerei, 1898.

Joteyko, Józefa. *La science du travail et son organization*. Paris: Librairie Félix Alcan, 1917. Translated into English as *The Science of Labour and its Organization*. London: G. Routledge; New York: E. P. Dutton, 1919. Reprint, 1974.

Kalisz online. History. "RYS HISTORYCZNY PARAFII EWANGELICKO-AUGBURSKIEJ W KALISZU." http://www.info.kalisz.pl/historia/ewangelicy.htm (Aug. 2014).

Karim, Abdul. "Muslin." In *Banglapedia: National Encyclopedia of Bangladesh*, 7:179–81. Dhaka: Asiatic Society of Bangladesh, 2003.

Kästner, Ingrid. "'Leider haben mich Familienverhältnisse abgehalten . . .': Leben und Werk der Ärztin Maria Michailowna Manasseina." In *Geschlecterverhältnisse in Medizin, Naturwissenschaft und Technik*, edited by Christoph Meinel und Monika Renneberg, 62–68. Bassum: Verlag für Geschichte der Naturwissenschaft und der Teknik, 1996.

Kearns, Michelle. "Reception to Mark Reopening of Reinstein Family Collection." *Buffalo News*, 16 Oct. 2010.

Kelly, Georgianna. "Adelaide Lukanina's 'A Year in America.'" MA thesis, University of Kansas, 1995.

Kennedy, Don H. *Little Sparrow: A Portrait of Sophia Kovalevsky*. Athens: Ohio University Press, 1983.

Klawe, Zofia Podgórska. "Kobiety-Lekarze Warszawy." *Archiwum Historii Medycyny* 28 (1965): 243–50.

Knight, Amy. "The Fritschi: A Study of Female Radicals in the Russian Populist Movement." *Canadian-American Slavic Studies* 9, no. 1 (1975): 1–17.

Koblitz, Ann Hibner. *A Convergence of Lives: Sofia Kovalevskaia: Scientist, Writer, Revolutionary*. Boston and Basel: Birkhäuser, 1983.

Komerov, V. L. "Olga Aleksandrovna Fedchenko." *Izvestia Rossiiskoi Akademii Nauk* 15, s. 6, no. 1 (1921): 247–48.

Kovalevskaya, Sofya. *A Russian Childhood*. Translated by Beatrice Stillman. New York: Springer Verlag, 1978.

Kovalzon, Vladimir M. "Some Notes on the Biography of Maria Manasseina." *Journal of the History of the Neurosciences* 18 (2009): 312–19.

Kulikowski, Jerzy. "Medycynierka." *POLITYKA.PL* 34, 21 Aug. 2007. http://www.polityka.pl/historia/226843,1 ,medycynierka.read (Aug. 2014).

Kuznetsov, N. IA. "Sofia Michailovna Pereiaslavtseva." *Revue Russe d'Entomologie* 6 (Dec. 1903): 422.

Kvasov, D. G., and A. K. Fedorova-Grot. *Fiziologicheskaia Shkola I. P. Pavlova: Portrety i kharakteristiki sotrudnikov i uchenikov*. Leningrad: Isdatel'stvo Nauka, 1967.

Lagnado, John. "Was the First Biochemist a Woman?" *Biochemist* 14, no. 5 (1992): 21–22.

Ledkovskaia-Astman, Marina, Charlotte Rosenthal, and Mary Fleming Zirin, eds. *Dictionary of Russian Women Writers*. Westport, Conn.: Greenwood Press, 1994.

Leier, Mark. *Bakunin: The Creative Passion*. New York: St. Martin's Press, 2006.

Le Lièvre, Audrey. "Fedchenkoana." *Garden* 112, no. 1 (1987): 65–68.

Lipinska, Mélanie. *Histoire des Femmes Médecins depuis l'Antiquité jusqu'à nos Jours*. Paris: G. Jacques, 1900.

Lipschitz, Sergei Iulevich, and Vladimir Nikolaevich Sukachev. *Russkie Botaniki: Biografo-bibliograficheskii slovar*. 4 vols. Moscow: izd-vo Moskovskogo obshchestva ispytatelei prirody, 1947–1952.

Loskutova, Marina. "Public Science as a Network: The Congresses of Russian Naturalists and Physicians in the 1860s–1910s." *Baltic Journal of European Studies* 1, no. 1 (2011): 153–69.

Maeder, Eva, and Peter Niederhäuser, eds. *Käser, Künstler, Kommunisten: Vierzig russisch-schweizerische Lebensgeschichten aus vier Jahrhunderten*. Zurich: Chronos, 2009.

Manacéine, Marie de. *Sleep: Its Physiology, Pathology, Hygiene, and Psychology*. New York: Scribner's, 1899. Reprint, Washington, D.C.: American Psychological Association, 2009.

Marek, Miroslav. "Genealogy of the Bobrinski Family." http://genealogy.euweb.cz/ascania/ascan13.html (Aug. 2014).

Masevich, Alla G., and Aleksandra K. Terenteva. "Zhenshchiny-astronomy." *Na rubezhakn poznaniia vselenna. Istoriko-astronomicheskie issledovanniia* 23 (1991): 90–111.

Matrikeledition der Universität Zürich, *1833–1924*. http://www.matrikel.uzh.ch/active/static/index.htm (Aug. 2014).

McCartee, Divie Bethune, and Robert Elliot Speer. *A Missionary Pioneer in the Far East: A Memorial to Divie Bethune McCartee*. Edited by Robert E. Speer. New York: Fleming H. Revell, 1922.

Mechnikova, Olga. *Life of Elie Metchnikoff, 1845–1916*. Translated by Sir Edwin Ray Lankaster. London: Constable, 1921.

Meijer, J. M. *Knowledge and Revolution: The Russian Colony in Zurich (1870–1873): A Contribution to the Study of Russian Populism*. Assen: International Instituut voor Social Geschiedenis, 1955.

Mir Slovarei Istoricheskii Slovar'-poisk. "Ziber-Shumova, Nadezhda Olimpievna." http://mirslovarei.com/content _his/ziber-shumova-nadezhda-olimpievna-17677.html (Aug. 2014).

Mongillo, Pasqualina. *Marussia Bakunin, una donna nella storia della chimica*. Sovaria Mannelli: Rubbettino, 2008.

Musabekov, Iusuf S. *Iulia Vsevolodovna Lermontova, 1844–1919*. Moscow: Nauka, 1967.

"Nadejda Alexandrovna Polovtzeva (1865–1920)—Genealogy." http://www.geni.com/people/Nadejda-Alexandrovna-Polovtseva/6000000007872512514 (Aug. 2014).

Nalivkin, D. V. *Nashi pervye Zhenshchinye-geologi*. Leningrad: Nauka, 1979.

Nassonov, N. [Memorial for Maria Iwanovna Pavlova]. *Anatomische Hefte, Abteil. 2, Ergebnisse der Anatomie und Entwicklungsgeschichte* 9, *Biographisches Geschichte* 11 (1900): 518–19.

Nazarska, Georgeta. "Bulgarian Women Medical Doctors in the Social Modernization of the Bulgarian National State (1878–1944)." *Historical Social Research* 33, no. 2 (2008): 232–44.

Neuima, Maria N. "Zhenshchiny-astronomy Pulkovskoi Observatorii." *Istoriko-astronomicheskie issledovanniia* 10 (1969): 229–39.

Nevskaya, N. I. "Leningrad Women-Astronomers in the Early Days of Soviet Power." In *Proceedings of the International Conference on the Role of Women in the History of Science, Technology and Medicine in the 19th and 20th C.*, Vezprém, Hungary, 15–19 August 1983, 1:139–44.

Nicolaus, Rodolfo Alessandro. "Maria Bakunin." *La Chimica e l'Industria* 42 (1960): 677–78.

———. "Maria Bakunin: Commemorazione." *Rendiconto dell' Accademia delle Scienze Fisiche e Matematiche* 28, s. 4 (1961): 15–21.

———. "Bakunin, Marussia." In *Dizionario Biografico degli Italiani*, 34:223–24. Roma: Istituto della Enciclopedia italiana, 1988.

Ogilvie, Marilyn, and Joy Harvey, eds. *The Biographical Dictionary of Women in Science: Pioneering Lives from Ancient Times to the Mid-20th Century*. 2 vols. New York and London: Routledge, 2000.

Paskalev, D., D. Radoinova, and B. Galunska. "Doctor Zaharina Dimitrova (1873–1940): Pioner v oblasti issledovania mikrostruktury epefiza." *Zhurnal Nefrologiia, Istoriia Meditsiny* 15, no. 2 (2011): 115–18.

Patuelli, Francesca. "Bakunin, Maria." https://scienzaa2voci.unibo.it/biografie/5-bakunin-maria (Aug. 2014).

Petkov, P., and A. K'nev. "130 godini ot rozhdenieto na d-r Zakharina Dimitrova: Izvestna b sveta, nepoznata b Bulgariia." *Asklepios* 17 (2004): 184–86.

Popov, Ia. "Introduction II." In Vera Evstaf'evna Bogdanovskaia, *Ocherki, stati, otryvki: Pamiati Very Evstafevny Bogdanovskoi, 1867–1896*, 1:iii–ix. St. Petersburg: Ia. Popov, 1898.

Prokhorov, A. M., ed. *Bol'shaia sovetskaia entsiklopediia*. Moscow: Izd-vo Sovetskaia Entsiklopedia, 1970–1981. Translated to English as *Great Soviet Encyclopedia*. New York: Macmillan, 1973–1982.

Rebiére, Alphonse. *Les Femmes dans la Science: Notes recueilles*. 2nd ed. Paris: Nony, 1897.

Rebisse, Christian. "Pierre-Augustin Chaboseau (1868–1946): Un Servidor Desconocido." http://www.biblioteca.amorc6.org/personajes/doc_download/513-pirre-augustin-chaoseau-1868-1946-un-servidor-desonocido (Aug. 2014).

Reshetina, N. S. "Liubov' Nikolaevna Zapol'skaia." Ryazan State University, Science Library. http://library.rsu.edu.ru/blog/wp-content/uploads/2013/01/zapolskaya.pdf (Aug. 2014).

Rohner, Hanny. *Die ersten 30 Jahre des medizinischen Frauenstudiums an der Universität* Zürich, *1867–1897*. Zurich: Juris Druck, 1972.

Rose, Barbara. *Tsuda Umeko and Women's Education in Japan*. New Haven: Yale University Press, 1992.

Roussanova, Elena. *Julia Lermontowa (1846–1919): Die erste promovierte Chemikerin*. Norderstedt: Books on Demand, 2003.

Siemak-Tylikowska, Alicja. "Maria Grzegorzewska (1888–1967)." *Prospects: The Quarterly Review of Comparative Education* 23, no. 3/4 (1993): 625–37.

Smorscek, E. "Le centenaire de la naissance de Jósefa Joteyko." *Bulletin de psychologie scolaire et d'orientation* 15 (1966): 45–50.

Sobolevskii, A. I. "P. S. Uvarova. 1840–1924. Nekrolog." *Izvestiia Rossiiskoi Akademii Nauk* 19, series 6, nos. 6–8 (1925): 141–44.

Sowerwine, Charles. *Sisters or Citizens? Women and Socialism in France since 1872*. Cambridge: Cambridge University Press, 1982.

Stadtarchiv Göttingen, Personen. http://www.stadtarchiv.goettingen.de/personen/schloezer_dorothea.htm (Aug. 2014).

Tee, Gary J. "Sof'ya Vasil'yevna Kovalevskaia." Paper presented at the Eleventh New Zealand Mathematics Colloquium, Palmerston North, 21–24 May 1976. *Mathematical Chronicle* 5 (1977): 113–39.

Tiburtius, Franziska. *Erinnerungen einer Achtzigjährigen*. 3rd ed. Berlin: C. A. Schwetschke & Sohn, 1929.

Tobies, Renate, ed. *"Aller Männerkultur zum Trotz": Frauen in Matematik und Naturwissenschaften*. Frankfurt/Main: Campus Verlag, 1997.

Todes, Daniel P. *Pavlov's Physiology Factory: Experiment, Interpretation, Laboratory Enterprise*. Baltimore: Johns Hopkins University Press, 2002.

Tollmien, Cordula. "Zwei erste Promotionen: Die Mathematikerin Sofia Kowalewskja und die Chemikerin Julia Lermontowa." In *"Aller Männerkultur zum Trotz": Frauen in Mathematik und Naturwissenschaften*, edited by Renate Tobies, 83–129. Frankfurt/Main: Campus Verlag, 1995.

Tuve, Janette E. *The First Russian Women Physicians.* Newtonville, Mass.: Oriental Research Partners, 1984.

Uglo, Jennifer S., comp. and ed. *The Northeastern Dictionary of Women's Biography*. 3rd ed. Revised by Maggy Hendry. Boston: Northeastern University Press, 1998.

University Archives, Bern University. http://www.unibe.ch/eng/about/portrait_archive.html (Aug. 2014).

University of Poznan, Main Library Publications. "Zakład Antropologii. Prof. fizjologii układu nerwowego Dr nauk przyrodniczych Michalina Stefanowska (20 XI 1855–15 XII 1942)." http://www2.bg.am.poznan.pl/wydawnictwa/_1919_39/115_135.pdf (Aug. 2014).

Valkova, Olga A. *Ol'ga Aleksandrovna Fedchenko, 1845–1921*. Moscow: Nauka, 2006.

———. "The Conquest of Science: Women and Science in Russia, 1860–1940." *Osiris* 23 (2008): 136–65.

Verein feministische Wissenschaft Schweiz. *Ebenso neu als kühn: 120 Jahre Frauenstudium an der Universität* Zürich. Edited by Katherina Belser, et al. Zurich: Verein feministische Wissenschaft Schweiz, Schriftenreihe, 1988.

Vinokurov, S. I., and R. V. Chagovets. "Mar'ia Manasseina i ee rol' b otkrytie beskletochnogo brosheniia." *Biokhimiia* 15, no. 6 (1950): 558–62.

Vogel, Charity. "Women to Remember." *Buffalo News*, 13 Nov. 2005.

Vucinich, Alexander. *Science in Russian Culture, 1861–1917*. Stanford, Calif.: Stanford University Press, 1970.

Wandycz, Piotr S. *The Lands of Partitioned Poland, 1795–1918*. Seattle and London: University of Washington Press, 1974.

Windholz, George. "Three Researchers in Pavlov's Laboratories." *NWSA Journal* 1, no. 3 (1989): 491–96.

Zemanek, Alicja. "Pro Memoria: 130 rocznica Gabrieli Balickiej-Iwanowskiej (1871–20 II 1962)." Cracow, Institut Botaniki, *Wiadomości Botaniczne* 45, no. 1/2 (2001): 74.

Index

About the Author and Contributor

Mary Creese (née Weir) was born in Orkney, Scotland, in 1935. After almost thirty years as a research chemist, she turned to the subject of women's contributions to scientific work, particularly in the nineteenth century. Since 1990 she has published more than forty articles on early women scientists. She holds a BSc degree from the University of Glasgow and a PhD from the University of California, Berkeley (both in chemistry), and has coauthored some twenty technical publications. From 1962 to 1964 she held research and teaching positions in Canada, first at the University of British Columbia, Vancouver, and then at the University of Calgary, Alberta. Subsequently she worked at the University of Kansas, most recently as an associate at the Hall Center for the Humanities. She is married to Thomas Creese, and they have two daughters.

Thomas Creese was born in New York in 1934 and spent his early years in New Jersey and Pennsylvania. He holds degrees in mathematics from the Massachusetts Institute of Technology and the University of California, Berkeley. From 1964 until he retired in 2007 he was a member of the Department of Mathematics at the University of Kansas and has coauthored both a textbook and a research monograph. He and Mary Creese have collaborated on projects in computational chemistry and the history of women in science.

www.ingramcontent.com/pod-product-compliance
Lightning Source LLC
LaVergne TN
LVHW062149200926
848484LV00007B/1888
9781442247413